Principles of
FABRIC
FORMATION

Principles of
FABRIC
FORMATION

Prabir Kumar Banerjee

CRC Press
Taylor & Francis Group
Boca Raton London New York

CRC Press is an imprint of the
Taylor & Francis Group, an **informa** business

CRC Press
Taylor & Francis Group
6000 Broken Sound Parkway NW, Suite 300
Boca Raton, FL 33487-2742

First issued in paperback 2017

© 2015 by Taylor & Francis Group, LLC
CRC Press is an imprint of Taylor & Francis Group, an Informa business

No claim to original U.S. Government works

ISBN 13: 978-1-138-83769-0 (pbk)
ISBN 13: 978-1-4665-5444-3 (hbk)

Visit the Taylor & Francis Web site at
http://www.taylorandfrancis.com

and the CRC Press Web site at
http://www.crcpress.com

Contents

Preface

Textile fabrics are intrinsically strong while being flexible, permeable, moldable, and drapeable. In modern times, this porous continuum is being progressively viewed as a useful engineering material in diverse applications as in civil constructions, such as roads, dams, or bridges; in vehicular constructions, such as aerospace ships or automobiles and trucks; in communication engineering as flexible conductive material for generation and transmission of signals or for digital display; in power generation in the form of flexible solar panels; in the medical field as biosensors or for construction of human body components; and so on and so forth. Arguably then, not only students of conventional textile disciplines, but also of civil, mechanical, electrical, computer, materials science, medical, and associated disciplines would, in the future, have more than a cursory interest in acquiring a basic understanding about the manner of formation of this unique material. The choice of topics covered in this book is based purely on the author's subjective perception of the needs of an undergraduate engineering student who is in the process of acquiring a basic understanding about general principles of fabric-formation systems to become capable of making a conscious choice of the type of fabric that would be most appropriate for targeted end uses.

An important dimension of a textbook is the depth of treatment of the topics handled. If a book is targeted at a very wide reader base, then the treatment and choice of topics have also to be appropriately extensive. While targeting this book at undergraduate engineering students, the author had in view the limited time that such students have in pursuing one particular subject during the period of graduation. Understandably, such students may also need in-depth insight into a particular aspect of a specific topic for pursuing a certain developmental work. The literature listed at the end of individual chapters should provide useful leads in this regard.

All major fabric-formation systems, namely weaving, weft knitting, warp knitting, braiding, and nonwoven systems, have been covered in this book, and it ends with a chapter on triaxial, multiaxial, and three-dimensional fabric-formation systems. In its character, the last chapter is apparently somewhat different from that of rest of the book as the focus here is on the type of product. It is felt that this novel product constitutes a new class of engineering material and therefore needs to be brought to the notice of future engineers.

The reader can discern a greater weightage accorded in this book to the topic of weaving, which is partly caused by the necessity of going into salient features of weaving preparatory processes, namely winding, warping, and sizing, and partly by the complexity, flexibility, wide range, and scale of weaving machines globally in use. Each type of weaving machine, whether it be an under-picking power loom, one of the four types of shuttleless looms, a narrow fabric loom, or a circular loom, is commercially important and engaged in the production of a specific range of woven fabrics. Moreover, each of these types of weaving machines, along with multiphase looms, has a unique technological dimension and hence merits inclusion in a book of this kind. This same argument could also be applied for flat-knitting and sock-knitting systems, both of which have been bypassed in this edition while dealing with the topic of weft knitting. It must be underlined, however, that this act of omission is guided by the author's perception that weft-knitted constructions are yet to attain desired dimensions needed for developing a range of engineering materials although some technical applications of weft knits, such as in compression garments, biosensors, or spacer fabrics, are fairly well established. Indeed, knotted constructions are, in this sense, more important, and it is hoped that the system of knotted fabric formation would find due place in future editions.

Quite clearly, decisions that the author had to make in terms of choice of topics and depth of treatment are open to scrutiny by the readers. It is hoped that this first edition would provoke a constructive and rational response that would provide guidance for future modifications and alterations. Moreover, in spite of sincere efforts to avoid mistakes, many must have remained undetected by a lone pair of eyes. I am sure that a larger body of readers would be able to spot such errors more easily and bring them to the notice of the author.

Prabir Kumar Banerjee

Acknowledgments

The author would like to use this opportunity to pay homage and respect to the late Prof. Dr.-Ing. Heinz Hollstein of TH, Karl-Marx-Stadt of the erstwhile GDR, currently TU, Chemnitz of Germany, who guided him into the world of academics during the doctoral program and whose two-volume book *Fertigungstechnik Weberei* has been extensively used in preparing a part of this book. The author is also grateful to Prof. Deepti Gupta and her postgraduate student B. Shanmugam as well as Prof. Kushal Sen of IIT, Delhi for the material support provided by them during preparation of the manuscript and to Khasti Pujari for taking immense pain in preparing the diagrams of this book. A special mention must also be made of Mriganka Sekhar Naskar who permitted the author a free run of his establishment for studying closely and collecting critical information on narrow fabric looms.

The author gratefully acknowledges permissions granted by M/S CCI Tech, Inc. of Hong Kong, M/S CRC Press of Florida, M/S ITEMA S.p.A. of Italy, M/S Karl Mayer of Germany, M/S Lohia Starlinger of India, M/S Savio Macchine Tessili S.p.A. of Italy, The Textile Institute of Manchester, and The Textile Machinery Society of Japan for use of copyrighted diagrams in this book.

And last, the author would like to put on record his appreciation of Dr. Gagandeep Singh of CRC Press for his patience and perseverance in following up preparation of the manuscript.

Author

Prabir Kumar Banerjee, born in 1946, obtained his bachelor's degree in textile technology from the College of Textile Technology, Serampore, West Bengal in 1966; his master's degree in textile engineering from IIT Delhi in 1973; and his doctorate from the erstwhile TH, Karl-Marx-Stadt of GDR, which is currently the TU, Chemnitz of Germany, in 1979.

He worked from 1966 to 1971 in fabric manufacturing sectors of the Indian textile industry, and between 1980 and 2011, he taught at the Department of Textile Technology of IIT Delhi, focusing primarily on fabric-formation systems and technical textiles. During 2002–2003, he worked as director of the Indian Jute Industries Research Association, and during 2007 to 2009, he headed the Department of Textile Technology of IIT Delhi. During 2014, he taught for one semester at EiTEX of the University of Bahir Dar, Ethiopia.

He has guided nine PhD students in the fields of weaving, knitting, braiding, nonwovens, and geotextiles, published 36 papers in refereed journals, developed eight new products and processes, and holds three patents. He has developed e-modules on *knitting technology* and *shuttleless weaving* on behalf of the NPTEL program of MHRD, Government of India, and has contributed a chapter on "Environmental Textiles from Jute and Coir" in *The Handbook of Natural Fibres,* published by Woodhead, UK, in 2012. He is a recipient of the lifetime contribution award from the International Geosynthetics Society (Delhi Chapter) and the Central Board of Irrigation and Power of India for contribution to the growth of geosynthetics in India.

1

Textile Fabrics: An Overview

A textile material is made from fiber or other extended linear materials, such as thread or yarn. Conventional textile fabrics are usually planar sheets displaying the unique character of drape, which is the result of simultaneous large deformation in bending and shear modes. Textile fabrics also exhibit controlled in-plane and cross-plane fluid flow and moderate to very high resistance to tensile deformation, primarily along its two principal directions. Such fabrics belong to a class of materials typified by their compressional properties, for example, bulk compression or planar compression associated with high compressional resilience, their thermal and electrical conductivities, and also texture. Texture of textile materials is a cumulative effect of some surface properties, namely roughness, light scattering, compressibility under low pressure, surface tension, and topology.

With the ever-widening application of textile fabrics from common apparel to building blocks of a flying machine or of a human body, new techniques of assemblage of fibers and yarns have evolved and are evolving. As a result, many characteristics of conventional textile materials are found wanting in such new fabrics. For example, the quality of drape would be neither measured nor possibly be present in a 3-D textile fabric meant for replacement of a part of the human skeleton. Indeed, more often than not, such novel fabrics are highly skewed in the sense that many attributes of conventional textile materials would be totally absent, and only the set of desired attributes would dominate. Textile fabrics in the modern age are fast evolving into multidimensional materials.

Textile fabrics (henceforth termed only as fabrics) are primarily produced from fibers, which are long-chain polymers. A textile fiber, the basic building block of a fabric, is anisotropic in nature, and its properties along its length direction decisively affect properties of the intermediate (fibrous mat, yarn) as well as of the final product (fabric). Many fabric formation systems—such as weaving, knitting, braiding, and netting—require the linear discontinuous fibers to be first assembled into a linear but continuous form (yarn), which is subsequently converted into fabric. In the case of continuous fibers, namely filaments, an assembly of a multitude of such units into a suitable multifilament form is usually necessary for said fabric formation systems. However, a host of other (nonwoven) systems, such as needle punching, spinlacing, spunbonding, or stitch bonding, require another form of fibrous assembly, known as mat, web or batt, for production of the final fabric. A mat or batt is planar in nature and forms a more logical intermediate stage than the linear yarn for production of a planar fabric. Indeed, there are examples of fabrics with a pronounced third dimension, termed 3-D fabrics, some of which are being produced commercially, such as 3-D knitted Spacer fabrics, and some are in various stages of development, such as the triaxial three-dimensional 3A–3D fabrics for replacement of cartilage or intervertebral discs. Logical thinking suggests that developing 3-D fabrics might be easier from a 2-D fibrous assembly than from a linear assembly of fibers. However, most of the 3-D fabrics that are currently being developed, somewhat oddly, employ yarns and not fiber mats as the feed material.

If a fiber or a yarn is assumed to be a right circular cylinder of diameter d microns (10^{-6} m) having density ρ g/cc, then the mass in grams of a 9000-m length of such a cylinder—which is its measure in deniers—can be expressed as

$$[\{\pi d^2/4\} \times 10^{-12}] \times 9000 \times \rho \times 10^{6} \tag{1.1}$$

Thus, a typical 1-denier polyester fiber of density 1.38 g/cc and a 30 Ne—approximately 180 denier—cotton yarn of density 0.7 g/cc would yield diameter values of about 10 microns and 190 microns (0.19 mm), respectively. Considering the entire range spanned on one extreme by microdenier fibers, such as those produced by the electrospinning method (100 to 1000 nm) to thick coir yarns of approximately 100,000 denier, it transpires that fabric formation systems are called upon to deal with raw materials as thin as 0.1 micron to ones as thick as 5 mm, a span of nearly four orders in magnitude.

The domain of fibers, yarns, and fabrics in terms of thickness in microns is illustrated in Figure 1.1. It is observed that there is some overlap between the 2-D and 3-D fabrics in so far as thickness is concerned. Indeed, a fabric such as terry cloth or a carpet is fairly thick on account of loop piles or cut piles staying erect on the fabric plane. Similarly, solid woven belts produced on a specially designed belting loom or, for that matter, a doormat made of

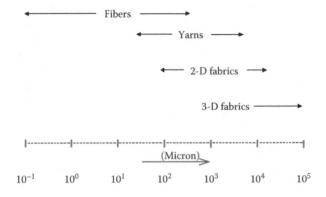

FIGURE 1.1
Range of thickness of commercial fibers, yarns, and fabrics.

coir yarns and produced on a handloom has a pronounced third dimension. Indeed, many nonwovens are also fairly thick although these are not classified as 3-D fabrics. A characteristic feature of a 3-D fabric is that it should be constructed from building blocks that are oriented in all three mutually perpendicular planes irrespective of the thickness of the respective fabric. Thus, the term 3-D refers more to directional orientation of the building blocks of a fabric and not to its thickness per se. The genre of commercial 3-D fabrics may also be expected to exhibit a limited range of width-to-thickness ratio although this aspect has not been explicitly spelled out anywhere. Moreover, the principles of formation of 3-D fabrics differ, at times radically, from the conventional ones, necessitating a completely separate classification of the products.

Schematic views of typical fabrics produced from yarns by weaving, weft knitting, warp knitting, braiding, and netting are shown in Figures 1.2 through 1.9. The path of yarns as well as their manner of interlacing/

FIGURE 1.2
Biaxial woven fabric.

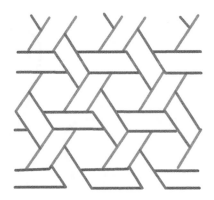

FIGURE 1.3
Triaxial woven fabric

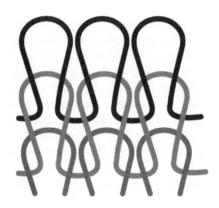

FIGURE 1.4
Plain weft knitted fabric.

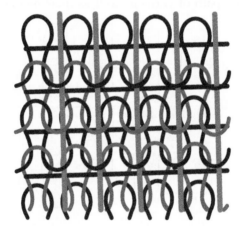

FIGURE 1.5
Biaxial reinforced weft knitted fabric.

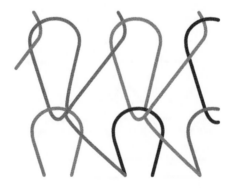

FIGURE 1.6
Tricot warp knitted fabric.

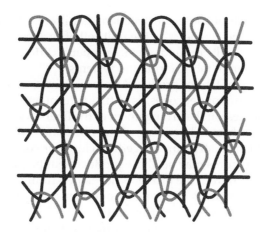

FIGURE 1.7
Biaxial reinforced Raschel knitted fabric.

FIGURE 1.8
Braided fabrics.

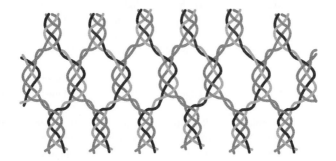

FIGURE 1.9
Bobbinet lace.

intermeshing in each of these constructions is different from those of the others. As a result, the geometry of the unit cell of each construction differs from the rest. This results in fundamental property differences between the products of the different fabric formation systems.

The cell geometry of fabrics produced from fiber mat, such as the needle-punched or the spunbonded ones, is difficult to quantify as there is an element of randomness in the arrangement of the fibers. Moreover, entanglements or fusion or chemical bonding of the fibers and even a combination thereof often hold such structures together. However complicated or indeterminate such structures may be, these products exhibit properties that are qualitatively different from the products developed from yarns. Indeed, it is often claimed that products developed from fiber mats can be easily engineered to suit specific end-use requirements.

Table 1.1 illustrates typical qualitative differences in some properties arising out of four typical commercial methods of fabric formation. Differences in properties arise not only out of the fabric formation method, but also from the raw material used. It is observed from columns 2 and 3 that weaving slit films results in the lowest porosity and thickness, and multifilament

TABLE 1.1

Comparison of Some Fabric Properties

	Woven (Biaxial)		Raschel Knitted	Needle Punched	Spunbonded
Property	Multifilament	Slit film (tape)	Yarns (biaxial)	Staple fiber	Filaments
Fiber orientation	Orthogonal	Orthogonal	Orthogonal	Distributed	Random
In-plane behavior	Orthotropic	Orthotropic	Orthotropic	Highly anisotropic to isotropic	Isotropic
Thickness	High	Low	Moderate	Highest	Lowest
Interfacial friction	High	Lowest	Moderate	Highest	Low
Initial modulus	Moderate	High	Highest	Lowest	Moderate
Breaking strength	High	High	High	Low	Moderate
Breaking elongation	Moderate	Low	Low	Highest	Moderate
Tear resistance	Low	Moderate	High	Higher	Highest
Bursting strength	High	High	High	Low	Moderate
Openings	Regular	Regular	Regular	Irregular	Irregular
Porosity	35%–45%	10%–15%	30%–40%	65%–90%	50%–80%
In-plane flow	Low	Lowest	Low	Highest	Moderate
Puncture resistance	Very high	High	Very high	Moderate	Low

weaving results in moderate porosity and reasonably high thickness. On the other hand, it is also possible to achieve similar fabric properties employing two very different formation methods. This is evident from the comparison of biaxial weaving and biaxial Raschel knitting products, listed in columns 2 and 4 of Table 1.1. In such an event, the choice of fabric formation system would be primarily governed by considerations of process economics. The important role that a different fabric formation system plays is highlighted by columns 5 and 6, which demonstrate that isotropicity, a very important requirement of material for engineered products, is easily achieved with both nonwoven methods cited here.

2

A Brief Outline of Various Fabric Formation Systems

2.1 Weaving System

Two sets of yarns, namely warp and weft, are needed for the formation of a woven fabric. Warp yarns are usually stronger and more compact than the weft or filling yarns. In view of its relative softness, the weft yarn is expected to flatten out in the woven fabric and fill up the open space that arises out of gaps between yarns. Hence, the weft yarn is also termed as the "filling" yarn. Warp is supplied in the form of a sheet, wrapped around a barrel that is known as a weaver's beam, to a weaving machine called a loom. A warp sheet is stretched along the length of a loom and taken through various guiding elements to the weaving zone. The weft yarn is supplied in a suitably packaged form, which is carried to the weaving zone by a shuttle. A shuttle passes through splits created in the warp sheet, dragging the weft yarn along.

The weft yarn is trapped in the warp sheet once all yarns in the sheet are brought together to a common plane. Subsequently, the warp sheet is split again but in a different manner compared to the previous cycle, and another segment of weft yarn is trapped. This process of sequential trapping of weft segments across the warp sheet is known as weaving. The fabric produced in such a manner is carried away from the weaving zone and wound onto a cloth roller.

Evidently, warp yarns remain aligned along the length of a woven fabric while weft yarns are oriented along its width direction. Accordingly, a woven fabric is basically a biaxial assembly of warp and weft yarns.

The weaver's beam, as also the weft package, is a product of a preparatory process, and yarns, the basic raw material of a weaving system, are a product of yarn formation systems, and they are usually packaged in the form of bobbins.

Each bobbin contains a limited length of yarn (for example, 2000 m) that contains many objectionable faults. Such faults need to be removed, and the resultant clean yarns from the supply bobbins need to be joined together to form a package of suitable dimension containing a sufficiently long length of yarn (for example, 100,000 m).

A very important preparatory process forms the interface between yarn formation and fabric formation systems, and it is known a winding. For stand-alone yarn and/or fabric formation units, winding is the final process of the yarn manufacturers, but for integrated units, quite often, winding machines are situated in the preparatory section of fabric formation departments. A commonly encountered wound package is known as cone because geometrically it is a frustum. Yarns from a number of cones are converted into a warp beam by assembling them first in a warping machine and then applying a layer of protective coating to each yarn on a sizing machine. This protective coating enables warp yarns to withstand weaving strain. The product of the sizing process is a sized beam, which is taken to the drawing-in or entering and knotting section for taking each warp yarn through the required guides. Fully equipped with these guiding elements, a sized beam of warp yarns, termed a weaver's beam, is taken to a weaving loom. Driving elements of a loom operate the warp beam as well as other yarn-guiding elements. The process of linking the driver and driven systems is known as beam gaiting.

A weft package is also produced from a cone containing the weft yarn. Such a cone may be taken directly to a loom, on which guiding elements unwind yarn from the cone and carry the same across the warp sheet. Alternately, small packages known as pirns may be made from the cone on a pirn-winding machine. A pirn is very similar to a bobbin in appearance. However, the geometry of yarn coils in a pirn is quite different from those encountered in a bobbin. Moreover, as opposed to yarn in a bobbin, the yarn in a pirn is devoid of all objectionable faults.

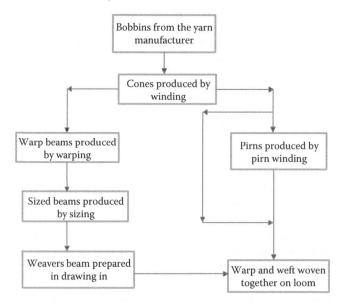

FIGURE 2.1
Material flow in the weaving process.

The material flow in a weaving process is illustrated in Figure 2.1. It is observed that, after yarn formation, an elaborate process of yarn preparation is necessary for weaving a fabric. Indeed, the yarn itself is, more often than not, assembled from discontinuous fibers in a still more elaborate yarn-formation process. Thus, a very large lag time is inherent in this textile system of conversion of fibers to yarns first followed by conversion of yarns to fabric. Moreover, the large number of operations contributes to a high cost of production.

2.2 Knitting System

For producing a knitted fabric, one usually needs either warp or weft yarns although both warp and weft yarns can also be employed simultaneously for developing some special knitted fabrics. The technique of producing fabrics by employing only yarns that resemble weft as used in the weaving process is known as weft knitting, and the technique of converting a sheet of warp yarns resembling a warp sheet of the weaving process into a knitted fabric is known as warp knitting. The manner in which a yarn is fed to knitting needles for producing a weft-knitted fabric of four wale lines A, B, C, and D is illustrated in Figure 2.2 whereas Figure 2.3 demonstrates the corresponding process for warp-knitted fabric. In both types of fabric, a loop of yarn is the basic building block, generated by needles and associated knitting elements. A knitted fabric is generated by an arrangement of loops in the form of a

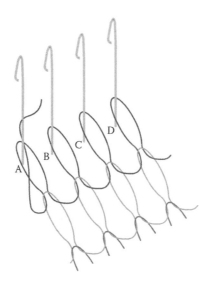

FIGURE 2.2
Yarn path in a weft-knitted fabric.

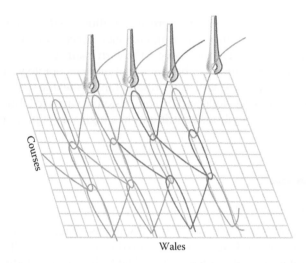

FIGURE 2.3
Yarn path in a warp-knitted fabric.

matrix. The geometry of a weft-knitted loop is different from that of a warp-knitted loop. An example of these two types is shown in Figures 2.4 and 2.5. A typical loop shown in Figure 2.4 is made up of one needle loop "a", two arms "b", and two legs "c". Over the contact zone "d" of its two legs a loop gets intermeshed with a neighboring loop in the same wale line.

It is important to note that the geometry of a loop and its manner of arrangement in the resultant matrix determines the structure of a knitted

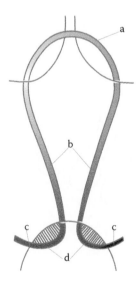

FIGURE 2.4
Geometry of weft-knitted loops.

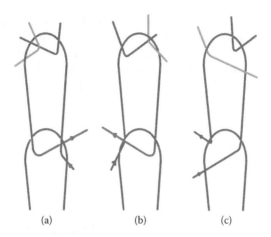

FIGURE 2.5
Geometry of warp-knitted loops. (a) Technical front side of a closed loop. (b) Technical back side of a closed loop. (c) Technical back side of an open loop.

fabric and therefore its properties. The difference in loop geometry of warp- and weft-knitted loops is caused by the manner in which yarn(s) is (are) fed to the needles, which are responsible for manipulating these yarns into loops and, at the same time, intermesh these to form the desired matrix.

In a weft-knitted matrix, shown in Figure 2.2, the path of yarn is across the length of fabric, a situation akin to the manner in which weft is supplied to a weaving loom. A warp-knitted fabric is, on the other hand, developed from a very large number of yarns, which are fed along the length direction of the fabric. This situation is analogous to feeding of warp sheet to a loom.

As a result, the material flow in a knitting system can be represented by Figure 2.6. A comparison of Figures 2.1 and 2.6 reveals that the yarn-preparation

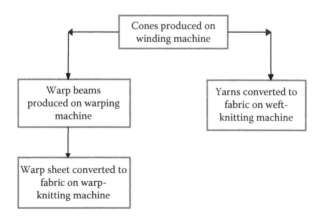

FIGURE 2.6
Material flow in the knitting process.

phase is less elaborate for the knitting system. One would expect, therefore, a reduction in lag time as well as in basic costs incurred for production of knitted as compared to that of woven fabric. Moreover, the weft-knitting system offers the unique facility of producing a fabric from only one strand of yarn. This facility is frequently made use of for characterizing the performance properties of yarn.

2.3 Braiding System

Braiding is probably older than weaving and is used primarily for production of narrow fabrics and tapes. A tubular braided fabric is produced on a circular machine in which spindles carrying strands of yarns move in a serpentine manner around the periphery of a circle (Figure 2.7). In order that the resultant fabric may have an acceptable cover, each strand may be made up of a number of yarns. In such an event, each yarn strand resembles a flat tape. In the simplest system, one half of the spindles rotate clockwise along the periphery of a circle, and the other half rotates counterclockwise. When

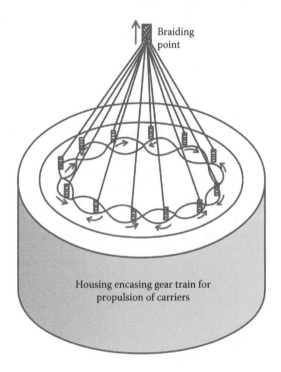

FIGURE 2.7
Braiding process.

a clockwise-moving spindle S_1 approaches a counterclockwise-moving S_2, both execute an additional radial motion whereby one, say S_1, moves away from the center of the circle while S_2 moves toward the center. Having thus avoided collision and crossed each other, they return to the periphery to continue their journeys in opposite directions. The interlacement between yarn strands carried by the spindles S_1 and S_2 result out of their relative rotation about each other at locations when planes of these spindles are at their farthest.

The braiding point, at which yarns from braiding spindles merge and get converted to fabric, is analogous to the fell of a woven cloth, at which warp and weft yarns merge. In the resultant fabric, at the braiding point, the strand of S_1 would appear above (or below) that of S_2 forming an interlacement very similar to that of a woven fabric with the difference that the crossing threads in a braided fabric might be highly acute or obtuse to each other. Moreover, such constituent yarns in a braided fabric are oriented in bias directions as opposed to along length and width directions in a woven fabric.

A braided fabric may also be produced in flat form by not permitting the spindles to execute a complete rotation about the braiding point. This would result in a discontinuity that manifests in a clean slit along the length of the tube. Indeed, if braiding spindles are also allowed to traverse along the hollow space enclosed within the circular serpentine path, a solid braided structure would result.

The simplicity and versatility of the braiding process as compared to the weaving or knitting processes is thus quite evident. Yarn from bobbins delivered by yarn-forming systems can be directly wound onto braiding spindles without need of any preparation whatsoever as braiding strain on yarns is very marginal. Hence, the lag time as well as production cost can be brought down to a bare minimum. In spite of its simplicity, this process is not employed for mass production of fabrics as the product is invariably a narrow strip. With the prevailing technology, the width ratio of product to that of the machine is extremely low. Hence, one would need an enormously large machine to produce fabrics of standard width. Moreover, the extremely large number of fairly massive moving machine elements employed in guiding the interlacing yarns slows down the production process appreciably.

2.4 Netting and Lacing Systems

The principle of making a net or a lace by the Bobbinet lace-making method is illustrated in Figure 2.8. As shown in the figure, 13 vertical threads, termed warp yarns, appear to be held under tension and thus act as columns, around which 13 weft yarns are twisted in a planned manner. All these weft yarns act simultaneously on the 13 warp yarns whereby no two weft yarns engage the same warp yarn at any instant. Each weft yarn is made to twist around

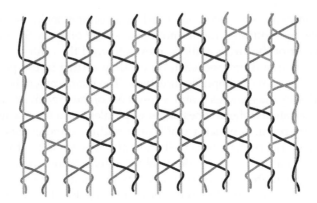

FIGURE 2.8
Principle of Bobbinet lace making.

the designated warp yarn once and then move laterally either to the right or to the left for crossing over to the neighboring warp yarn. This cycle is repeated until a weft yarn reaches the edge of the warp sheet. Here, the weft yarn would twist one and a half times around the boundary warp yarn and start moving in the opposite direction, that is, toward the other edge of the warp sheet. At any given cycle, one weft yarn out of the 13 is in the process of reversing its direction of travel while half of the rest travel toward the left edge of the warp sheet and half to the right. This leads to six crossover points of the weft yarns across the sheet of 13 warp yarns. Moreover, such crossover points alternate between gaps of the neighboring warp yarns.

When the resultant lace is taken off the machine, the vertical columns of warp yarns collapse into a zigzag shape on account of being pulled alternately to the right and to the left by weft yarns. The resultant lace material exhibits well-defined and stable hexagonal cells, illustrated in Figure 2.9.

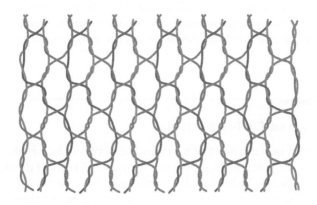

FIGURE 2.9
Relaxed lace.

In order to carry out the lacing process, therefore, two sets of yarns, namely warp and weft—very similar to the weaving process—are required. The warp yarns are unwound from a supply beam and stretched along the length of the machine into two layers. The weft yarns are individually wound onto flanged bobbins, which, in turn, are mounted on carriages. These carriages can be swung along the arc of a circular path while executing limited lateral movements through the gap between two layers of warp sheet, thus enabling them to change their relative positions. By selective transferring of carriages across a section of warp yarns and by changing this selection in a planned manner, the twisting sequence of weft yarns around the warp yarns is realized.

This process is extremely slow, complex, and energy-consuming, primarily owing to the compulsion of having to provide one carrier for each weft yarn and ensuring that a very large number of such carriers are moved in each cycle of operation through gaps between and around selected warp yarns. Hence, such systems are not encountered in commercial practice anymore, and commercial nets and laces are manufactured much more elegantly by weaving, knitting, and braiding systems.

2.5 Nonwoven Systems

The underlying philosophy in development of this system of fabric production is to bypass the large number of operations involved in assembling fibers first into yarns and then preparing the yarns suitably for the fabric-formation process. Rather, if one could just assemble the fibers or filaments into an orderly 2-D collection of a specific width and mass per unit area, termed mat, and then reinforce this intermediate product suitably, then one could logically generate fabrics of desired performance properties in a more straightforward manner. The reduction in the number of processes, which is achieved thereby, is remarkably high, which should lead to a drastic reduction in lag time as well as in the production cost. A fiber mat can be made to any thickness—as thin as a few microns, which a spider prepares, or many centimeters thick—by choosing suitable fibers and arranging them into the desired number of layers.

The orientation distribution of fibers in a mat needs to be paid particular attention to so that the resultant product has the desired performance properties. Thus, a product that should be strong in a particular direction should have most of the fibers oriented along that direction only. Similarly, for a product that should permit transmission of fluid along a particular direction, the fibers should be oriented accordingly. On the other hand, a product that should enable a fluid drop to spread very quickly over the entire fabric surface should have fibers arranged in a random manner. A mechanism that takes care of arranging the fibers in this desired manner as well as distributing this mat uniformly over the desired width continuously is the primary requirement of any nonwoven process. The nonwoven process thus renders yarn formation and yarn preparation redundant.

Some mat-forming processes have rendered even a separate stage of fiber formation redundant. Synthetic fiber-spinning systems have been modified in such a way that the resultant filaments, instead of getting wound onto packages, get randomly distributed onto a conveyor belt. In such a case, the end product of the spinning system itself is a mat and not packages of filaments. Filament-spinning processes have also been modified in such a way that staple fibers are produced by breaking the filaments into discrete segments as they emerge from the spinning systems. These fibers emanating continuously from the spinning system are subsequently deposited on a mat-forming surface.

In the pursuit of engineering new nonwoven products, interesting innovations have been taking place in the filament-spinning system itself as the same has become integrated in the production of nonwovens. In electrospinning, nanofibers with fiber diameters in the range of about 100 nm to 10 μm are being spun from a polymer solution through electrostatic force. When a droplet of polymer solution is subjected to a high electrical voltage, as shown in Figure 2.10, the charges drag the solution to form fibers, overcoming the surface-tension forces of the solution.

Fibers, both natural and synthetic, also coming in staple form, have finite length. Such fibers can be converted to a mat by mechanical and hydraulic means. In the hydraulic method, the fibers are initially made into an aqueous suspension similar to that employed in papermaking. Water is subsequently

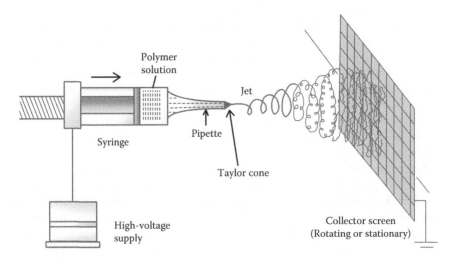

FIGURE 2.10
Electrospinning process.

drained from the slurry, and the fiber mat is generated. The mechanical method of mat formation involves processes similar to those employed in spun yarn formation up to the stage of carding.

Once the fiber mat is formed in one of the many ways described here, it has to be reinforced. Three methods, namely mechanical, chemical, and thermal, are employed either individually or in combination to form a nonwoven fabric. For example, a controlled heating of the mat might cause some of the constituent fibers to melt or become tacky at certain locations. Upon cooling, these spots would harden and thereby bond the neighboring fibers. These bonded joints impart strength to the entire assembly of fibers. One may also apply suitable adhesives to the mat and thereby bond the fibers suitably. The properties of such bonds as well as their distribution in the mat affect properties of a nonwoven fabric.

3

Yarn Winding

3.1 Objectives

Yarn produced either during spinning of staple fibers or during extrusion of filaments from polymeric materials needs to be wound into a suitable package form. Such a process of winding is not the focus of this chapter. The technology of suitably rewinding packages already formed during the spinning process—such as a ring bobbin as shown in Figure 3.1—into packages—such as a cone as shown in Figure 3.2—of desirable shape, build, and quality that would be compatible with the subsequent processes is going to be addressed in the following.

A typical ring bobbin of cotton yarn can measure 180 to 360 mm in length and be up to 72 mm in diameter containing 2000 to 4000 m of yarn weighing approximately 80 to 120 g. A cone, on the other hand, can be 90 to 150 mm tall with a base diameter of nearly 300 mm and conicity varying between 4°20′ and as high as 11°, carrying approximately 2 to 3 kg of yarn of 50 to 100 km in length.

The objectives of winding can be briefly stated as

- Removing objectionable faults present in yarn contained in the supply package
- Building a package of suitable dimensions compatible with the downstream processes

Additionally, one would also like to ensure that the surface of the yarn is not damaged in any way during the process of winding (e.g., increasing hairiness) and, if possible, even improved (e.g., through waxing). The winding of yarn must be done in such a way as to permit unwinding in the following processes with a minimum of difficulty at the required speed. Moreover, the package shape, size, and build must be most suitable for the specified end use.

FIGURE 3.1
Ring bobbin.

FIGURE 3.2
Cross-wound cone.

3.2 Package Build

Packages may be broadly categorized into two groups, namely (1) flanged and (2) flangeless. A flanged package (Figure 3.3) is equipped with two discs mounted on a hollow tube. These discs are meant to provide lateral support to the layers of yarn wound in the intervening space. The flangeless package, such as a cone (Figure 3.2), does not exhibit any such support. The build of the package, that is, the exact manner in which yarn is wound around and along the tube is therefore very critical in the case of the flangeless packages as the yarn mass has to be self-supporting. In the following, the build of a package is considered from the point of view of flangeless packages only.

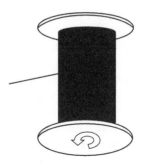

FIGURE 3.3
Flanged bobbin.

Flangeless packages permit yarn unwinding at very high speed through over-end withdrawal.

Flangeless packages can, in turn, be classified into two groups, namely (1) parallel wound and (2) cross-wound. Parallel-wound flangeless packages are invariably tapered at the two ends. As a result, a coil of yarn at the boundary layer of any diameter is supported against lateral slippage by yarn coils lying beneath this layer and just ahead of it along the package axis. In the process, however, some precious space is lost, leading to a lowering of yarn content in the package.

Parallel-wound packages may be built in three ways, namely (1) cop, (2) roving, and (3) combination build.

Cop-built packages, such as a pirn, are constructed from one end of the tube with a very short yarn traverse ℓ (Figure 3.4). After a few layers are wound over a short length of the tube and a specific thickness a is built up, the yarn-traversing mechanism is shifted by the amount b and winding continues, partly on a stretch of bare surface of the tube and partly on the previously constructed layers.

By programming a diameter-sensing system, it is possible to control the taper angle at the beginning, which is at the base of a cop. Because of the taper at the beginning, each layer of yarn wound subsequently on the body of the tube has to exhibit a slope inclined in the opposite direction. This

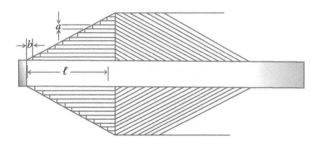

FIGURE 3.4
Initial phase in pirn building.

slope can be varied throughout the construction of the package or may be kept constant. The length measured along this slope and projected on the axis of the package is termed the chase length. A cop-built package is thus built up by continuously adding small bits of elements from the base toward the nose. The shape of the typical hump of such a package can take different shapes depending on the contour of the bare tube.

In roving-built packages (Figure 3.5), yarn is traversed at the beginning along the entire length of the bare tube. Once a certain diameter is built up, the total traverse length is reduced by a definite amount at both ends. This process is continued until the package attains the required diameter.

A combination-built package is a hybrid of the cop and roving build. In this case (Figure 3.6), the traverse at the beginning is much greater than that in the case of the cop build (Figure 3.4) although not extending the whole length of the tube. Thus, each yarn layer covers a long length of tube so much so that even while unwinding from the last few layers, the nose of residual yarn layers on the package is quite close to the tip of the tube. Incidentally, the nose of a package is closest to the unwinding device, and its base is farthest.

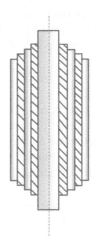

FIGURE 3.5
Roving-built package.

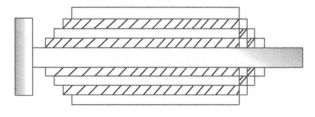

FIGURE 3.6
Combination build.

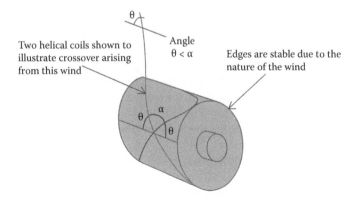

Two helical coils shown to illustrate crossover arising from this wind

Angle $\theta < \alpha$

Edges are stable due to the nature of the wind

FIGURE 3.7
Coils on a cross-wound package.

Commercially produced cross-wound packages are of two types, namely cheese and cone. All cross-wound packages have very sharp walls at the package base and nose (Figure 3.7). These walls may either be normal to the plane of the package axis or may even be curved or slightly inclined. Cross-wound packages exhibit a very large angle of wind (α) and very low number of coils per traverse. The length of traverse and the traverse speed are usually constant throughout the package build although, in pineapple-shaped cones, the traverse is slightly reduced as package diameter increases. A cheese is built on a hollow cylindrical core whereas a cone is constructed on a frustum. The taper of cone, that is, its conicity may remain constant throughout a package or may gradually rise as in packages with accelerated taper.

3.3 Unwinding Behavior

During over-end unwinding, fluctuation in unwinding tension is primarily caused by the phenomenon of ballooning. A typical profile of unwinding tension as a function of time arising out of unwinding yarn from a bobbin and rewinding it onto a cone employing a common cone-winding machine would exhibit an envelope of periodic short-term fluctuation the mean value of which rises slowly at the beginning and fairly sharply as the bobbin nears exhaustion. The unwinding tension at any instant is directly proportional to the square of balloon height and inversely proportional to the square of unwinding radius.

Considering the three types of flangeless parallel-wound packages, it can be inferred that short-term fluctuation in unwinding tension, that is, within a traverse, is primarily caused by change in the radius of unwinding point between the nose and shoulder of a cop-built package whereas change in balloon height would be the source in a roving-built one. Obviously, then the short-term fluctuation in tension in a pirn would be much lower compared to that in a tube of filament yarn. However, the problem in the unwinding of a pirn becomes progressively critical as the package nears exhaustion. On one hand, the balloon height has, by then, increased to a very high value, leading to a nearly exponential rise in tension, but on the other hand, there is a severe problem of licking of the tube by the balloon (Figure 3.8), causing an unwarranted addition in tension.

A roving-built package would also exhibit a continuous rise in unwinding tension caused by the gradual fall in unwinding radius. However, the corresponding rise in tension with a roving build would be much less than in the cop-built one. Thus, the roving-built package is advantageous from the point of view of long-term tension increase whereas the cop-built one is better in terms of the short term. Evidently, a hybrid one should provide a good compromise if designed properly. The combination-built package therefore provides a much steadier average value of unwinding tension.

Of the two cross-wound packages discussed, a cone provides a better solution insofar as unwinding is concerned. In a cone, the short-term tension variation is caused by continuous changes in both unwinding radius and balloon height whereas in a cheese the change in balloon height only should be the cause. In reality, however, there is considerable licking by yarn around the nose of a cheese, leading to sharp peaks in unwinding tension. Hence, the taper in the cone proves beneficial in suppressing these sharp peaks in short-term tension fluctuation. The long-term tension fluctuation in both cheese and cone is caused by a gradual reduction in package diameter. Hence, in this respect, there is not much to choose between the two.

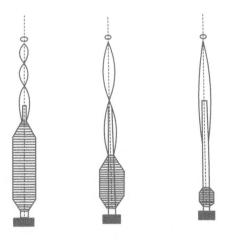

FIGURE 3.8
Licking of tube during unwinding from cop-built package.

Cross-wound packages exhibit a very large wind angle (α in Figure 3.7). During reversal of a traverse, there is an abrupt change in this angle as well. The wind angle quickly drops to a value of zero and then again sharply increases in magnitude—accompanied by a change in sense of the coil angle (θ)—over a distance of only a few millimeters near the two package walls. This too has the effect of causing a sudden and large periodic tension fluctuation in cross-wound packages.

3.4 Precision and Random Winding

A yarn package can be generated by rotating the same about its own axis and imparting a lateral to-and-fro motion to yarn at the point of winding. The traversing motion leads to reciprocation of yarn between two extreme edges along the package length. Continuous rotation of the package as well as traversing of the yarn may be achieved in a number of ways. A package may be rotated by either of two methods, namely

- By pressing the package onto a rotating drum, commonly termed surface drive (Figure 3.9)
- By mounting the package directly on a rotating spindle, commonly termed spindle drive (Figure 3.10)

In the case of surface drive, the drum rotates at a constant rpm; hence, its surface speed remains constant. Assuming no slippage between drum and package, it can be inferred that the package would have the same surface speed as that of the drum. Referring to Figure 3.9, it is observed that the frictional force $\mu.N$, acting tangential to package surface, is the prime mover of the package, where

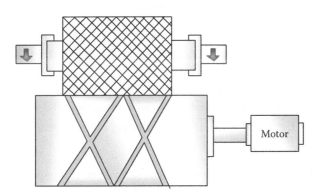

FIGURE 3.9
Principle of surface drive.

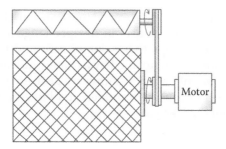

FIGURE 3.10
Principle of spindle drive.

μ is the coefficient of friction between the surface of the package and drum, and N is the normal force acting on the drum package contact line. Hence, the coefficient of friction between drum surface and yarn should be high enough to enable such a system to function; alternatively, the normal pressure should be very high. This implies that yarns of smooth and delicate surface should not be wound on surface-driven machines as the unavoidable slippage between the drum and layers of yarn would not only damage the yarn surface, but also disturb the winding process, leading to defective packages. It is to be noted that irrespective of the type of yarn, the rpm of a surface-driven package would keep on diminishing in a hyperbolic fashion with increase in package diameter.

Spindle-driven packages, however, rotate at constant rpm. Obviously, yarn surface character does not play any role here in the building up of the package. However, as the package diameter keeps on increasing, the surface speed would also rise, leading to a rise in winding speed and, hence, in yarn tension. This problem can be circumvented by either restricting the maximum package diameter to a moderately low value or by designing a closed-loop system that ensures a smooth and gradual reduction of package rpm with an increase in the package diameter. The latter solution is expensive whereas the former may prove uneconomical.

Yarn traverse on package can be carried out by

- Reciprocating a yarn guide along the length of a package, commonly termed reciprocating traverse
- Guiding the yarn by helical grooves designed on the surface of a rotating drum, commonly termed rotary traverse (Figure 3.11)
- Guiding the yarn by a series of rotating blades, termed multipede traverse (Figure 3.12)

Reciprocating traverse, encountered, for example, on rotor-spinning machines or on slow-speed nonautomatic winders as well as on pirn-winding machines, is subject to inertial problems and therefore is unsuitable for high traverse speed. The repeated accelerations and decelerations of a reciprocating mass

FIGURE 3.11
Rotary traverse by grooved drum.

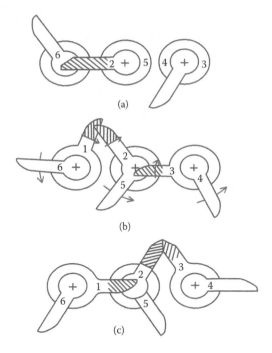

FIGURE 3.12
Principle of multipede traverse. (a) All pairs of blades in overlapped state at extreme left reversal point of traverse. (b) Opposing direction of rotation of each pair of blades. (c) Pairs of blades approaching extreme right reversal point of traverse.

not only result in high consumption of energy, but also in wear and tear of the driving elements. A typical traverse speed on a 12,000-rpm pirn-winding machine would be in the range of 40–50 m/min.

A rotating drum, on the other hand, results in rotary traverse, which should, in principle, be devoid of any jerk. The speed of rotation of the drum determines the traverse speed, and its maximum value would be governed by factors such as design of the groove, the material of the drum, the rigidity of the driving elements, etc. A typical traverse speed on a 4000-rpm drum of 1.5 scrolls with a traverse of 150 mm would be 400 m/min.

Multipede traverse systems combine the principles of reciprocating and rotary traverse, leading to the possibility of a much higher traverse speed. In such a system, rotating blades (Figure 3.12), which are synchronized properly, guide the yarn to and fro. Very little mass is involved here, and there is no change in direction of rotation either. Such systems are usually employed in very high speed winders needed, for example, on filament-spinning machines.

A winding machine based on either of the two principles, namely surface drive or spindle drive, can be equipped with any of the three traversing systems discussed.

Considering now two successive layers of yarn on a package (Figure 3.13), it is observed that the coils of one layer cross the coils of the other layer. The included angle between two crossing coils (α in Figure 3.7) can be termed as the angle of wind or wind angle. The normal range of angle of wind varies between 30° and 55°. Minimum package density is reached when the wind angle is 90° or the two crossing coils are perpendicular to each other. Under such a condition, a very large air gap would exist between adjacent crossing yarns. This degree of openness is not required even for dyeing packages for which an angle of 55° is quite adequate. Packages for warping and shuttleless weaving are usually wound with an angle of 30°. As the wind angle changes from 30° to 55°, there is normally a reduction in package density by 20% to 25%. Cotton packages for dyeing usually exhibit a density of 0.30 to 0.35 g/cc whereas those for warping and weaving have values between 0.40 to 0.50 g/cc. Besides affecting density of package, the angle of wind also affects its stability.

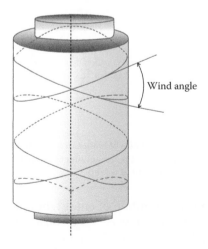

FIGURE 3.13
Laying of coils in a double traverse.

A cross-wound flangeless package owes its stability to the axial components of yarn tension, directed opposite to each other in two neighboring layers (Figure 3.14). In the event of any impactual axial disturbance, these oppositely directed forces counter the tendency of any layer to slip out of the package (slough off). Evidently, a higher angle of wind would result in a larger axial component and, hence, a greater stability. A very low hypothetical angle of wind of say 5° to 10° would result in near parallel coils, resulting in very low stability. Consider now a pirn. During shuttle checking, a pirn of 30 g yarn content can be subjected to an axial impactual force of over 50 N. This can disintegrate an incorrectly built pirn. A typical yarn layer on a pirn may contain 13 coils within a space of 3.75 cm. If winding were carried out throughout the pirn in this fashion then a wind angle of about 3° would result on the shoulder and 9° on the nose. The wind angle of a coil on a package can be calculated from the formula

$$\text{wind angle} = 2 \tan^{-1} (\ell/\pi Dw) \tag{3.1}$$

In this expression, D, w, and ℓ represent package diameter, number of coils per traverse, and the length of traverse, respectively. Reducing the number of coils by a factor of two would raise the angles to 5° and 14°, respectively. It is inferred from these figures that coils on the shoulder of a pirn are more prone to sloughing off than those near the nose. However, this sloughing-off tendency can be suppressed by periodically switching over to a smaller number of coils per traverse, aiming at a compromise between maximizing the capacity of the pirn and minimizing loss due to sloughing off. This is realized in practice by resorting to winding and binding layers.

A pirn winder is a spindle-driven machine in which the rotational speed of the pirn and the traverse speed of the yarn are constant. Cylindrical packages, such as cheese, can also be wound on spindle-driven machines,

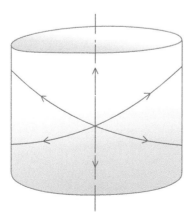

FIGURE 3.14
Axial components of yarn tension in succeeding coils.

especially when yarns with a delicate surface have to be prepared. On such systems, the wind angle goes down with an increase in package diameter, but the number of coils wound per traverse remains constant (Figure 3.15). Owing to a continuous drop in wind angle and, hence, in stability of the package with increasing diameter, spindle-driven systems are not suitable for large-diameter packages.

Surface-driven winders, on the other hand, are usually characterized by a combination of constant surface speed and constant traverse speed. This single difference in basic character results in an exactly opposite behavior of the system, namely a constant wind angle but a continuously diminishing number of coils per traverse (Figure 3.16). Thus, very large packages can be constructed on surface-driven winders without worrying about instability.

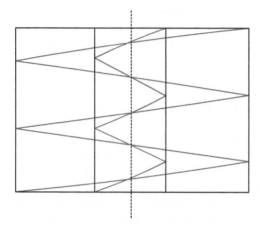

FIGURE 3.15
Winding variables as a function of package diameter on a spindle-driven package.

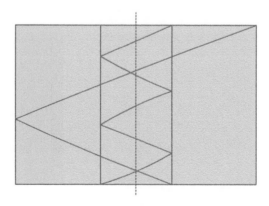

FIGURE 3.16
Winding variables as a function of package diameter on a surface-driven package.

The continuously diminishing number of coils per layer with increasing package diameter leads, however, to two critical problems, namely patterning and nonuniformity in package density.

The wind angle and, hence, the density of the package produced on a surface-driven winder with rotary traverse, that is, employing a grooved drum, is controlled by varying the drum scroll. Drums are usually made with 1.0, 1.5, 2.0, and 2.5 scrolls. With 1.0 scroll, the drum makes one revolution for one complete traverse of yarn, resulting in the lowest density and highest angle of wind. A 2.5-scroll drum makes 2.5 revolutions for one traverse of the yarn. For a drum of diameter d, length of traverse ℓ, and scroll s, the theoretical crossing angle of the grooves is

$$\text{crossing angle} = 2 \tan^{-1} (\ell/\pi ds) \qquad (3.2)$$

The corresponding expression of wind angle on the package would be

$$\text{wind angle} = 2 \tan^{-1} (\ell/\pi Dw) \qquad (3.3)$$

In this expression, w represents the number of coils per traverse, and D the package diameter. For surface-driven winders, the product Dw is constant and is exactly equal to the product ds.

The number of package revolutions in one traverse equals the number of coils in the corresponding layer. Hence, the quotient K between package rpm and the number of double traverses per minute yields the value of the number of coils in two layers that is in one double traverse. On a spindle-driven winder, the package rpm remains constant; hence, the number of coils in every layer remains the same. But on a surface-driven winder, the package rpm diminishes continuously as the package builds up; hence, the number of coils in a layer also goes down with an increase in package diameter. Therefore, the number of crossings between yarns of two adjacent layers would remain constant in a package built on a spindle-driven winder whereas the same would keep on falling in a package built on a surface-driven winder.

If the total number of coils in two consecutive layers, that is, the ratio K, is a whole number, then yarn returns to the same winding point at the start of the next double traverse, leading to patterning. In Figure 3.13, this number is equal to four, which is a whole number, and the starting point of the next double traverse coincides with that of the one just wound. By carefully choosing this number, it is possible to shift the winding point at the beginning of each double traverse and introduce a phase shift among pairs of yarn layers around the package axis, distributing the crossing points in the process evenly over the package surface, provided this number can be maintained at a constant value. This is possible in spindle-driven winders, which are therefore also termed precision winders. The term *gain of wind* is employed to indicate the amount by which wind per double traverse should differ from the whole number so that not only a pattern is avoided, but each yarn coil is also laid at a precise distance from its neighbors. In the event that this distance is chosen to be equal to yarn diameter, then

$$\text{gain of wind} = \text{yarn diameter}/(\pi D \sin \theta)$$

In this expression, θ is coil angle, that is, the angle subtended by a yarn coil to the package axis (Figure 3.7).

In surface-driven winders, such a control is not possible, as a result of which the yarn coils are located randomly on a package surface.

The problem of patterning on random winders is countered by taking recourse to anti-patterning motions. Of the many commercial solutions, the concept of a computer aided package (CAP) and step precision winders are quite novel. Keeping in view the fact that anti-patterning is basically aimed at avoiding critical values of K during package build-up and that the continuous drop in value of K is due to a continuous drop in package rpm, the obvious solution would be to reduce the traverse speed continuously and proportionately so that a particular value of K can be maintained over a period of time during which the winding system passes through critical values of K (Figure 3.17). Subsequently, the traverse speed may be raised back to the original value in one step, thereby climbing down quickly from one convenient value of K to another. Such winders would, however, need to be equipped with microprocessor-controlled inverters for the traversing system.

The computer-aided package-building principle is based on the observation that the effective contact point between a cylindrical drum and a conical package, which is located approximately at one third the traverse distance from

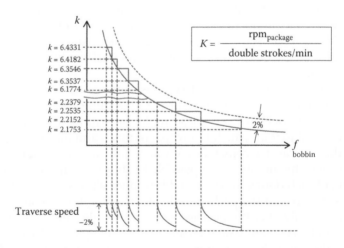

FIGURE 3.17
Wind per double traverse as a function of package diameter on a step-precision winder.

the package base, determines instantaneous package rpm. As and when the ratio of the package rpm and the traverse speed reaches a critical value favorable for patterning, the cradle holding the package against the drum is slowly rotated, resulting in a gradual shift of the contact point from the base to the nose. This process continues during the critical phase of pattern formation, holding thereby the instantaneous value of K at a noncritical level. Once this phase passes by, the cradle is swung back to its normal position, thereby reestablishing the original contact point between drum and package.

Uniformity of package is another criterion, which is influenced by the nature of the winding system. If a constant yarn length per unit of surface area is decided as the measure of uniformity, then it can be shown that the following relationship between a continuously diminishing wind angle and an increasing package diameter can satisfy the criterion.

$$D \sin (\alpha/2) = \text{constant}$$

This condition is totally violated in random winders, in which the wind angle is constant. Such packages would get progressively softer from the core to the surface. In precision winders, the wind angle does go down with increasing diameter, but the relationship

$$D \tan (\alpha/2) = \text{constant}$$

does not quite fulfill the requirement. It does appear therefore that building a large and uniform package still necessitates conceptually new solutions.

3.5 Features of a Modern Surface-Driven Cone-Winding Machine

Typical winding units of a modern cone- or cheese-winding machine are shown in Figures 3.18 and 3.19, respectively. The supply package, namely the ring bobbin, is housed in a magazine in Figure 3.18 whereas, in the other case (Figure 3.19), it is mounted on a peg. Replenishment of the magazine is a manual process whereas complete automation in feeding is achieved in the other system, which involves a conveyor system that removes the peg carrying an exhausted bobbin by one carrying a full bobbin.

An important component of the yarn supply unit is the booster/balloon controller (Figure 3.20). This device adjusts itself continuously and maintains a constant distance from the package nose. The balloon is thus kept under control, limiting fluctuation in unwinding tension. The central unit of a winding unit carries (a) tensioner, (b) splicer, (c) electronic yarn clearer, and (d) waxing unit.

FIGURE 3.18
Cone winder with magazine bobbin holder.

FIGURE 3.19
Cone winder with bobbin supplied by conveyor belt, Savio machine manual.

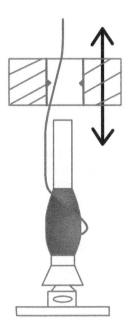

FIGURE 3.20
Booster for controlling yarn ballooning.

3.5.1 Tensioner

The tensioner of a modern machine senses yarn tension continuously and, employing a feedback loop, maintains the same at a constant level. For a simple disc tensioner, the relevant controlled variable would be spring pressure. The discs of a tensioner are also driven slowly by a motor against the running yarn, thus preventing deposition of dirt particles, wear, or even cutting of the disc surface by the yarn.

3.5.2 Yarn Splicer

In textile processes, yarns break and supply packages come to an end. There is, hence, a need to join yarns. The natural way to join yarns is to knot them, but yarn knots affect the quality of a finished fabric. Knots can also cause breakages during fabric production because a lumpy knot can get caught up in gaps between various guides through which yarns have to pass during fabric manufacturing processes. There are ways to join yarns other than knotting. One may join yarns by using wrapper fibers, glue, heat, etc. One can also very effectively splice two yarn ends and join them.

An important element in modern winding machines is the splicer. Yarn splicers come in varied forms for different types of yarns. A splicer may function on pneumatic, hydraulic, thermal, or mechanical principles. Splicers can be employed for joining single yarns spun from staple fibers or from continuous filaments. Splicers can also be employed for joining broken plied or even cabled yarns.

Pneumatic splicing is a very commonly encountered method: interlacing fibers of yarn ends by employing a blast of compressed air. In such a process, two yarns are layed side by side into a part of the splicer called the *splicing chamber*. The chamber is closed, knives trim off the waste ends, a blast of air enters the chamber, the fibers get intermingled, and a yarn joint is established. The most important reason for splicing is improving overall quality. It takes longer to make a splice than a knot, but the improvement in joint quality and in overall production efficiency can be startling.

3.5.2.1 Ends-Opposed Pneumatic Splicing

Ends-opposed splicing is the form of joint that was used when splicing technology was first developed. It makes the best-looking splices.

- The two yarns to be joined are placed into the splicer from opposite sides—one from the right and one from the left.
- When the splicer operation is activated, the two yarns are joined together by an air blast, rather analogous to that of linking the fingers of two opposing hands.
- Once the waste ends have been trimmed off, the resulting splice has a very flat form.
- The operation takes several seconds to complete.

3.5.2.2 Ends-Together Pneumatic Splicing

Ends-together splicing is a very simple form of joint. It was first used to splice yarns, which had proved to be completely resistant to normal ends-opposed splicing.

- The two yarns to be joined are placed into the splicer from the same side.
- When the splicer operation is activated, the two yarns are joined together by an air blast, rather like the fingers when two hands are clapped together, palm to palm.
- Once the waste ends have been trimmed off, the resulting splice has a distinct "tail" at approximately 90° to the thread line.
- The operation is very quick and simple, but the results are only satisfactory for some noncritical applications.

3.5.2.3 Principles of Pneumatic Splicing

The principles of pneumatic splicing are simple to explain although the process is quite complex. Direct observation of splice formation is difficult because the process occurs extremely quickly. Even high-speed photography has proved to be of limited use. Most of the conclusions about splicing performance have come from indirect observations—for example, by using tracer filaments, which can be observed and photographed. Understanding of the pneumatic splicing process is therefore based on interpretation rather than theoretical analysis.

The process of splicing is easiest to understand in the case of a simple transverse splicing chamber with a single central blast hole, splicing a continuous filament yarn with a low twist level.

The yarns to be joined are placed into the splicing chamber. This is simplified by providing the chamber with a hinged cover (a chamber pad or pad), which is normally open. When the pad is closed and the waste ends are cut to length by the integral knives, an air blast is introduced into the splicing chamber at very high speed. The air is highly turbulent, and the violent small-scale disturbances radically disrupt the arrangement of the fibers in the splicing chamber. Those fibers, which happen to lie across the opening of the air-feed hole, are separated by the direct blast. Those that lie elsewhere in the chamber are subjected to a chaotic pattern of vortices downstream of the entry point, which produces twisting and intermingling.

The resulting splice has a characteristic form. The central section, which corresponds to the air entry point, appears essentially unchanged with the fibers lying largely parallel. To either side of this central section, the fibers lie in dense clusters, highly twisted and intermingled together. Each cluster usually terminates in a small tail in which the extreme tips of the spliced yarns have not been fully bound into the structure. When a load is applied to the splice assembly, the fibers in the clusters slip very slightly until the entire structure stabilizes as the inter-fiber frictional forces take the load.

Turbulence is random, so no two splices are identical on the microscale. Nevertheless, the whole splice is much longer than the scale of the intermingling, so splices are, for all practical purposes, identical. With continuous-filament yarns, using even a very simple splicing process, very high splice strengths can be achieved, typically 90%–95% of that of the parent yarn.

Staple yarn comes in a multiplicity of forms: fiber length may be short or long; the spinning system may take one of many forms; the twist level can take almost any value; yarn may be singles, twofold, or manifold with almost any level of folding twist; and yarn may be assembled from one or many fiber types.

A splicing process acts at the level of the individual fibers. It is therefore clear that splicing of spun yarns, although conforming in general terms to the simple account, must be more difficult. Much of the development effort required for making pneumatic splicing more universal in its application has gone into the generation of specific solutions for staple yarns.

For spun yarns with their more complex construction it is necessary to present the yarns to the splicing chamber in a form that facilitates the splicing action. To this end, some splicer manufacturers have developed end-preparation techniques that subject the fibers to a preliminary treatment. Others have used novel chamber forms. Some have modified the splicing process itself.

As a general rule, the appearance of splices made in staple yarns is rather less compact than those made in continuous-filament yarns. If the yarn construction is particularly difficult, splices may be fully acceptable in terms of mechanical properties but may have fluffy "tails."

3.5.2.4 Non-Pneumatic Splicers

For certain types of yarns made from natural vegetable fibers, such as highly twisted cotton yarns, denim and open-end yarns, linen yarns, and plied cotton yarns, the strength of the spliced joint can be improved by the use of an injection splicer. In such systems, a small quantity of distilled water is added to the splicing air. The added moisture improves pliability and enhances cohesion of the constituent fibers. Thermosplicers are normally used for yarns made from animal fibers or blends of these with man-made fibers with which the added heat energy serves a similar purpose.

For splicing of short staple compact yarns and yarns blended with elastomers—of both S and Z twists—mechanical twinsplicers can be employed. The principle of functioning of this system is depicted in Figure 3.21. In the first phase, two pieces of broken yarns are placed between a pair of discs along their diameters. Rotating one disc in a direction opposite to that of the other disc leads to generation of oppositely directed torques at the two ends of the imaginary diameters. This means that, depending on the direction of twist in the yarn and the directions of rotation of the two discs, the two ends of the two yarn segments would either be twisted or untwisted. By choosing the rotational direction of discs conforming to the direction of yarn twist, the two pieces of yarn are, in the first phase, untwisted and straightened. In the second phase, the prepared central part is moved in and their tails pulled out. In the third phase, the yarns are matched and re-twisted. Subsequently, the discs are opened and the splice extracted.

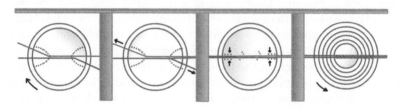

FIGURE 3.21
Principle of functioning of a mechanical splicer, Savio machine manual.

Such splicers work on purely mechanical principle. Accordingly, yarn segments that are jointed remain under positive control during all three phases. The untwisting–re-twisting action of yarns is done between two interfaced discs with self-compensating properties so that they can adapt to yarn diameter. Three critical settings, namely untwisting, re-twisting, and straightening, influence the quality of the joint.

3.5.3 Yarn Clearers

After being taken through a splicer, yarn is scanned by an electronic clearer in which it gets checked for quality of splice, long defects, repetitive defects—such as thick and thin places—and foreign fibers as well as vegetative matters. Yarn passing through a clearer at a speed of 1000–1500 m/min is scanned either for its dielectric properties by a sensor working on capacitance principle or for its optical width and reflectivity by optical sensors.

Capacitance is the property of a capacitor to store energy in the form of an electric field between two conducting plates. The dielectric property of the medium in the intervening space between conducting plates affects the capacitance. All other factors being equal, greater permittivity of the dielectric gives greater capacitance. Glass, for instance, with a relative permittivity of seven, has seven times the permittivity of pure vacuum. Consequently, glass will allow for the establishment of electric field flux seven times stronger than that of vacuum. Capacitance of homogenous dielectrics in a uniform dielectric field is generally proportional to the mass but is also a function of frequency and, for materials like cotton, additional substances (e.g., moisture) and irregular geometry and shape may affect capacitive measurement results.

Optical sensors employ suitable light sources, reflectors, and receivers for recording the extent of light scattered and absorbed with and without the yarn in its path. Such a system can measure the width of the yarn and, from the nature of light reflected, can detect the presence of a foreign substance in the body of the yarn.

Assuming a circular cross-section of yarn and a certain degree of homogeneity of the body of twisted fiber assembly, it can be deduced that mass per unit length of yarn is proportional to the square of its diameter. Hence, the sensitivity of the capacitance system, which measures variation in mass per unit length, is significantly higher than that of the optical system, which measures the variation in the optical diameter of yarn. For example, the doubling of yarn mass (100% increase) would result in a diameter increase of 42%, all other factors remaining the same.

In spite of its higher sensitivity, the capacitance system suffers from two drawbacks, namely the sensitivity to moisture and inability to isolate signals for hairs/loops and foreign matters from that of the core yarn body. In this regard, the optical system has been found to be superior.

Yarn fault can be grouped into two categories: deviation of the mass or diameter per unit length from a desired value and deviation of the light reflectance from a desired level. The first type of deviation is caused by slubs, neps, short and long thick and thin places, and hairs or loops sticking out of the main yarn body, and the other type results primarily out of contamination. The desired quality level of wound yarn would depend upon requirements of the end use, and the same can be set at the control panel of the winding machine. Whenever the threshold is crossed, a cutter comes into action, and the faulty segment of yarn is removed, the broken ends spliced, and then winding is restarted. A spliced joint whose quality has also to conform to the overall quality requirements thus replaces every objectionable yarn fault. Hence, the clearer is located after the splicer in the path of the moving yarn such that a bad splice can also be removed and replaced by a proper one.

3.5.4 Yarn Waxing

Waxing is required for yarns meant for weft knitting and similar applications in which the package of wound yarn is directly converted to fabric and yarn coefficient of friction plays a major role in the fabric-formation process. The yarn-waxing unit should logically be the last element in the yarn path as only the cleaned yarn needs to be waxed to the required degree before being wound onto the package. However, on some winding machines, the waxing unit is located even before the splicer!

In the waxing unit, a controlled pressure is applied on the block of wax, which is made to bear against the moving yarn while being slowly rotated against the direction of yarn flow for ensuring even pickup. The added wax has to be removed later on from yarn and, hence, should be emulsifiable in a normal scour bath employing common detergents. Usually waxes used for yarns have a melting point varying between 48°C and 62°C. The closer the ambient temperature of the winding section is to the melting point of the wax, the greater would be the percentage of wax picked up by the yarn. The coefficient of friction of yarn does not, however, keep on falling with an increasing amount of wax picked up. Indeed, as depicted in Figure 3.22, after

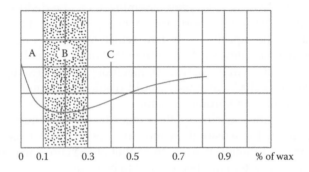

FIGURE 3.22
Variation of coefficient of friction of yarn as a function wax pickup.

an initial drop (zone A), the value of coefficient of friction remains steady over a range of wax pickup (zone B) followed by a continuous rise (zone C). The wax percentage at which the minimum value of friction coefficient is reached depends on the type of yarn, type of wax, the angle of wrap at the yarn–wax contact point, and the yarn tension as well as winding speed. For instance, a 100% increase in winding speed (from 700 to 1400 m/min) may cause a 50% rise in wax pickup (from 1.2% to 1.8%).

3.5.5 Drum Winding

3.5.5.1 Drum Drive

Each grooved drum of a winding machine is driven directly by a servomotor. Drums might be mounted directly on the motor shaft or be driven by toothed belt. A drum lap guard and drum lap brush are provided at the back of a drum to prevent development of lappers, that is, coils of yarn getting wound on the drum in the event of a thread break. Deflector plates cover the drum at the front either partially or fully. After a completed yarn joining cycle, the deflector plates guide yarn toward the center of the drum so that the jointed yarn can slide securely into the drum groove. Moreover, during bobbin runout or yarn break, the lower contours of the deflector plates guide the free upper yarn end toward the drum center.

Servomotors driving the drums are programmed to vary winding speed in such a manner that the unwinding tension in yarn being withdrawn from the supply package remains within desirable tolerance limits. As mentioned in Section 3.3, long-term yarn unwinding tension rises steadily as the unwinding point shifts from the tip of the bobbin being unwound toward its base. A servomotor can be programmed to gradually reduce the yarn-unwinding speed in a desirable manner so that the long-term tension variation is suppressed to the extent required. This feature supplements the controlled disc tensioning system, which suppresses short-term tension fluctuation.

3.5.5.2 Effects of Drum Groove Geometry

Yarn ultimately passes through a yarn trap, shown at the bottom of Figure 3.23, before it enters the yarn-traversing zone. The trap is equipped to retain the lower end of yarn in the event of a thread break. It also functions as a thread guide and forms the apex of the winding triangle, shown in Figure 3.23. The base of this triangle is formed by the projection on the vertical (say the XZ) plane of the locus of yarn element entering the drum groove. The projection of the actual path of yarn from the yarn trap to the drum groove on the three mutually perpendicular planes exhibits shift. The shift is caused by the helical groove of the drum, which forces yarn to execute a to-and-fro motion. As the path length along two sides of the triangle is the longest and along the altitude the shortest, one would expect this shift to cause a periodic tension

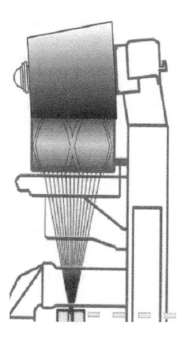

FIGURE 3.23
Passage of yarn from trap to drum.

fluctuation in the yarn segment beyond the yarn trap. The locus of yarn end entering the drum groove also traces a complex path along the XZ and YZ planes, caused by periodic changes in depth and inclination of drum groove. These shifts contribute to additional tension fluctuation, and yarn abrasion with the base and walls of the drum groove results in a harsh treatment of the yarn surface.

Beyond the drum groove entry point, a yarn segment follows a narrow curved channel within the groove before landing on the body of the package. Barring the path along the altitude of the winding triangle, this channel is always inclined opposite to the direction of movement of yarn along the base of the winding triangle. Thus, there is generally a lag between the entry point of the yarn in the drum grove and its winding point on the package body. The projection of yarn path on the XZ plane between the yarn trap and the winding point on the package thus exhibits kinks of varying extent. Moreover, wall angle and depth of groove keep on varying from one end of the drum to the other and is accompanied with variation of wrap angle between the drum and the yarn, which fluctuates between 90° and 120°. These measures are aimed at preventing yarn from jumping out from within the drum groove during the traversing motion, thereby meting out harsh treatment to the yarn. Thus, for most of its journey from the yarn trap to the package, the yarn path exhibits a highly complex configuration and is subjected to a fairly harsh treatment by the drum groove. It can therefore be stated that, in spite of the tension-controlling systems before the yarn trap, a less-ordered condition prevails at the final winding stage insofar as the nature of the yarn path and the treatment of the yarn are concerned, which constitutes serious limitations of rotary traverse systems.

The helix angle of the drum groove, that is, the angle between the instantaneous tangent to the drum groove and the drum axis, may exhibit a continuous change from one edge of a drum to its opposite edge. The helix angle near the drum end on which the nose of a cone rests is lower than the angle near the drum end–supporting base of a cone, and this difference increases with package conicity. This variation is to compensate for the gradually increasing surface area of a conical package and, hence, the necessity of slowing down the traverse speed as one moves from the nose to the base of a cone for maintaining constant yarn length per unit surface area. The helix angle also exhibits a gradual rise in the neighborhood of the two edges of a drum. This is required for allowing a smooth transition of the clockwise spiral of the groove into the counterclockwise one and vice versa during the change in the direction of traverse. Such high angles cause a considerable rise in the amount of yarn wound near the nose and near the base of a cone, which makes them harder than the rest of the package body. To counter this feature, the drum itself is applied a traverse motion, which results in a slow lateral shift of drum relative to the package, thus spreading the extra yarn length over an area near the base and nose of a package.

3.5.5.3 Package Cradle

The package being wound is held between two adaptors of a cradle as depicted in Figure 3.24. The right-hand adaptor can be shifted laterally against spring pressure to allow the package sleeve to be gripped firmly. A sensor in the adaptor records the number of package rotations. In conjunction with a sensor recording the number of drum rotations during formation of a package, the computational system of a modern winding machine can keep yarn length variation and diameter variation in a package within a tolerance range of 2% and 1%, respectively. The cradle can be adjusted to deliver packages of conicity varying between 0° to 5°57′. Usually, lower package conicity is associated with shorter traverse. The conicity can, on requirement, be increased even up to 11°.

The right-hand package adaptor is provided with a pneumatic braking system for bringing the package to a quick halt in the event of a yarn break or bobbin runout. The cradle pressure on the package is adjustable for producing soft, medium-density, and hard packages. Increasing cradle pressure due to increasing cradle weight of the package is compensated, keeping the operational cradle pressure nearly constant throughout the package build. The cradle linkage is oil-cylinder damped, thus eliminating vibrations. The line of contact between package and drum is also maintained nearly undisturbed during the entire build of a package.

FIGURE 3.24
Side view of yarn package held by cradle on a winding drum, Savio machine manual.

3.5.6 Peripheral Features

A modern winding machine, in addition to having a number of winding units, is also equipped with a large number of support systems, which make possible winding high-quality packages consistently. These are

- A central computing station through which all important input variables are fed via touch screen and which commands functioning of the different elements, such as of the various servomotors in a synchronized manner.
- The conveyers of full and empty ring bobbins as well as of the full and empty cones and sleeves.
- The full package doffing system that patrols on tracks located above winding units. The doffer winds a nose tail on the completed package and doffs the same. The doffer also places a new cone/sleeve from the magazine into the package cradle and winds a transfer tail.
- A traveling cleaning system, which patrols the entire frame and sucks away loose fibers or yarns from various locations and deposits in a centrally located chamber.

4

Warping

4.1 Objectives

The purpose of warping is to prepare a warp sheet of desired length containing a specific number of yarns. During this process, the warp sheet is wrapped on a flanged barrel in such a manner that tension in each yarn and density of yarn mass in the cylindrical assembly are maintained within a given tolerance level throughout the body of the product known as a warp beam.

4.2 Warping Systems

A cone produced on a winding machine forms a basic supply package, which can be used

1. To produce pirns of weft for a shuttle weaving machine or spindles of yarn for a braiding machine
2. To produce beams of warp for weaving, warp knitting, or some non-woven fabric-manufacturing systems
3. Directly on a shuttleless weaving machine as weft or on a weft-knitting machine as a feed package

   Discussion in this chapter is restricted to conversion of cones of warp yarn into a warp beam of given specifications.

   Let us assume that a piece of woven fabric of 5 m width and 1000 m length is to be produced from warp yarn of 10 tex arranged 20 yarns per cm in the fabric. This would necessitate production of a beam in which 10,000 yarns are arranged parallel to each other in a sheet form. The length of this sheet has to be somewhat higher than 1000 m to account for the crimped path of warp in the final fabric as well as wastages during weaving and contraction of fabric during subsequent relaxation processes.

Let contraction of grey fabric during relaxation process = 6%

Let warp crimp in fabric = 5.5%

Let wastage of warp during conversion of beam to fabric = 0.001%

Based on the above figures, it is found that a length of 1060 m of gray fabric needs to be produced that would contain 10,000 warp yarns each of 1118 m of straight length to which an additional length of approximately 1.12 m of warp sheet needs to be added to account for a total of 1119.2 or approximately 1120 m of warp sheet in the warp beam. Such a sheet made of 10 tex yarns would weigh 112 kg only. A 2 kg cone of 10 tex yarn contains 200,000 m of yarn. Therefore, a minimum of 56 such cones would be required to produce this sheet without, of course, accounting for the wastage during conversion of these cones into the sheet.

However, 56 cones can have only 56 ends. In order to generate 10,000 ends from 56 cones, one has to take out 1120 m from each cone in the form of a sheet and then break the yarns from the cones and take out another 1120 m in the form of a sheet and repeat this process 178.57 times. The sheets produced each time, when superimposed or placed side by side, would yield the sheet of desired specifications. However, the trouble is that 178.57 is not a whole number, and therefore, the solution is not feasible. The number 10,000 has to be therefore broken up into combinations A and B that would give a realistic indication of the number of cones (A) and the number of times (B) the yarn has to be broken from a cone with each cone containing ℓ m yarn weighing g grams. The possible combinations {A, B, ℓ, and g} are {10,000, 1, 1200, 12}, {5000, 2, 2400, 24}, {2000, 5, 6000, 60}, {1000, 10, 12,000, 120}, {500, 20, 24,000, 240}, {250, 40, 48,000, 480}, {100, 100, 120,000, 1200}, and {50, 200, 240,000, 2400}. The first four combinations are impracticable as a very large number of cones each containing a very small quantity of yarn would be required. The underlined fifth and eighth combinations deserve to be considered as both have relative advantages. If the fifth solution is adopted, then the number of times the yarn has to be broken is reasonably small (20), and the eighth solution requires use of a reasonably small number of cones (50) each of commercially standard dimensions. The fifth solution can be chosen in case cones of 10 tex yarns can be employed either for production of a sufficiently large number of such fabric rolls, say 10 or more, or for production of other beams for which the chosen combination applies. In such eventualities, one does not have to keep an inventory of partially consumed cones as cones containing only 240 g material are not normally produced in the manufacturing sector. If, however, neither of these conditions applies, then the eighth solution has to be adopted. The two solutions necessitate completely different methods of warping, namely

- Beam warping for solution 5
- Sectional warping for solution 8

The use of yarns of different colors in warp may also dictate the choice between the two warping systems. A faultless warping of a number of stripes, each of a complex pattern of colors, is more easily accomplished on a sectional warping system. However, beam warping is also employed for colored warping when the designs have a substantial portion of gray yarns.

Sectional warping is also used for direct preparation of a weaver's beam when sizing is unnecessary. This applies to a high proportion of twofold colored towel pile beams.

4.3 Elements of Warping Systems

Block diagrams showing the elements of a beam-warping and of a sectional warping machine are shown in Figures 4.1 and 4.2. Although the final product of both beaming systems is known as a warp beam, the beam from a sectional warping system is complete in all respects whereas the number of warp yarns in a beam from a beam-warping system is a fraction of that targeted in the final weaver's beam. Hence, in the beam-warping route, a number of such warp beams have to be assembled during the subsequent sizing process for producing the final beam that can be sent to a weaving machine. Thus, with the fifth combination, 20 beams have to be assembled in the sizing machine. In the sectional warping system adopted for the eighth combination, 200 sections of 50 yarns each are, at first, wound side by side on a sectional winding drum, and then yarns from these 200 sections are

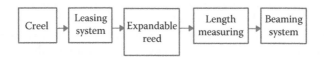

FIGURE 4.1
Elements of a beam-warping system.

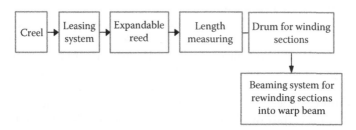

FIGURE 4.2
Elements of a sectional warping system.

unwound in a sheet form from the drum and rewound onto a beam. The beam-warping system is also known as direct warping as warp beams are produced directly from cones arranged in creel. Consequently, the sectional warping system is labeled as indirect as the warp beam is produced not from the cones in the creel but from the sections wrapped onto a winding drum.

Individual elements of the two beaming systems are considered in some detail in the following.

4.3.1 Creel

A creel is a three-dimensional assembly of pegs. Each peg is designed to grip the inside wall of the shell of a cone securely and be strong enough to support a cone in space. As yarn from a cone is withdrawn around its nose, a balloon is formed that is made to pass through a yarn guide, a tensioning system, and a thread detector (broken thread stop motion), all mounted on the frame of a creel. A creel that can house 250 pegs must therefore have 250 sets of thread guides, tensioners, and thread detectors. As each peg occupies a different location in creel, the yarn path between the expandable reed and the peg is different for each yarn. This path difference results in a tension profile of the yarns, which varies with the design of the creel.

A creel in its plan view is usually V-shaped with the yarns emerging from outside of the two arms of V and converging at the expandable reed. Each arm of the V is, in fact, shaped like a rectangle. Thus two rectangular blocks are arranged in V form in such a creel, which contains a very large number of cones supported by pegs and is therefore suited for beam-warping machines. On the other hand, one rectangular block is sufficient to serve as a creel for a sectional warping machine. However, a creel of a sectional warping machine must be equipped with a driving system for imparting a traversing motion to the whole creel during the winding of a section on the drum. Similarly, it should be possible to relocate the creel to a new position at the beginning of the winding of a new section. Indeed, the creel of a sectional warping machine has to be mounted on tracks that would permit it to be moved from one end of the winding drum to the other.

Basically, four types of creels are available for a beam-warping system. These are truck, swiveling, continuous-chain, and magazine creels. Truck creels employ reserve creels, which can be pushed in place of an exhausted creel to save on creeling time, that is, the time required to pull out empty cone shells from pegs and push the new cones in place. In case the creel size is very large and maneuvering the trucks becomes difficult, then swiveling peg assemblies are employed that can be swiveled in and out in groups. Each such assembly carries two sets of pegs, one set carrying cones from which yarns are being withdrawn by a warping machine and the other set carrying reserve cones. On exhaustion of cones, one has to swivel the pegs and bring the loaded pegs into working position, thus saving on time. During running of the machine, the exhausted cones are replenished. In continuous-chain creels, the frame of pegs carrying exhausted cones is moved by a chain drive away from the outer walls of arms of the V to its inside position while the reserve framework carrying full cones is driven simultaneously from the inside position to the outside. In all the three examples cited, time is saved in replenishing the supply packages. However, yarns from new packages have to be tied (knotted or spliced) with yarns unwound from previous packages, the tail ends of which should be around respective thread guides in the creel. This additional time is saved in magazine creels, in which individually swiveling pegs are employed and tail ends of yarns emanating from cones being unwound are tied in advance to leading ends of reserve cones. Upon exhaustion of cones, the pegs are swiveled, and the warping process is restarted immediately.

The swiveling, continuous-chain, and magazine creels carry twice the number of cones required for warping. Hence, space and accessibility become constraints if a large number of cones have to be used in the warping process. The truck creel, on the other hand, needs space for the reserve truck.

Automation has taken place in creels in terms of the threading-in of new yarns through tensioners and thread detectors. In the event of an end break, automated ladders move into the relevant position on the creel to assist the operator and lessen the machine downtime. Modern tensioners employ feedback control systems.

4.3.2 Leasing System

Leasing is a method of segregating yarns from neighboring ones and maintaining their location in a warp sheet (Figure 4.3). This assists the operator (warper or weaver) in easily locating the broken yarn and preventing crossed yarns in a warp sheet. Simple end-and-end leasing involves splitting a warp sheet into two layers of odd- and even-numbered ends. On a warping machine, this is carried out with the help of a leasing reed. A band or a

FIGURE 4.3
Simple leasing system.

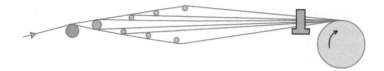

FIGURE 4.4
Multiple leasing system.

rod inserted between these two layers would show either the odd-numbered or the even-numbered yarns passing above the band or rod. In order to create a neat crossover line, another rod or band is subsequently introduced between the two layers with the order reversed. Thus, neighboring yarns are forced into complete wave cycles of 180° phase difference. Identifying or locating yarns at the crossing points of these waves is relatively simple. Leasing is also required for another purpose, namely separating yarns from each other after the warp sheet has been sized and dried. A sized warp sheet is completely encased in size film and cannot be used in the subsequent process unless yarns are individualized. If one attempts to split this sheet into odd and even yarns at one go, then the size film would be broken into segments equaling the number of yarns in the sheet. Apparently, a large energy would be consumed suddenly, and because of uneven distribution of the same across the warp sheet, stress concentrations become a distinct possibility, leading to potential rupture of yarns and peeling-off of size film. Hence, multiple leasing is resorted to in which the warp, depending on the closeness of yarns in the sheet, is split into multiple layers, say six or eight or 10. Accordingly, every sixth or eighth or 10th yarn is separated from the main body of the warp, and a lease band is introduced in the gap (Figure 4.4). After sizing and drying of this sheet has been carried out, gentle tugging at such bands would help segregate the individual layers in steps, thereby individualizing the yarns again. In case the warp needs no sizing, then a multiple leasing is not required and end-and-end leasing would suffice. Leasing has to be carried out at the beginning of the warping process, but the bands for multiple leasing are introduced in the sheet either at the beginning (sectional warping, beam to be sized) or at the end (beam warping, beam to be sized). In either case, the leading edge of the sheet of a warped beam would exhibit the multiple lease (Figure 4.5). Automatic leasing systems enable quick and fault-free leasing, especially for multiple leasing.

4.3.3 Expandable Reed and Length-Measuring System

A warp sheet from a creel converges into the expandable reed through the dents of which warp yarns pass. As the reed wires can be moved either closer to or away from each other, the distance between dents can be adjusted to the desired value of yarn spacing in the warp sheet. Thus, the desirable density of the warp yarn sheet is set at the expandable reed.

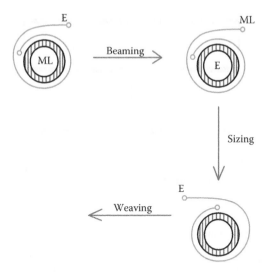

FIGURE 4.5
Leasing on a sectional warping system.

The length of warp sheet in a beam is crucial from the point of minimizing wastage, and the length-measuring unit forms therefore a very important component of the beaming system. The measurement methods and their resolutions vary over a wide range. In its simplest form, the moving warp sheet is taken around a freely mounted roller whose angular displacement can be translated both into yarn speed and yarn length by a suitable electronic counting system.

4.3.4 Drum for Winding Sections

The drum on a sectional warping machine is a hollow cylinder with a conical flange on one end. The leading edge of a section of warp sheet is hooked on the drum surface at an appropriate location, and then, as the drum rotates, the section starts getting wrapped on the drum surface. As the winding continues, the number of layers of warp on the drum starts increasing, raising steadily the effective instantaneous diameter of the drum at which a layer is being wound. The diameter of a bare drum being around 0.8 m (for circumference = 2.5 m) to 1 m (for circumference = 3.14 m), a substantial length of a warp section can be wound on a drum without causing a significant change in the effective drum diameter and, hence, in the speed of winding even if the drum rpm is kept constant. The speed of winding in a modern warper can be as high as 1000 m/min.

If the layers of a section were allowed to be wrapped on the same location of the drum, then, after a while, the sectional view of warp sheet would appear like a rectangle, and the two edges of the warp sheet would have a tendency to collapse in the absence of any lateral support (Figure 4.6). To prevent this

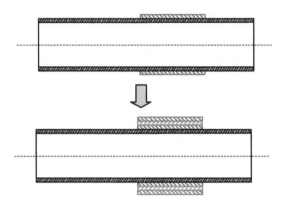

FIGURE 4.6
Section winding without a flange.

from happening, every section is given a lateral shift or a traverse during the winding process such that the cross-section of warp sheet wound on the drum appears like a parallelogram instead of a rectangle (Figure 4.7). The conical flange of the drum supports one edge of the parallelogram, and the other edge appears tapered. This tapered edge needs no support and indeed provides support to the inclined edge of the next section. The conical flange of the drum therefore plays a vital role in sectional warping systems.

Winding drums are manufactured either with fixed conical flanges or with flanges that can be adjusted for variable conicity. The conicity angle of the fixed flanges is offered in the range of 7° to 14° while the conicity

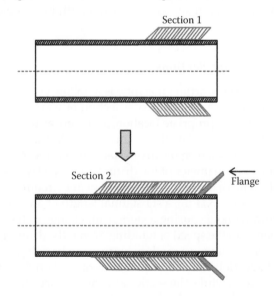

FIGURE 4.7
Section winding with flanged support.

of the other type can be varied over a much wider range. In the variable-conicity drums, plates hinged at one end on the drum body are arranged along the circumference of the drum. By raising the free ends of the plates simultaneously away from the drum by an equal amount, the conical surface is created. This surface, however, would not be as smooth as the surface of a fixed conical flange. Moreover, as one moves away from the hinge of the plates toward their tips, the gaps between the edges of neighboring plates keep on increasing. These two factors result in stress concentration on yarn segments—explained subsequently with the help of Figure 4.9—in contact with the discontinuous flanged surface as well as in a difference in length between yarns wound on the flange and those on the main body of the drum at the corresponding diameters. As a consequence, fixed-conicity flanges are preferred for delicate and sensitive yarns. However, the variable-conicity system is more versatile insofar as processing a wide range of yarns is concerned as would be obvious from the following deduction.

In Figure 4.8, each parallelogram represents the cross-sectional view of a section of warp sheet built on the winding drum. The area enclosed by any parallelogram is given by

$$s[\{x^2 + b^2\}^{1/2}]\sin \alpha \tag{4.1}$$

In this expression, s represents section width, x the traverse imparted to sections, b the thickness of the resultant yarn sheet on the drum, and α the flange conicity.

The length of a warp sheet of thickness b wrapped on a drum of bare diameter d can be expressed as $[\pi(b + d)]$. Hence, the volume of yarn sheet contained in a section can be expressed as

$$[s[\{x^2 + b^2\}^{1/2}]\sin \alpha] [\pi(b+d)]$$

$$= [sx\{1 + \tan^2 \alpha\}^{1/2} \sin \alpha] [\pi(b+d)] \tag{4.2}$$

$$= \pi s x (b+d)\tan \alpha$$

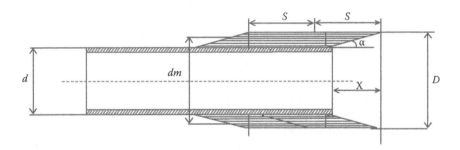

FIGURE 4.8
Geometry of a sectional view of a drum with sections of warp.

The width of a section *s* is given by warping requirements (desired number of ends per inch and total number of ends in a section) and cannot be treated as a variable while adjusting the settings of a winding machine. Moreover, for obvious reasons, it is desirable to maintain as small a value of *b* as possible. Thus, with fixed-conicity flanges, one has only one variable, namely the traverse to be adjusted for adjusting the volume to be wound, while with variable conicity drums both traverse and the cone angle can be varied.

The factors affecting traverse are yarn diameter and the required volume of warp sheet on the drum. Clearly, thicker yarns require a section to be traversed by a higher amount. Thus, with fixed *b* and α, the volume constraint may yield a value of *x*, which may not be in agreement with the thickness requirement of the yarn. Hence, if a wide range of yarn thickness is to be processed, then the variable-conicity flange is advantageous.

The value of *b* is governed by thickness of yarn and yarn spacing as well as tension applied to a section during winding. Higher yarn thickness, lower yarn spacing, and lower yarn tension result in a higher value of *b* for the same length wound. Thus, a database is needed to relate a warp length of given specification of warp and winding tension with *b*. Nevertheless, the relation $b = x \tan \alpha$ implies that, in the event of a fixed value of α, the traverse has to be accurately adjusted to get the exact value of *b*. In the event of flexibility in choosing α, the accuracy of setting would naturally be better.

The winding tension, as discussed previously, affects the value of *b* and, therefore, the values of *x* and α. In sectional warping, the importance of winding tension has another dimension. Normally, the first section is wound from cones having the highest diameter, and with every section, the diameter of cones keeps on diminishing. Because of the ballooning effect, the unwinding tension keeps on increasing as the cone diameter diminishes. Thus, the successive sections on a drum would be wound under progressively higher tension, resulting in a progressively diminishing *b* from the first section to the last. As the value of *x* is the same for all sections, different taper angles of the supporting side of each section would be the consequence, creating instability in the warp sheet on the drum. Moreover, during the next process of beaming-off in which all yarns of the assembled warp sheet are wound onto a beam, a large tension difference across the sheet would be encountered. Such a beam would be difficult to size or weave. Hence, tension-leveling systems employing feedback-control loops are installed on creels of modern sectional warping systems. Such systems employ servomotor-controlled systems and maintain unwinding tension in yarns within tolerance limits. Thus, during winding sections on a drum, the length of yarn as well as yarn tension are monitored closely so that these two variables remain within tolerance limits for each section.

As mentioned earlier, drums with fixed conicity are preferred for winding delicate yarns, that is, yarns with low modulus and high extensibility. This point is elaborated with the help of Figure 4.9.

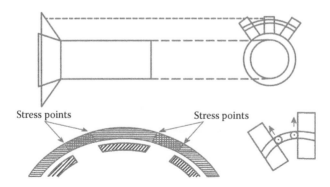

FIGURE 4.9
Variable winding conditions with a variable-conicity flange.

The conical flange of a variable-conicity drum is made up of fin-like bars, each of which is hinged along the periphery of the corresponding drum end. Rotating these bars about their hinges changes the flange conicity. Two such positions are shown in the figure. As the bars are swung up for increasing the conicity, the distance between adjacent edges of any two neighboring bars grows as one moves from the base to the tip of a bar. The segments of yarn lying on these bars would follow a curved path, and those lying in the gap between adjacent bars would be straight. Hence, the net circumferential length along the bars would be always less than the one corresponding to a circle of equivalent diameter. Accordingly, within each yarn layer wound above the base of the bars, a length difference would be observed between yarns wound on the flange and those wound on the solid drum surface. Moreover, this length difference would grow as one moves to a higher diameter. This length difference between yarns of the same layer would result in permanent deformation of the shorter yarns during the subsequent beaming process. Over and above this deformation, the curved segments of a yarn are supported along their lines of contact with the bar surface while the intervening straight segments are not supported at all, causing pressure points to develop at the edges of the bars where the curvature changes abruptly from a finite value to zero. The larger the gap between adjacent bars, the more pronounced would be these zones of stress concentration.

The quality of sections wound on the drum is also affected by the direction of rotation of the drum. The drum in Figure 4.10 rotates in a counterclockwise direction, and the yarn sheet coming from the left is layed to the winding-on point on the drum surface by a guide roller. The guide roller and the winding-on point are quite close, and therefore, the intervening segment of yarn sheet is constrained to follow a definite path. The top of the sections being wound is also fully visible and accessible for any intervention, which would not have been the case had the drum been rotating clockwise. Furthermore, the automatic insertion of a size-splitting element during

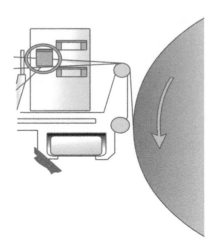

FIGURE 4.10
Rotational direction of drum.

the multiple leasing process is facilitated by the location of the guide roll. If the drum were to rotate in the clockwise direction, then the corresponding guide roll would have needed shifting during the leasing process, causing a disturbance to the threads.

A critical aspect of warping relates to the interdependence among the distance between the creel, the speed of winding, and the braking torque that can be applied on the drum. When yarn breaks during the winding process, the drum has to be brought to a stop within less than one revolution as otherwise the broken end would get embedded in the subsequent layer. Reversing the drum for finding out the broken end in the absence of any accumulating system would create entanglements among yarns and, hence, is not feasible. Thus, the larger the winding drum and the higher its speed, the more efficient the braking system has to be. Although a higher distance between the creel and the winding drum would ease the situation, it would also mean a higher space requirement.

Having wound the first section to the desired length, the yarns are cut after introducing the lease band and the end of the section securely tucked in within the warp sheet. The other end of the yarns is securely gripped in front of the reed assembly. Subsequently, the reed assembly and the creel are shifted to the starting position for the next section, the lease band inserted into the tip of the new section, and winding on the drum repeated.

4.3.5 Beaming Systems

In the indirect system of beaming, the ends of sections wound on a drum are collected in a sheet form and then wrapped onto a flanged barrel. The barrel is given a spindle drive, and the resultant tension created in the warp sheet rotates the winding drum, unwinding all the sections wound on the drum simultaneously. The process of winding the assembled sections in layers of yarn sheets onto the barrel results in the desired warp beam.

The torque required to rotate the drum is made up of the moment of inertia of the drum as well as other resistances, such as braking force applied on the drum, friction in its bearings, etc. The beaming operation is therefore carried out at a fairly low speed of around 300 m/min. Proper density of the resultant warp beam is regulated with the help of press rolls, and the angular velocity of the beam is gradually reduced with a build up of diameter. Warp beams of diameter in excess of 1 m are produced on a beaming system. As the layers in each section are wound with a traverse, a matching traverse in the opposite direction has to be imparted to the winding drum so that the edges of the warp sheet coming out of a drum remains always parallel to the flanges of the warp beam.

In a direct-beaming system, a reasonably small number of yarns, varying, for example, between 250 and 500, is pulled out of the creel and wound on a flanged barrel. As a very small resistance of unwinding is encountered in this process, the beaming speed can be in excess of 1000 m/min. The beam can be surface-driven or spindle-driven. The surface drive is supposed to result automatically in a constant winding speed, and the spindle drive necessitates an additional control system for reducing the angular velocity of the beam with a rise in beam diameter. Press rolls are employed on the beam for generating the required hardness. In the event of a thread break in a surface-driven machine, a brake must be applied individually on the beam, the driving drum, and the press rolls as well as on the measuring roll. This is more easily achieved with a thyristor-controlled driving system as prevails in a spindle drive. The modern direct warpers are therefore spindle-driven. A surface drive is also harsh for the warp, especially during starting and stopping.

Further Reading

Ormerod A (2004). *Modern Preparation and Weaving Machinery*, Woodhead Publishing Limited, UK, ISBN: 1 85573 998 4.

5

Yarn Sizing

5.1 Objectives

The main objective of sizing is to form a uniform layer of protective coating over warp yarn and lay down protruding fibers that project out of its surface (Figure 5.1a and b).

Sizing material should be easily removable during desizing. The size material should penetrate the body of yarn to such an extent that the size film gets firmly anchored. Excess penetration of size not only means an excess material consumption but also reduced yarn flexibility. A well-sized yarn has higher work of rupture than an unsized one.

The film of size surrounding a warp yarn has to be elastic in tension and in repeated flexing, and its surface should be well lubricated and possess high abrasion resistance. It should not delaminate from the yarn body during the weaving process but should come away easily during the desizing process.

5.2 Importance of Sizing

A film of size not only protects neighboring warp yarns on a weaving machine from getting entangled through formation of globules of fibers while rubbing against each other, but it also improves the work of rupture of yarn, which is crucial for withstanding weaving strains. A size recipe depends not only on the nature of warp yarns, but also on weaving conditions. A modern sizing machine can process yarns coming from approximately 25,000 ring spindles and feed 150 projectile looms of 3 m width. It is therefore evident that a typical commercial weaving unit must operate with a limited number of sizing machines, the maintenance and proper operation of which is very critical to the efficient functioning of the entire weaving shed.

From the foregoing, it can be inferred that an understanding of the nature of sizing materials as well as of the crucial process variables is central to ensuring a well-sized warp.

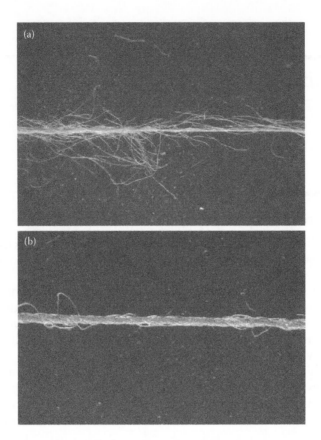

FIGURE 5.1
Views of (a) unsized and (b) sized yarn.

5.3 Sizing Material

The principal components of a sizing material are

- Adhesive
- Lubricant
- Antimicrobial agents

The components are mixed into a slurry in a preparatory section and sup-plied to the sizing box of a sizing machine in a controlled manner. Preparation of the sizing slurry varies according to the composition of the sizing material and may or may not involve heating (Ormerod, 2004). Adhesive is the most critical of the three components.

5.3.1 Desirable Nature of Bonds between Adhesives and Fiber Material

In order to hold down protruding fibers onto the yarn body, the number of bonding sites between the sizing material and the fibers constituting the yarn must be sufficiently high. The weaker the bond strength between the sizing material and the fibers, the larger the number of bonding sites must be. A sizing material should ideally have a low bond strength as it has to be disposed of after sized yarns have been woven into a fabric. Strong bonds not only cause problems during size removal, but also involve unnecessary expenditure. So a suitable optimization between the adhesive power of sizing adhesives and ease of removal has to be worked out.

5.3.1.1 van der Waals Forces

van der Waals forces are very weak forces of attraction arising from

- Momentary dipoles occurring due to uneven electron distribution in neighboring molecules as they approach one another
- The weak residual attraction of the nuclei in one molecule for the electrons in a neighboring molecule

The higher the number of electrons present in a molecule, the stronger the van der Waals forces will be. van der Waals forces are the only type of intermolecular forces operating between nonpolar molecules. For example, van der Waals forces operate between hydrogen (H_2) molecules, chlorine (Cl_2) molecules, carbon dioxide (CO_2) molecules, dinitrogen tetroxide molecules (N_2O_4), methane (CH_4) molecules, etc.

5.3.1.2 Dipole–Dipole Interactions

Dipole-dipole interactions are intermolecular forces that are stronger than van der Waals forces. They develop between molecules that have permanent net dipoles as found in polar molecules. Dipole–dipole interactions occur between SCl_2 molecules, PCl_3 molecules, and CH_3Cl molecules. If the permanent net dipole within the polar molecules results from a covalent bond between a hydrogen atom and either fluorine, oxygen, or nitrogen, the resulting intermolecular force is referred to as a hydrogen bond (see Section 5.3.1.3). The partial positive charge on one molecule is electrostatically attracted to the partial negative charge on a neighboring molecule.

5.3.1.3 Hydrogen Bonds

Hydrogen bonds occur between molecules that have a permanent net dipole resulting from hydrogen being covalently bonded to fluorine, oxygen, or nitrogen. For example, hydrogen bonds develop between water (H_2O) molecules;

TABLE 5.1

Strength of Different Types of Bonds

Type of Bond	Bond Strength (kJ mol^{-1})
van der Waal's forces	$\ll 50$
Dipole–dipole interactions	~50
Hydrogen bond	~100
Single covalent bond	~300

ammonia (NH_3) molecules; hydrogen fluoride (HF) molecules; hydrogen peroxide (H_2O_2) molecules; and alcohols, such as methanol (CH_3OH), molecules as well as between carboxylic acids, such as acetic acid (CH_3COOH), and between organic amines, such as methylamine (CH_3NH_2). Hydrogen bonds are stronger intermolecular forces than either of van der Waals forces or other forms of dipole–dipole interactions because the hydrogen nucleus is extremely small and positively charged and fluorine, oxygen, and nitrogen are highly electronegative so that the electron on the hydrogen atom is strongly attracted to the fluorine, oxygen, or nitrogen atom, leaving a highly localized positive charge on the hydrogen atom and a highly negative localized negative charge on the fluorine, oxygen, or nitrogen atom. This means that electrostatic attractions between these molecules will be greater than for the polar molecules that do not have hydrogen covalently bonded to fluorine, oxygen, or nitrogen.

Bond energies of the above-mentioned bonds along with that of a covalent bond are given in Table 5.1.

5.3.2 Starch as a Suitable Material for Sizing Cotton

Before hunting for a suitable adhesive material for sizing cotton, the chemistry of cotton fiber needs to be understood first. The bonding of chemical groups of a cellulose unit of cotton is shown in Figure 5.2. It is clear from the figure that cotton has a number of −OH groups, which are potential groups

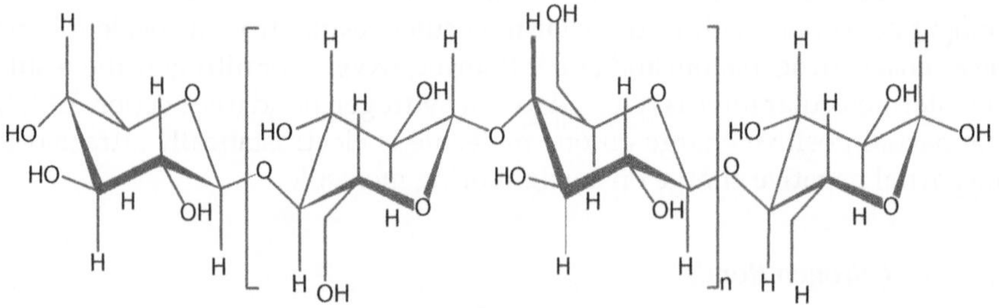

FIGURE 5.2
Organization of chemical groups in a cellulose molecule of cotton.

for H-bonding. Hence, an adhesive material for sizing cotton should have the following characteristics:

- High degree of polymerization
- Large number of –OH groups
- Ease of removal

Considering natural materials, it is observed that some polysaccharides available in abundance satisfy these criteria. These are, namely, cellulose, gums, and starches. Among these, cellulose (cotton linters and wood pulp) can't be used directly as it has to be modified to make it emulsifiable for application, and natural gums may not be cost-effective. Hence, this leaves starch as the commercial alternative.

Starch is usually composed of two components, a straight chain polysaccharide of glucose and a branched chain polysaccharide of glucose. Amylose, the straight chain component, is relatively low in molecular weight, water soluble, and makes up to 20%–30% of starch, and amylopectin, the branched chain component, is relatively high in molecular weight, water insoluble, and makes up 70%–80% of starch.

Starch alone suspended in cold water is essentially unable to act as an adhesive because it is tightly bound in granular form. The granules consist of crystalline regions of straight chain molecules and straight chain sections of aligned branched chain molecules. The crystalline regions are linked together by more amorphous areas in which the molecules are not aligned. It is these granules that must be opened up to obtain adhesive bonding.

The various processes to break up starch granules are boiling, acid treatment, alkali treatment, and oxidation. These processes basically break the interpolymer hydrogen bonds and make starch soluble.

5.3.2.1 Boiling of Starch

The simplest method of breaking up starch granules is boiling. During boiling, the starch granules first swell and then burst with coincident thickening of suspension. The temperature at which this thickening occurs is called gelation temperature. For starches in pure water, this temperature is between 57°C and 72°C. Basically, the intermolecular hydrogen bonds are broken as water enters the starch granule. The crystallinity of the starch solution is lost during gelation.

5.3.2.2 Acid Treatment (Thin Boiling Starch)

Thin boiling starches are made by adding a small amount of acid to a starch suspension that is held just below its gel point. The acid cleaves the polymer at the glycoside linkage, thereby lowering the viscosity of a solution made from it. Hydrolysis occurs within the granule without breaking the granule as such.

5.3.2.3 Alkali Treatment

Strong bases activate starch. Although starch granules can be completely gelatinized in aqueous alkali, the degree of granule swelling depends upon the nature of the starch; the nature of the alkali; the relative amounts of starch, alkali, and water; the temperature, and the presence or absence of neutral salts. In aqueous alkaline slurries, starch granules absorb most of the alkali. An increased reactivity due to alkali absorption is a major factor in commercial manufacture of starch derivatives. A further increase in reactivity of starch in aqueous alkali is obtained by addition of neutral salts, especially sodium sulfate. These salts shift the starch–alkali absorption equilibrium such that alkali absorption is increased.

5.3.2.4 Oxidation of Starch

Starch granules are oxidized with sodium hypochlorite, which converts the 2–3 hydroxyls into –COOH groups breaking the ring at that point. Sodium bisulfite is added to destroy excess hypochlorite. The granular structure is retained, and films from oxidized starch are better than those formed from thin boiling starch.

5.3.2.5 Typical Behavior of Starch as Aqueous Suspension

Typical traces for variation of viscosity of corn and potato starch suspensions in water with temperature are shown in Figure 5.3. Initially, as water enters starch granules, starch swells up and then bursts. This causes a rapid increase in the viscosity of the suspension. On further heating, more interpolymer H-bonds break. This is shown by a decrease in viscosity of the

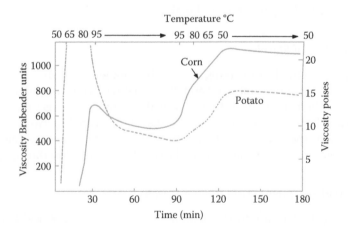

FIGURE 5.3
Temperature-dependent variation of viscosity of corn and potato starch suspension in water.

suspension. Finally, on cooling, there is again formation of H-bonds, and the paste thickens—a process known as retrogradation of starch.

5.3.3 Sizing of Synthetic Fibers

Synthetic fibers and yarns pose a very different problem in sizing. The basic problems associated with sizing of synthetic yarns arise out of the following issues:

- Absence of –OH groups
- Hydrophobicity of material

Typical examples of these fibers are polyester and polyamide, whose structures are shown in Figures 5.4 and 5.5, respectively. It is clear from the figures that there are no –OH groups, which are mainly responsible for H-bonding. The C=O group is also capable of forming H-bonds, but the ease of formation of H-bonds is less than that of C–OH group. This can be attributed to the fact that it is easier for the O atom to draw an electron across a single bond rather than a double bond. Hence, for sizing of these hydrophobic yarns, one has to primarily depend on the following forces:

- van der Waals forces
- Dipole–dipole forces

Hence, the following requirements have to be satisfied for sizing yarns made of polyamide and polyester:

- Adhesive and fiber have to be brought into close proximity.
- Adhesive should not contain steric hindering groups.

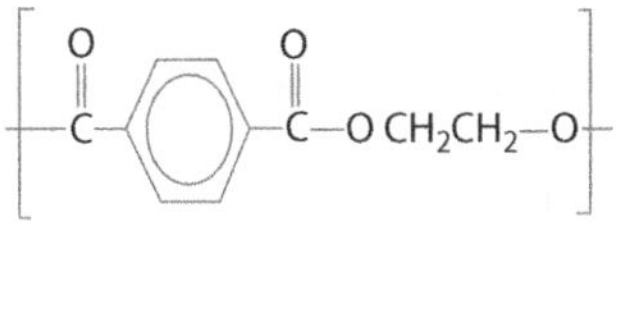

FIGURE 5.4
Molecular structure of a polyester.

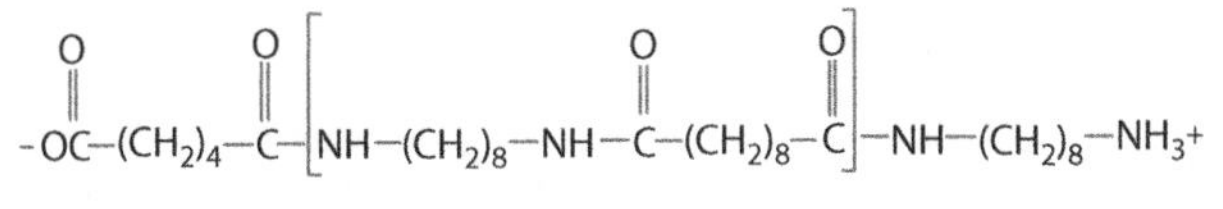

FIGURE 5.5
Molecular structure of a polyamide.

TABLE 5.2

Sizing Agents for Some Important Fibers

Type of Fiber	Chemical Nature of Size Adhesive
Polyester	Vinyl copolymers
Polyamide and polyacrylonitrile	Polyacrylic acid
Acetate	Polyvinyl alcohol

- Adhesive molecules must be as linear as possible.
- Adhesive material should have favorable dipole–dipole interacting (polar) groups.

Adhesives used for some common synthetic fibers are listed in Table 5.2.

5.3.4 Hot Melt Adhesives

These are specially formulated materials that come in the form of bricks. When heated suitably with the aid of an applicator, the brick melts and flows onto the yarn sheet. Such a sizing process is carried out in tandem with warping on a sectional warping machine. Sizing with hot melt adhesives is dealt with in some detail in Section 5.5.3.

5.4 Sizing Machine

The block diagram in Figure 5.6 illustrates material flow in a typical sizing machine.

5.4.1 Creel

Warp beams are arranged in a creel in such a manner as to facilitate

1. An easy withdrawal of warp sheet from each beam
2. Proper assembly of each sheet into the final form in which sizing of the assembled yarn sheet would take place
3. Easy access of the operator to each element of the creel

FIGURE 5.6

Block diagram of material flow in a sizing machine.

Creels are designed for efficient utilization of space while maintaining the lowest possible tension variation in yarns of an individual beam during unwinding from full beam to the empty state as well as between yarns of different beams. Ideally, all yarns across the entire assembled sheet leaving the creel section should be under the same tension throughout the sizing process.

Irrespective of the arrangement of beams in a creel, the rotation of each beam is controlled by a brake. A brake may typically consist of a rotor and multiple pneumatically operated calipers. The quantity of calipers is based on tension requirements.

5.4.2 Size Box

The assembled yarn sheet is taken to the size box and immersed in size liquor by an immersion roller. The size box, shown in Figure 5.7, is usually equipped with heating coils linked to a temperature-control system for maintaining temperature and thereby the viscosity of the size liquor. The depth of the immersion roller below the level of liquor is adjusted in conjunction with the speed at which the yarn sheet is taken through the size box so that the net amount of liquor picked up by the yarn is kept at a desired value. The higher the speed of the yarn, the less the dwell time of yarn within the liquor would be for any specific setting of the immersion roller. Reduced dwell time would be expected to result in less size liquor pickup if the viscosity of the liquor is kept constant. On the other hand, a higher speed of the yarn in the liquor would also mean a higher drag force, proportional to the square of the yarn speed, which would result in a larger amount of liquor being carried away by the yarn. There is therefore equilibrium between dwell time, viscosity, and pickup. It is important that the level of size liquor, which affects dwell

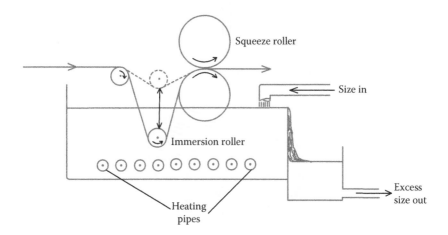

FIGURE 5.7
Typical size box.

time, also be held constant during the entire sizing process. This necessitates a level-control system that maintains a steady inflow of size liquor to the box.

The yarn sheet emerging from the size liquor is next taken through squeeze rollers from which some amount of liquor carried by the yarn sheet is squeezed out and flows back into the box. The effect of nip of the squeeze rollers is illustrated in Figures 5.8 and 5.9.

During its passage through the squeeze roller nip, the wet yarn section gets flattened, causing a temporary reduction in the volume of yarn body under the control of the nip (Figure 5.8). Upon emerging from the nip, the yarn section tends to recover its original form. During this process of recovery of void, an inward radial flow of size liquor into the yarn body takes place (Figure 5.9). A controlled squeezing not only results in the right amount of wet pickup, but also in the desired extent of penetration of size into the yarn body. A low penetration of liquor within the yarn body results in a coating that can easily flake off, and an excessive penetration reduces yarn flexibility.

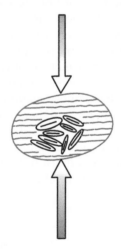

FIGURE 5.8
Squeezing pressure temporarily flattening sized yarn.

FIGURE 5.9
Sucking of sized material into yarn body.

A pair of rollers, one with a hard surface and the other with a relatively compressible one, is usually employed for creating an effective nip of squeeze rollers. For example, a combination of steel and nitrile rubber–covered roller could be a suitable combination for generating pressure in the range of 1 to 5 tons per linear meter. The hardness of the two surfaces as well as the smoothness and concentricity of the two rollers are very critical for a desired level of squeezing pressure on all yarns uniformly across the entire width of a warp sheet. A setup with two pairs of squeezing rollers and two immersion rollers results in a double dip–double squeezing system, which is more effective than single dip–single stage squeezing if the viscosity of the liquor is high and/or the speed of sizing is high. A single dip–double squeezing can also be effected through suitable arrangement of squeeze rollers. In such an arrangement, the compressible roller is located between two harder ones.

The relationship between concentration, viscosity, and wet and dry pickup of the size material depends considerably on the nature of the size material and the squeezing pressure applied. Dry and wet pickup of size materials are defined as follows:

- Dry pickup is the ratio of mass of bone-dry size material picked up to that of the corresponding mass of bone-dry yarn expressed as a percentage.
- Wet pickup is the ratio of mass of size paste picked up by the corresponding mass of yarn expressed as a percentage.

To exploit the benefits of high-pressure squeezing, a high concentration of size solution is required. This ensures that, even if a greater portion of size liquor picked up by the yarn sheet from the size box is removed by squeezing rollers, the amount retained in the yarn sheet still contains a sufficient quantity of sizing ingredients for providing necessary protection to the yarn. But the maximum achievable concentration in sizing liquor is limited by its viscosity because an increase in viscosity retards flow and therefore penetration of size into the yarn. Hence, use of a size formulation that results in a low viscosity at high concentration can be effectively employed in conjunction with high-pressure squeezing to ensure that wet pickup is reduced for the same dry pickup. Such a measure is tantamount to having a lesser amount of water in the sized sheet, which, in turn, means lower consumption of heat energy during drying.

In the event of a very high density of yarn in a warp sheet with neighboring yarns touching each other, splitting the sheet into two halves and sizing each sheet in two size boxes and assembling them after completely drying each sheet can be resorted to. This is termed split sizing, and it results in a well-sized warp produced at much higher speed than if the entire sheet were sized together in the conventional manner. The layout of a split sizing system differs, of course, from that of a conventional system.

5.4.3 Drying Zone

Drying of the wet yarn sheet is usually carried out by physical contact with a hot surface. As the yarn sheet moves at a certain speed, the hot surface also must move at the same speed in order that no abrasion between the two may take place. Traditionally, multiple cylinders are employed for this purpose. The cylinders are driven at a desired angular velocity such that the surface speed equals the translational speed of the yarn sheet. As the cylinders rotate, the yarn sheet in contact with their curved surfaces also moves along a sinuous curved path, coming in contact with a cylinder and leaving it after a wrap greater than π but less than 2π radians before climbing on to the next cylinder, which rotates in a direction opposite to that of the previous one. In this manner, the two sides of a sheet are dried alternately in multiple steps. Drying cylinders come in diameters of 30" to 34" with working widths of 60" to 72".

These cylinders are heated from inside by injecting superheated steam and taking out the condensate through suitable traps. Alternatively, gas also can be employed for heating.

In steam-heating applications, the most efficient heat transfer occurs when high quality (100%) steam at saturation temperature is condensed in the heat exchanger. The majority of thermal energy in steam (latent heat of vaporization) is transferred when steam condenses to water. However, the steam discharge from most steam boilers contains water molecules or mist that has not evaporated. This is called "wet steam," and it has a lower thermal transfer efficiency, undesirable in many commercial applications. To improve steam quality, wet steam can be superheated to create 100% quality or "dry steam" using a circulation heater.

Part of the heat energy of steam/gas is spent in heating the metallic cylinder up to the required temperature. Direct contact between a size film–coated yarn sheet and the hot cylinder causes a localized instantaneous drying of the portion touching the cylinder. A moisture gradient develops in the yarn section (Figure 5.10) that leads to a flow of wet size across the yarn section from its wet side to the drier one. To prevent this migration, it is desirable that moisture be removed at an equal rate over the entire section of yarn sheet. Otherwise, one has to dry the two sides of yarn alternately in many small steps, a strategy adopted in multicylinder drying systems. In one typical layout of such a system, eight cylinders are arranged in two blocks of four. The wet yarn sheet is initially split into two layers while entering the first block. This not only breaks the continuous size film encasing the warp sheet, but sufficient space is also created between neighboring yarns. The two layers of yarn are dried to an extent—termed predrying—before they are assembled again and dried together in the next block of four cylinders.

As a result of wet splitting, a continuous coat of size film develops around each yarn. If the yarns were fully dried in this state, then the subsequent splitting zone would become redundant. Sufficient space between two

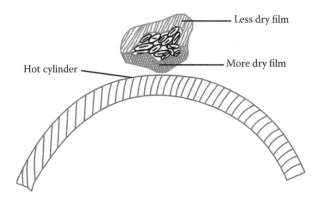

FIGURE 5.10
Size migration on yarn during drying.

adjacent yarns on a drying cylinder is also conducive to removal of steam from the drying yarns. Indeed, an efficient removal of steam from the surface of drying cylinders not only lowers the vapor pressure, thereby facilitating evaporation, but also prevents the formation of a pad of insulating steam layer between the yarns and cylinder. Such a pad, if allowed to form, would lower the thermal efficiency of the system. Owing to wet splitting, the load on each cylinder is also halved, resulting in higher machine speed.

At the instant of detachment of the warp sheet from the first drying cylinder, the size film surrounding each yarn would still be wet and soft. In order that part of the size film does not peel off from the yarn and remain stuck on the hot cylinder during the process of detachment, the first few drying cylinders are coated with PTFE (polytetrafluoroethylene). The very low coefficient of friction of PTFE as well as its high thermal stability permits the process of drying in steps without sacrificing speed of sizing (ca. 50 to 150 m/min). However, this additional layer of PTFE also lowers the thermal conductivity of the cylinder. (Thermal conductivity of PTFE at 273K is in the range of 0.26 to 0.44, and the corresponding value of air is 0.025, of water 0.56, and of stainless steel 45 to 65 W/K.m). Indeed, the thermal energy of steam injected into the cylinder is wasted considerably in overcoming the combined resistance of a layer of condensate on the inner wall of the cylinder, the wall of the cylinder itself, and the layer of PTFE. The thermal efficiency of the system is quantified by the ratio of the weight of water evaporated per unit time to the weight of the steam used. Efficiency of 50% to 60% would be typically realized commercially. Utilization of thermal energy involving multicylinder drying also depends on the extent of wrap of the yarn sheet around individual cylinders. This, in turn, depends on the relative arrangement of cylinders.

Hot-air drying and infrared-ray drying have also been attempted in the past. However, they have not proved to be commercially successful.

5.4.4 Splitting Zone

Yarns in a warp sheet coming out of the drying section adhere to each other to a degree that depends on the efficiency of the predryer. In the absence of a predryer, the entire yarn sheet would be encased in a continuous coat of sizing material. As yarns need to be individualized for the subsequent operation, the coat of size film needs to be broken at many places. Lease bands introduced during warping are made use of in splitting the sized sheet in steps. Rods are inserted in place of the lease bands as depicted in Figure 5.11 and held in position by suitable brackets. During the process of splitting, some amount of size material flakes off as waste. Simultaneously, however, a large number of relatively long fibers, protruding from the surface of yarns and bridging neighboring yarns also get broken into smaller pieces. Hence, splitting is considered to have some beneficial effect insofar as reduction of hairiness is concerned. However, this process results in an uneven and possibly even a ragged and damaged film around each yarn. This is illustrated in Figures 5.12 and 5.13. The neighboring yarns in Figure 5.12 are covered by a continuous film, and a long fiber bridges the space between yarns at the center. The arrows indicate locations where splitting forces act on the continuous size film, which, after being accordingly ruptured, leave very rough surfaces around each yarn as shown in Figure 5.13. If, however, the wet split yarn sheets are sufficiently dried prior to being assembled into the final sheet, thereby requiring no splitting of films, each yarn would be nicely and uniformly coated all around by size films as if

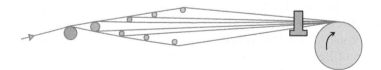

FIGURE 5.11
Splitting zone.

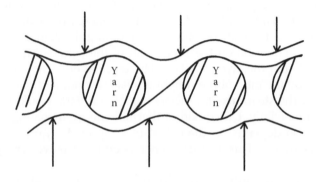

FIGURE 5.12
Size film–encased yarn sheet.

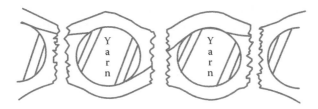

FIGURE 5.13
Effect of splitting on surface of sized yarn.

each yarn had been individually and separately sized. This forms the concept of single-end sizing.

5.4.5 Beaming Zone

The sheet of sized and individualized yarns is finally wound on beams suitable for mounting on looms of a desired width. As one sizing machine caters to a very large number of looms, and as looms may be of varying widths, the headstock of a sizing machine must have the capacity to take in the widest as well as the narrowest weaver's beam. The beam is driven at one end by a spindle to which a beam pipe is fastened by clamping the same in a three-jaw chuck. The other end of the beam pipe is provided as support to permit free rolling. This supporting end can be moved in and out to accommodate beams of various widths as well as to permit doffing of the full beam. The width of warp sheet must conform to the width of beam on which it is to be wrapped. Hence, the sized warp sheet is, at first, passed through an expanding reed/comb by means of which thread spacing is adjusted to the desired value.

The angular speed of a spindle-driven beam needs to be reduced continuously with the increase in beam diameter so that a constant yarn speed is maintained. This can be achieved by monitoring the yarn tension and controlling the transmission ratio between driving the motor and spindle by means of a PIV box. The hardness of the beam is controlled by maintaining the yarn tension at a desired level and by applying a press roll on the beam being wound. A soft beam is detrimental for weaving as the parallel wound yarns may collapse under fluctuating weaving tension. Moreover, a soft beam contains a lesser amount of material, thereby causing frequent interruptions to the subsequent production process. On the other hand, a very hard beam may not only raise the strain level in yarn layers close to the beam pipe beyond a critical level, thereby causing permanent deformation and changes in yarn property, but it can also cause a rise in warp yarn breakages during the weaving process owing to loss of resiliency of the warp beam. Hence, a precise control of yarn tension as well as of pressure on the beam throughout the beaming process is crucial to its quality.

After passing through an expandable comb, the yarn sheet is sensed by a moisture meter. The measured value of moisture content is compared against a reference value, and depending on the type of error, the speed of the entire machine is either raised (moisture level below the lower limit) or lowered (moisture level above the upper limit).

A modern sizing machine is equipped to produce beams up to a width of 6000 mm and a diameter of 1600 mm. The power required by such a machine can be around 30 kW.

5.4.6 Controls on a Modern Sizing Machine

A modern sizing machine is equipped with following types of control systems.

5.4.6.1 Stretch Control

During the process of sizing, yarns are subjected to stretch between the creel and size box, between the size box and drying cylinders, between the drying cylinders and draw rolls, and between the draw rolls and final beam. Depending on the type of yarn being sized, the net stretch has to be kept under control so that yarn does not lose elasticity while being sized. For example, a cotton warp sheet is expected to undergo a net stretch not exceeding 1.5%. A precise control on the speed of the driven elements is a prerequisite for such a control. A stretch meter can indicate the percentage of stretch between various zones.

5.4.6.2 Tension Control

Tensiometers are employed to measure tension in yarn as yarn tension is a direct outcome of yarn stretch. If the yarn property is known exactly, then stretch can be kept under control by controlling tension. Tension control is used to directly regulate the angular velocity of the beam.

5.4.6.3 Moisture Control

Moisture meter reading just prior to beaming is employed to control the overall speed of sizing without disturbing the speed ratios of critical driven elements. As pointed out, a moisture level higher than the desired one can be neutralized by speeding up the machine, and a moisture level lower than the desired level can be rectified by slowing down the machine.

5.4.6.4 Pressure Control

Pressure in the nip of the squeezing rollers affects wet pickup and hence the cost of drying. A better anchorage of size film in the yarn body reduces the size of the flake off, thereby ensuring better protection. On the other hand, too high a pressure may flatten the yarn, affecting its appearance and subsequent behavior. The amount of pressure to be applied depends on the material being sized as well as on the density of the sheet.

5.4.6.5 Temperature Control

Temperature of the PTFE-coated cylinders is lower than those of the succeeding ones when filament yarns are sized although no such difference is maintained while sizing spun yarns. Similarly, cotton yarns are dried at higher temperatures than polyester-cotton blended yarn. Indeed, the temperature of each cylinder needs to be controlled and maintained at a different level for different materials. This depends on the amount of water to be evaporated by each cylinder and the undesirability of overdrying a yarn surface and causing excessive migration during the initial stages of drying. Indeed, the migration of size in a filament yarn, which lies flatter on the cylinder in a wet state than a spun yarn, might even be across the section instead of being along the periphery.

5.4.6.6 Level Control

This system ensures a constant depth of immersion of the yarn sheet within the size paste and contributes to uniformity of wet pickup.

5.5 Special Sizing Systems

5.5.1 Dye Sizing

Colored warp threads are employed for weaving certain types of fabrics of which blue jeans/denim is a typical example. The warp yarns are colored blue through an indigo dyeing process.

A dyed warp sheet may be produced from cones containing colored yarns. To this end, one can dye yarn cones by first winding yarn on a perforated conical shell under low tension and pressure so as to create a soft package suitable for dyeing. Such packages can then be put in a closed chamber under pressure in which dye liquor is circulated. However, such dyed cones need to be rewound before being warped and sized. To circumvent such a tortuous process, sizing machines have been designed to dye a warp sheet first and then size the same at one go. Clearly, only a very limited range of materials and dyeing processes can be handled in this fashion. Dyeing cotton yarns by reactive dyes needs curing at an elevated temperature (ca. 130°C) with the help of hot air or steam, and the curing is a relatively slow process. Indigo dyeing, on the other hand, requires oxidation at room temperature and does not need a long dwell time. Hence, a dye sizing machine for denims needs many vessels in series and guide rolls, and dye-sizing cotton or polyester-cotton yarns with reactive and disperse dyes needs additional provision of curing chambers, occupying a considerable space for ensuring the required dwell time.

5.5.2 Single-End Sizing

This process basically involves a high-density package creel from which yarns are withdrawn and formed into a sheet, which is then sized straight-away. As a creel can hold a very limited number of yarn packages, the resultant yarn density in the sheet is very low such that distance between adjacent yarns in the sheet is a multiple of yarn diameter. Hence, the resultant size film forms securely around each yarn, doing away with the necessity of splitting the dried sheet. Single-end sizing is highly beneficial for filament yarns as individual filaments may rupture during the splitting process. However, the beam coming out of such a sizing machine would contain only a limited number of yarns, insufficient for a woven fabric. Hence, a separate beaming machine subsequently assembles the required number of such sized beams into the final weaver's beam.

In a modified version of this process, yarns from the creel are warped to a beam, and the beams are individually sized and assembled by a separate beaming process. This modification takes care of interruptions in a sizing process caused by filament/yarn breakages as a result of yarn unwinding from the creel. This modified process is easy to implement with a conventional machine setup.

5.5.3 Hot Melt Sizing

Hot melt sizing is carried out during the warping process by applying a molten, 100% active size material to the warp yarns. The sizing applicator is located between the warper creel and the warper. Warp yarns drawn from a warper creel pass over the top of an internally heated rotating cylinder. Deep and closely spaced parallel grooves, parallel to the yarn path are formed on the surface of this roller, each groove being meant for one yarn strand only, and allowing good surface contact with the yarn. The sizing agent is applied by pressing a block of solid size against the grooves of the hot cylinder. From the point of application, the molten size is carried by rotation of the cylinder to a zone where each moving yarn strand passes briefly through a corresponding size-filled groove, essentially tangentially to the cylinder, and then moves away coated with an optimum amount of size. The method of application is unique in its capacity for placing a quick-setting, high molecular weight, film-forming melt size on yarn at high speed.

The warps yarns travel at a surface speed much greater than the surface speed of the grooved roll. As a result, when the molten size reaches the yarns, each groove gets wiped clean of size. The wiping effect not only provides for application of size to the yarn, but also causes hairy staple ends in the yarn to lie down to an acceptable level. As yarn ends continue past the applicator and roll into the cooler surrounding air, the size sets up very quickly, bonding the fibers together. The accumulator positioned between the hot melt applicator and the warper allows the operator to properly locate and repair any broken ends.

A groove depth of three to five times the diameter of the yarn being sized is considered optimal, but deeper grooves, although harder to machine and maintain, can be used if desired.

Hot melt adhesives contain hydrogen bond–forming (O) groups, which improve bonding with fibers. The size can be a melt blend of ethylene/vinyl acetate copolymer with microcrystalline paraffin or hydrogenated tallow wax. In the case of ethylene vinyl terpolymers, the maleic anhydride group improves adhesion of size to the fiber. It is applied as a melt to warp yarn, desized with a hot petroleum solvent, and the extracted size and unrecovered solvent burned to nonpolluting carbon dioxide and water. These sizes have the unique advantage that their adhesive property can be doctored as per requirement by suitable selection of chain lengths of each group.

Alternately, a composition made of an intimate mixture of a water-soluble film-forming thermoplastic polymer and a melt-miscible, solid modifier selected from the class consisting of aliphatic and aromatic carboxylic acids, partial esters thereof, non-polymeric polyhydric alcohols, phenolic acids, and polyhydric phenols, wherein the proportion of polymer to modifier is between 90:10 and about 50:50 on a weight basis, can also be used as the said size material. The size can be removed from yarns by aqueous or alkali extraction.

The advantages of hot melt size are listed as follows:

- *Low energy consumption*: Because there is no water to be removed from sized yarn, the energy required to size by this method is about 80% less than that needed for aqueous sizing.

- *Low investment cost*: The investment cost for a hot melt sizing unit and for a machine required for assembling the warper beam will be lower than that for a conventional sizing machine and the size cooker.

- *Elimination of cooking area and its dumping*: There is no size solution to prepare and dump. A hot melt size is available in ready-to-use form from the supplier.

- *Superior quality sizing of warp yarn*: Due to the separate end sizing of each yarn, sizing quality is superior to that achieved by conventional sizing. Also the method provides for fiber lay.

- *Greater speed of size application*: Because there is no water to be removed, the speed at which size is applied is not limited by drying capacity. The rate at which the molten size sets up to a non-tacky state determines the sizing speed. A setup rate of one second or less is achievable without any assistance of any type of auxiliary equipment, allowing application speeds equivalent to conventional warping speeds.

- *Resistance to shedding*: Resistance to shedding is a result of the excellent elastic properties of the advanced size materials.

The disadvantages of this method are

- *Size recycling*: The projected cost of aqueous hot melt size recovery does not justify size recycling.
- *Operator training required*: Because the hot melt applicator and size material are unconventional, operator training is necessary. However, the method is less complicated and easier to operate than the conventional sizing method.
- *New equipment requirement*: Investment in new equipment is required to change over from conventional to hot melt sizing.
- *Different size*: The size material is not a wax and is unlike any conventional aqueous sizing materials.

5.5.4 Solvent Sizing

Increasing energy cost and pollution control legislation have resulted in development of solvent sizing. In this system, polystyrene resin is used as a sizing material, and the solvent is a chlorinated hydrocarbon, such as 111 trichloroethane, perchloroethylene, and trichloroethylene, etc. Because a chlorinated hydrocarbon has very low latent heat of vaporization, the energy saving will be enormous.

A warp sheet is passed through a size box, and the size material is dissolved in solvent at room temperature. The warp sheet coming out of the size box is wet split before being taken to a cylinder for drying. The wet warp sheet comes in contact with heated solvent vapor, which partially dries it. The whole unit is completely enclosed. Two condenser coils condense the solvent vapor and take it to a storage tank for further use. The size is mixed in a simple tank at room temperature using a high-speed propeller-type mixer.

The advantages of this method are listed as follows:

- *Reduction in energy cost*: Because the energy requirement for evaporation of solvents (chlorinated hydrocarbon) is low, the energy required to dry a warp sheet is also very low. Besides the obvious savings in production costs, this also results in higher production from a compact unit, showing savings in space and capital costs.
- *Versatile in terms of multiproduct weaving*: Solvent sizing is very versatile; the same machine can be used for sizing a wide variety of yarns made out of synthetic and natural fibers, which is due to the capability of solvents to wet all kinds of fibers easily.
- *Suitable for fibers of low wet strength*: Regenerated fibers have lower wet strength in aqueous medium compared to their dry strength. But when viscose or viscose-blended yarns are impregnated with solvents, there is no loss in strength.

- *No detectable size migration*: In solvent sizing, no detectable size migration onto drying cylinders occurs, which is due to the low surface tension of the solvents.
- *High production rate*: High production rate is possible due to quick drying of the warp sheets.

The disadvantages of this process are

- *Toxicity of chlorinated solvents*: Chlorinated solvents are highly toxic, causing environment-related problems (pollution).
- *Higher investment cost*: Due to special safety regulations, the investment cost is high. Also, additional energy is required for the vacuum system.

5.5.5 Cold Sizing

Cold sizing agents are blends of polymers based on PVA with high adhesive power and antistatic agents dissolved in water. These are available in different viscosities, concentrations, and adhesive powers.

Application of a cold sizing agent is carried out with a waxing system located between the warping drum and the beaming machine of a sectional warping machine. The size is applied by an indirect, two-sided tangential applicator. Through rotation of the rolls and simultaneous application of pressure, a finely adjustable size film is realized on the warp. By adjusting the difference in speed between the warp sheet and the applicator roll, pickup can be adjusted. Infrared radiators dry the warp sheet without contact. The advantages of this drying method are

- High and uniform drying performance.
- Fast reaction time when switching the radiators on and off; hence, no overdrying at machine stop or at low speed.
- Precise temperature control.
- Contactless drying promotes a smoothening effect of the warp threads by the applicator rolls.

The advantages of this sizing method are

- Accurate and controlled liquor application.
- Gentle sizing of warp, virtually without loss in elongation.
- Extremely highly viscous and highly concentrated size could be used.
- Saving in labor and space.
- No size cooking required.
- Capital expenditure for a lubrication device is low.

5.5.6 Features of New Sizing Methods

Some features of new sizing methods are listed as follows:

- Low energy consumption.
  - Size cooking consumes a lot of energy, which can be saved if cooking can be avoided or made more efficient.
  - Normally, 70%–80% of total energy consumption is required for drying. With the new methods, energy is saved during drying.
- Low sizing material requirement. With improved size material in terms of viscosity, concentration, and adhesive power, the sizing material requirement is less. Moreover, because size material cost is the main component of total sizing cost, any saving in sizing material improves the economy of sizing.
- Superior sizing quality. Better sizing quality is achieved through improved size-application techniques and drying, if at all required.
- Less space requirement. With hot melt and cold sizing, no separate machine for sizing is required as the process is incorporated with warping. Moreover, space for size cooking is also not required in any of these methods (hot melt, solvent, and cold sizing).
- Increased sizing speed. In applications in which the sizing speed is not limited by drying, sizing speed goes up considerably.
- Reduced workload on operative. The new methods are less complicated, easy to operate, and are user friendly.

Reference

Ormerod A (2004). *Modern Preparation and Weaving Machinery*, Woodhead Publishing Limited, UK, ISBN: 1 85573 998 4.

Further Reading

Goswami B C, Anandjiwala R D, and Hall D M (2004). *Textile Sizing*, Marcel Dekker, Inc., New York, ISBN: 0 8247 5053 5.

6

Basic Weaves and the Process of Drawing In

6.1 Basic Concepts

Warp and weft yarns, prepared in the manner described in the previous chapters, become, in the process, ready for the actual weaving process. Weaving is the process of interlacement of warp and weft yarns according to a specific plan for constructing a fabric. If a specimen of woven fabric (Figures 1.2 and 1.3) is closely observed, it is found that yarns are aligned in two or more directions. In the simplest of woven fabrics, yarns are found only in two directions: warp yarns along the length and weft yarns along the width. It can also be observed that warp yarn rides at a certain location above the weft, and at certain other locations, weft yarn rides above the warp. This pattern is repeated all over a woven fabric.

The pattern of interlacement of yarns in a woven fabric is known as weave. Weaves of different kinds, ranging from simple and plain to extremely complicated and elaborate damasks and brocades, can be employed to interlace warp and weft yarns. For such interlacements to occur, one needs to group warp yarns in a certain order and lift yarns of specific groups away from the main body of the warp sheet for forming an opening, that is, a shed through which a weft yarn can be inserted. Subsequently, upon lowering of these lifted yarns back to the main body of the warp sheet, the inserted weft yarn becomes trapped, forming thus the edge or the fell of a woven fabric. The sequence of lifting of warp yarns is determined by the weave in question. Healds are used for maintaining a specific order of grouping of warp yarns. Warp yarns belonging to a certain group are passed through openings in heald wires controlled by the respective heald frame. This frame is lifted and lowered, thereby moving the respective heald wires and, therefore, the corresponding warp yarns away from the neutral plane of the warp sheet.

Thus, healds are used for splitting the warp sheet into various groups and for lifting and lowering the individual groups in a preplanned manner during weaving. The plan of distributing the individual warp yarns to respective healds is known as drafting. A woven fabric materializes on a loom out of interaction between the drafting plan of the warp yarns and the lifting plan of the healds. Hence, before the warp yarns can be sent to the weaving loom, they have to be sent to a drawing-in section, where, according to the desired drafting plan, warp yarns are drawn in through wires of respective heald frames. A weaver needs to be fully aware of the drafting and lifting plans required for realizing a certain fabric weave on a loom out of a given warp sheet. Conversely, if a certain weave needs to be realized on a loom, the weaver should be able to formulate appropriate drafting and lifting plans for the warp sheet.

6.2 Identification of Warp and Weft

When a fabric sample is analyzed, the first step is to identify its warp and weft yarns. If a portion of a fabric is observed through a pick glass (a magnifying glass, which can be focused onto a small area of the fabric) and the number of yarns covered within the area under focus is counted separately for the two directions, then the direction in which the larger number occurs is usually that of the warp. The number of yarns per inch is known as the yarn (or thread) density, and for warp, it is termed ends per inch (epi), and for weft, it is termed picks per inch (ppi). In general, epi is greater than ppi. Another method of identifying warp yarns is to compare twist in yarns of the two directions. Warp yarn is usually twisted with a higher twist multiple than weft, and at times, warp yarns even exhibit a plied structure, and weft yarns are usually softer and bulkier. Weaving strains that warp yarns have to withstand require that they are hard and strong whereas the softer weft yarns, once embedded in a fabric, are meant to spread and flatten out easily, thus blocking out the pinholes that would otherwise occur in a fabric owing to interlacement between adjacent pairs of warp and weft yarns. Flattening of weft yarns improves fabric cover.

6.3 Introduction to the Basic Weaves

Once warp and weft directions have been identified from a fabric sample, the weave of the sample, that is, the interlacement pattern of yarns has to be found out and recorded on a piece of paper. In order to execute this task, one has to focus on a particular warp yarn and count the number of times it remains consecutively under and above the interlacing weft yarns. For transferring this observation onto paper, the convention of filling up or leaving blank the spaces in a square paper is usually adopted. Accordingly, a filled-up square is meant to represent the corresponding warp yarn as floating above the interlacing weft yarn. Similarly, a blank space indicates that the corresponding warp yarn passes below the weft yarn at the particular location.

In general, due to interlacements between warp and weft yarns, the fabric shrinks in both length and width directions. This is due to the fact that yarns cannot remain straight after interlacements and are forced into a wavy configuration. A wavy or crimped form requires a greater length than the straight form. Hence, fabric length and width would be less than the corresponding lengths of warp or weft yarns embedded in the fabric. This length difference gets manifested as fabric shrinkage. As the number of interlacements is highest for a plain weave, the maximum shrinkage is also observed in plain-woven fabrics. Weaves such as twill result in less fabric shrinkage than plain. Therefore, suitable allowances should be kept at the warping stage to counteract shrinkage loss due to crimp in warp threads.

In Figure 6.1, a square paper exhibits four columns and four rows that have been drawn and each row and column numbered. The four vertical columns represent four warp yarns, and the four horizontal rows stand for four picks of weft. The filled-up squares correspond to warp yarn being above, and the blank shows it to be below the corresponding weft yarn.

The construction shown in Figure 6.1 is known as plain weave. Each alternate yarn in either direction behaves similarly insofar as the pattern of interlacement is concerned. This weave therefore repeats on two ends and two picks. Nearly 70% of all fabrics manufactured commercially are made of plain weave construction. A plain weave exhibits the maximum number of interlacements among the yarns, resulting in maximum strength realization of yarns employed in the resultant fabric. Moreover, both sides of a plain-woven fabric have similar appearance and surface properties.

The construction shown in Figure 6.2 is that of a 2/2 twill weave. If the first warp yarn is considered, then it is observed to be consecutively below the picks two and three while it rides above the first and fourth picks. It is also noticed that all four warp yarns exhibit a similar pattern of interlacement although there is an upward shift by one pick for every successive warp yarn. As a result, a ladder with a positive slope is formed. This would get manifested in the fabric in the form of a diagonal line, typical of

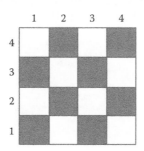

FIGURE 6.1
Convention for representation of weaves.

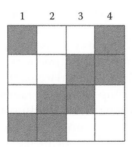

FIGURE 6.2
Four-end twill weave with positively sloped diagonal.

twill weaves. Slope of the diagonal line can be reversed if successive warp yarns are shifted down by one pick (Figure 6.3). The diagonal lines with their smooth floats of weft yarns reflect incident light uniformly in a certain direction, and therefore, the fabric appears to be more lustrous than one made of plain weave. On account of floats, a twill fabric also appears smoother to the touch. Because of the reduced number of interlacements, it is also possible to pack in a higher number of yarns in unit length of a twill-woven fabric. Hence, a twill weave allows construction of a denser and therefore heavier fabric. Moreover, owing to greater mobility of yarns caused by a lesser number of interlacements as compared to plain weave, twill-woven fabric drapes better.

At this stage, it would be appropriate to differentiate between the terms weave, construction, and structure of a woven fabric. The weave of a fabric represents the pattern of interlacement between warp and weft yarns. Its construction is represented by the set {weave, count of warp and weft yarns, epi and ppi}, and its structure is represented by the set {crimps in warp and weft, fabric thickness, and, in some cases, the pore size distribution}.

A 2/2 twill repeats on four ends and four picks. Instead of a 2/2 twill, it is also possible to weave a 3/1 twill (Figure 6.3) or a 1/3 twill, both repeating on four ends and four picks. Such weaves are termed unbalanced as the number of warp yarns that are above weft yarns at any pick are not equal to the number that passes below. A 2/2 twill fabric exhibits a similar surface on both sides of fabric, and 3/1 and 1/3 twills have warp floats predominantly

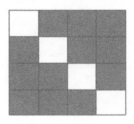

FIGURE 6.3
Four-end twill weave with negatively sloped diagonal.

on one surface, the other surface showing predominantly weft floats. A blue jean is a typical example of such a weave, showing blue warp yarns on one face and gray weft yarns on the other.

A satin weave exhibits long stretches of warp floats, that is, warp remains over weft across a large number of weft yarns. This results in a fabric surface having very high luster and smoothness. If, instead of warp, weft yarns float over the warp yarns in a similar manner, then the construction is known as sateen. The fabric woven with these types of constructions are very thin and delicate due to the very limited number of interlacements. Typical examples of satin and sateen weaves are shown in Figures 6.4 and 6.5, respectively.

Apparently, the constructions 3/1 twill and 1/3 twill should result in similar fabrics with the surfaces simply reversed. Similarly, a 7 up, 1 down satin or, its reverse side, a 1 up, 7 down sateen should logically result in the same fabric, and differentiating between them may not appear to be necessary. However, due to differential tensioning of warp shed lines on a weaving loom (see Section 7.2.2.6), one could expect some difference in crimp distribution in the interlacing yarns and, therefore, some difference in the properties of the resultant fabrics. More important, however, is the necessity

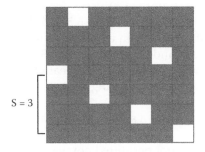

FIGURE 6.4
Seven-end satin weave.

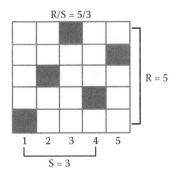

FIGURE 6.5
Figurative representation of a sateen weave.

to differentiate between the two weaves when a combination of different weaves is employed in complex constructions. As, for example, a construction employing plain weave over the first six warp yarns followed by six yarns of 2/1 twill and six yarns of 3/1 twill would be quite different from the one employing plain weave over the first six warp yarns followed by six yarns of 2/1 twill and six yarns of 1/3 twill. Insofar as satin and sateen weaves are concerned, the nomenclature used may be indicative of the pattern of shifting, and in this sense, a 7/3 sateen is structurally different from a 7/3 satin. This is explained in Section 6.4.

6.4 Repeat and Shift

Repeat and shift are the two basic factors that are considered in describing a sateen (satin) weave. R denotes repeat, and S denotes shift. The generalized representation is R/S. Repeat R stands for the number of picks required for repetition of the pattern, and shift S denotes the step across the warp (for sateen) or along the warp (for satin) yarns for interlacement in the subsequent pick. An example is given in Figure 6.5. This R/S representation is particularly useful for satin and sateen weaves as is explained in the following.

For such weaves, $R \geq 5$, and S is such that $1 < S < (R - 1)$. Suppose that we wish to construct a sateen weave with a repeat of 7, that is, $R = 7$. Now the shift can be determined in the following way:

Seven can be represented as a sum in many ways: $1 + 6$, $2 + 5$, and $3 + 4$.

The shift of 1, which is not permitted anyway, would result in a 1/6 twill as shown in Figure 6.6. For $S = 2$, the sequence of the number of warp yarns being lifted above the successive picks would be first, third, fifth, seventh, second, fourth, and sixth (Figure 6.7). Thus, the value of S decides the lateral

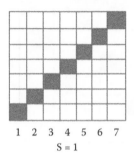

1 2 3 4 5 6 7
S = 1

FIGURE 6.6
Seven-end sateen weave with step of one.

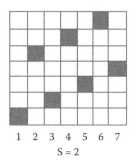

1 2 3 4 5 6 7

S = 2

FIGURE 6.7
Seven-end sateen weave with step of two.

shift in number of warp yarns that would be lifted up for interlacement with successive picks. It is observed from Figure 6.7 that the interlacement points are spread out in such a manner that long floats of weft surround every interlacement point. These weft floats would effectively drown the small segment of warp that floats above, and therefore, the surface of fabric would virtually exhibit only weft yarns, a typical feature of weft-faced sateen. The sequence of interlacement for the other possible values of S is worked out in the following, and the resulting constructions are shown in Figures 6.8 through 6.11.
   For

   S = 3: first, fourth, seventh, third, sixth, second, and fifth warp yarns to be lifted sequentially (Figure 6.8)

   = 4: first, fifth, second, sixth, third, seventh, and fourth warp yarns to be lifted sequentially (Figure 6.9)

   = 5: first, sixth, fourth, second, seventh, fifth, and third warp yarns to be lifted sequentially (Figure 6.10)

   = 6: first, seventh, sixth, fifth, and so on leading to a left-hand twill weave (Figure 6.11)

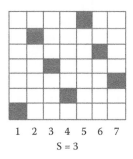

1 2 3 4 5 6 7

S = 3

FIGURE 6.8
Seven-end sateen weave with step of three.

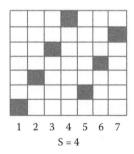

FIGURE 6.9
Seven-end sateen weave with step of four.

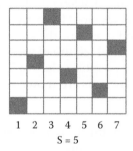

FIGURE 6.10
Seven-end sateen weave with step of five.

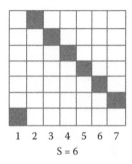

FIGURE 6.11
Seven-end sateen weave with step of six.

Another example for repeat on eight threads can be considered as follows:

$$8 = 1 + 7, 2 + 6, 3 + 5, \text{ and } 4 + 4$$

For

S = 2, the sequence would be 1, 3, 5, 7, 1...

  = 3, the sequence would be 1, 4, 7, 3, 6, 2, 5...

  = 4, the sequence would be 1, 5, 1...

  = 5, the sequence would be 1, 6, 3, 8, 5, 2, 7, 4...

  = 6, the sequence would be 1, 7, 5, 3, 1, 7...

Hence, S = 2, 4, and 6 lead to unviable weaves as quite a few warp yarns would not get interlaced at all. It is thus observed that, in addition to the inequality condition stated in the foregoing, the sufficient condition for a viable weave would be that the common divisor between R and S should be equal to one. Thus, for R = 9, one can state that the viable sateen weaves would be generated by choosing steps of 2, 4, 5, and 7 but not steps 3 and 6 as illustrated in the following:

For

S = 2, the sequence would be 1, 3, 5, 7, 9, 2, 4, 6, 8

  = 3, the sequence would be 1, 4, 7, 1—hence, not viable

  = 4, the sequence would be 1, 5, 9, 4, 8, 3, 7, 2, 6

  = 6, the sequence would be 1, 7, 4, 1—hence, not viable

  = 7, the sequence would be 1, 8, 6, 4, 2, 9, 7, 5, 3

For R = 6, S can have values 2, 3, 4, 5, none of which satisfies the conditions stipulated. Therefore, no sateen weave is possible on R = 6.

The satin weave results in a fabric with warp floats visible on the face side of the fabric. The face side of a fabric is meant to be the exposed side. The weave is expressed as a fraction with the numerator showing the number of picks required for the repeat and the denominator representing the vertical shift in interlacement point. The difference in the distribution of interlacement points of a 7/3 satin and a 7/3 sateen is illustrated by Figures 6.4 and 6.8.

Plain, twill, and sateen (satin) are the three basic weaves. All woven constructions are based on derivatives of these three weaves.

6.5 Drafting and Lifting

For the above weaves to materialize in fabric, warp yarns have to be selectively lifted or lowered in such a manner that the weft passes over or below these yarns as required by the woven construction. The movement of warp yarns in the vertical plane is controlled by healds. A heald is made up of many similar heald wires, which come in different forms, shapes, and materials. These wires are either directly fastened to two parallel bars fastened to a heald frame or are loosely mounted on narrow metallic strips, which, in turn, are fastened to the parallel bars. Either way, an assembly of heald wires in a heald frame is termed simply as a heald or, on occasion, as a heald shaft.

Warp yarns, which behave in similar fashion throughout the weaving process, are drawn through the same heald. Thus, for a plain weave, warp yarns can be split into two groups whereas for a 1/3 twill warp yarns have to be split into four groups. For a plain weave then, two healds are required whereas for a 1/3 twill one needs four healds. Yarns belonging to the same group are drawn through the eyes of heald wires mounted on the same heald. Hence, if all healds with warp yarns drawn suitably through their respective heald wires are stacked close to each other along parallel vertical planes and each given the requisite amount of vertical motion, some in an upward and others in a downward direction, then the warp sheet would be split open suitably for insertion of the relevant pick. Working out a drafting plan involves choosing the appropriate number of healds and assigning warp yarns of a weave repeat to wires of the proper healds. A lifting plan involves working out the proper sequence of raising corresponding healds over the number of picks in a weave repeat and is a result of the weave and drafting plans.

The principle of drafting and lifting plans is illustrated with the help of an example shown in Figure 6.12. The number of healds available is four, and the weave repeats over six ends and eight picks. Box A denotes the weave, B the drafting plan, and D the lifting plan. Box C is used to convert B into D with the aid of A. A close observation of fabric weave shows that warp yarns 2 and 6 as well as 3 and 5 behave similarly. Therefore, warp yarns 2 and 6 can be drawn through wires of the same heald, and the warp yarns 3 and 5 can be drawn though wires of another heald. The warp yarns 1 and 4 behave dissimilarly to the rest of the yarns and therefore need to be drawn through wires of separate healds. Hence, for the case at hand, a total number of four healds is sufficient. Based on this observation, the drafting plan is worked out in the manner shown in box B. Warp yarn 1 is passed through heald number 1, warp yarns 2 and 6 are passed through heald number 2 as they behave similarly, warp yarns 3 and 6 through heald number 3, and warp yarn 4 through heald number 4. The markings are done by filling up the respective square with a hollow circle, and thereby the drafting pattern is obtained.

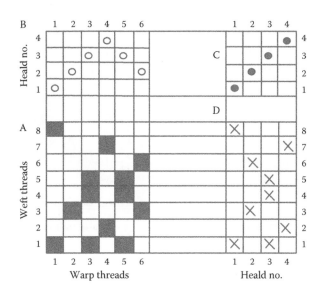

FIGURE 6.12
Principle of drafting and lifting.

The lifting plan depicts the sequence of lifting of healds with respect to successive pick insertion (weft insertion). An intersection of fabric weave and drafting plan is used to develop this plan. For example, to insert pick number 1, the first, third, and the fifth warp yarns need to be lifted. These yarns are incidentally drawn through healds 1 and 3. Hence, in the bottom row of the lifting plan, corresponding to the first pick shown in box A, crosses have been inserted in the squares numbered 1 and 3. The rest of the crosses in the lifting plan are inserted following the same logic. The total lifting plan has eight rows and four columns, corresponding to the eight picks and four healds in question.

The drafting plan depends on warp density. If the number of warp yarns per unit of linear space (yarn density) is very high, that is, warp yarns are packed very close to each other, then more than one heald can be used for lifting similar warp yarns as it leads to reduced inter-warp friction. These healds can be bunched together, or they can be spaced out. Such a draft is known as a skip draft, shown in Figure 6.13 along with some other common forms of draft. In this particular case, six heald shafts have been employed for a plain weave although only two are required. However, the drafting method permits bunching of the first three healds as one and the last three healds as another. The odd-numbered yarns are drawn through the first three, and the even-numbered yarns are drawn through the last three.

A point draft is employed for constructing waved twill, and as the figure shows, only six healds are required to generate a weave repeating on 10 warp yarns. Such weaves exhibit symmetry about the vertical axis.

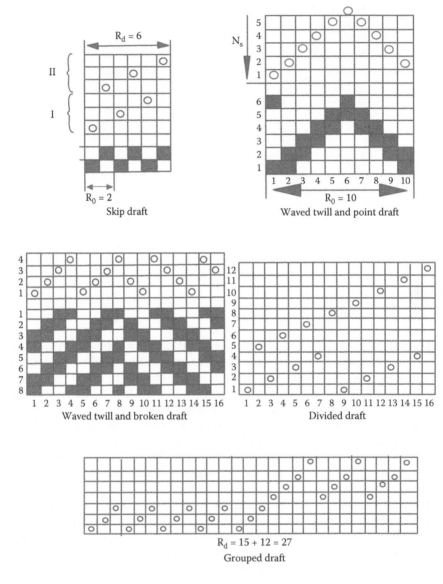

FIGURE 6.13
Some common forms of draft.

A broken draft exhibits a shift vis-à-vis the point draft. The divided draft is meant for simultaneous weaving of two different warps. The first four healds in the figure take care of the odd-numbered yarns coming from one beam, and the other eight healds take care of the even-numbered yarns coming from the second beam. Straight draft has been employed for each set of warp coming

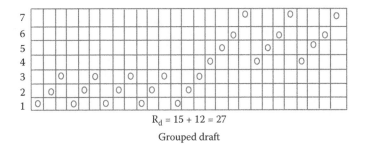

$$R_d = 15 + 12 = 27$$

Grouped draft

FIGURE 6.14
Grouped draft.

from the two beams. Grouped draft (Figure 6.14) is employed for separately handling different groups of warp yarns coming from the same beam, encountered when fabrics such as those with stripes are required to be produced.

6.6 Methods of Generating Weaves

Weaves can be classified as regular and irregular. Regular weaves are those in which all warp ends interlace in the same manner and the interlacing patterns of consecutive warp ends are displaced relative to each other in equal steps. Moreover, all picks of weft also interlace in the same way as the warp ends. Hence, in regular weaves, step number and float arrangement are constant in one complete repeat. Some examples of a regular weave repeating on three ends and three picks are shown in Figure 6.15a through c. Irregular weaves do not exhibit such similarity in pattern of interlacement as exemplified by the three weaves shown in Figure 6.16a through c. However, all weaves, regular and irregular, can be developed systematically from basic weaves through certain operations.

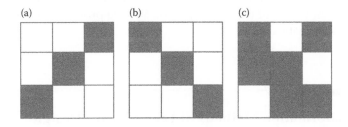

FIGURE 6.15
Some regular weaves. (a) 1/2 right handed twill. (b) 1/2 left handed twill. (c) 2/1 left handed twill.

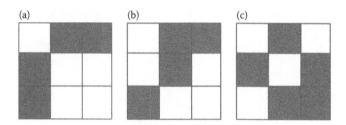

FIGURE 6.16
Some irregular weaves. (a) A combination of one 2/1 and two 1/2 weaves. (b) Another combination of one 2/1 and two 1/2 weaves. (c) Combination of two 2/1 and one 1/2 weaves.

Milasius and Reklaitis (1988) developed a method for automatically generating 2-D and 3-D weaves represented by 2-D weave matrices. The 2-D weaves are grouped into regular and irregular weaves, and all regular weaves are treated as one whereas the irregular 2-D weaves and the 3-D weaves are treated as individual types. When the parameters of a type of weave are specified, a 2-D binary weave matrix is generated automatically, which can be directly used as is or be easily converted into a lifting plan. Attempts to formulate 2-D weaves were made by using the principle of weave coding. The authors explained the need for a computer-aided weave coding system arising from increasing use of computers in the field of textiles (CAD and electronically controlled dobby). In each case, it is necessary to store weave patterns in a computer memory. Direct design input through selection of all individual cells is time-consuming and could also be a source of error in a large weave pattern. Hence, initially, the interlacing sequence of the first column, which is different for each pattern, could be created, and then the remaining part of the design can be created by shifting previous columns by a constant step. Weave-designing methods were classified according to four levels of complexity. The first and second levels are used to create both simple and complex weaves (plain, twill, satin, etc.), and the third and fourth level codes are usually employed to develop complex weaves (crepe, diamond, honeycomb, etc.) in which Boolean matrices serve as the operands.

Chen et al. (1996) presented an approach to describe weaves as mathematical functions. Accordingly, all regular weaves may be described by using a simple mathematical model, and all irregular weaves are exclusive of regular weaves. When values of the weave parameters are provided, the weaves are generated automatically.

6.7 Transformation Methods of Fabric Weave Design

The science of fabric weaves involves studying interlacing patterns of either warp or weft floats as weave is an important crucial factor affecting both appearance and properties of resultant fabrics. Ping and Lixin (1997) studied and tried to formulate a mathematical model for fabric weave design. In the process of designing fabric weaves, the key is to establish an appropriate mathematical model and provide a suitable algorithm. Accordingly, a mathematical model for fabric weave design was established, which is a set of mapping (F) that functions on the set. The paper describes the elements of F, their characteristics, and their definition in mathematical terms. This model could serve as a basis for CAD in fabric weaves. The mathematical model for fabric weave design, which is one type of nonassociative algebra in a two-element field composed of the basic transformations in the two-element field, was established in the light of mapping between {0,1} matrices of fabric weaves. The transformation methods described in the paper (Figure 6.17) are summarized in the following:

1. *Negative interchanging*: In weave design, warp-floating points are changed into weft-floating points and vice versa.

2. *Symmetrizing*: Taking the x-axis or y-axis as the axis of symmetry, the weave is rotated around the two axes for creation of a new design.

3. *Overlapping*: The intersection points of two weave designs are overlapped for creation of a new design. Overlapping two warp-floating points produces one warp-floating point, overlapping two weft-floating points produces one weft-floating point, and overlapping one warp- (weft-) floating point and one weft- (warp-) floating point produces one warp-floating point.

4. *Rotating*: Rotating a weave design in a clockwise (or counterclockwise) sense through $\pi/2$, π, or $3\pi/2$ radians results in new weave designs.

5. *Transposing*: A new construction is created by transposing between the warp and weft yarns.

6. *Order changing*: By changing the orders of the warp (weft) yarns in a weave, a new construction is created.

7. *Adding (reducing) warp (weft) yarn*: By inserting or removing a warp (weft) yarn in a weave design, a new construction is created.

8. *Expanding in any direction of above, below, right, or left*: By joining another weave in any direction of above, below, right, or left of a weave construction, a new one can be created.

9. *Putting weave blocks together*: Filling up the selected weaves according to a given pattern results in a new woven construction.

10. *Accumulating warp (weft) yarns one by one after cyclic removal of the intersection points of the warp (weft) yarns*: Based on one warp (weft) yarn, the intersection points are removed in a cyclic manner, and then the resultant warp (weft) yarns are accumulated one by one for obtaining a new woven construction.

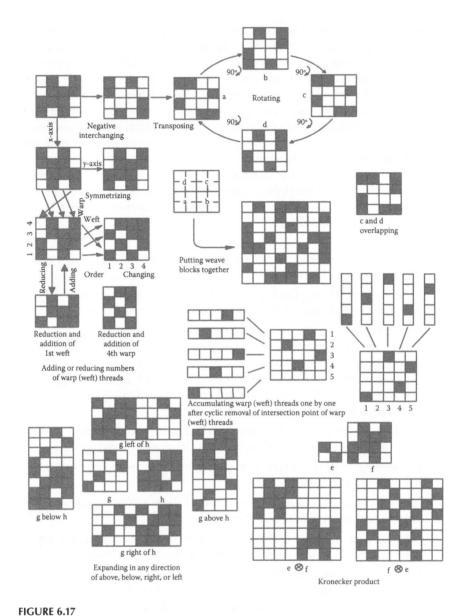

FIGURE 6.17
Principles of transformation of weaves. (From Ping, G., and Lixin, D., *Journal of the Textile Institute*, 88, I, 3, 265–381, 1997, CRC Press.).

11. *Kronecker product*: According to the requirement, a multiplicand or a multiplier is assumed, and a Kronecker is allowed with a certain weave matrix, thus changing the weave's arrangement to form a new construction. (The concept of a Kronecker product is explained in the Appendix.)

6.8 Process of Drawing In

Once the drafting plan has been decided, one needs to choose a reed of the appropriate count and width. This depends partly on the width of the targeted fabric and partly on the loom that is going to be employed for production.

Warp yarns, coming either from a single beam or from a pair of beams, need to be drawn through drop wires, heald eyes, and the reed, in that order. If drop wires are of the open type, then they may be mounted on a warp sheet after the yarns have been drawn in. Otherwise, the drawing-in sequence needs to be maintained.

In order that yarns from a warp sheet are held under tension and kept in properly aligned condition, warp beams are mounted on suitable frames and prepared. This frame with a warp sheet is then transported to the drawing-in frame.

The drawing-in process may be manual or even fully automatic. The manual process is labor-intensive and time-consuming. The modern concept of flexibility in production systems and low lead time cannot be put into practice with a manual drawing-in system. It is not uncommon to find a weaving department with very modern preparatory and weaving machines although the drawing-in section is entirely manual. Such a bottleneck makes the production system very rigid, needing a very high response time.

The drawing hook, shown in Figure 6.18, is central to this process. In a manual system, the drawing hook is operated by a human being whereas it would be machine-operated in the event of an automatic system. The figure shows a drawing hook pulling in a yarn through the eye of a drop wire and

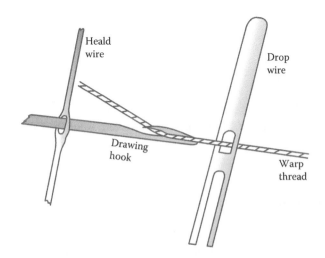

FIGURE 6.18
Drawing hook.

about to pull it through the heald eye. Evidently, size of the drawing hook and its surface properties are crucial in ensuring a good performance.

Subsequent to completion of the drawing-in process, the edge of a warp sheet sticking out in front of the reed needs to be properly secured so that no thread can slip back. This can be done by splitting the sheet into small groups and tying individual groups into suitable knots. This series of knots of warp yarns in front of the reed can be opened up, one by one, during the subsequent beam-gaiting process and tied across the width to a cloth strip already mounted on the loom. Alternately, as practiced in more modern systems, a biodegradable plastic sheet can be welded to the edge of the warp sheet extending beyond the reed, and this very sheet of plastic can be directly secured to the fabric take-up system of a loom. Subsequently, this sheet may be discarded and allowed to degrade in nature. This system obviates the time-consuming tying-in process and also reduces wastage of yarns.

Warp beam with yarns drawn in through drop wires, heald eyes, and the reed are mounted on a truck and transported to a loom for the final weaving process.

References

Chen X, Knox R T, McKenna D F, and Mather R R (1996). Automatic generation of weaves for the CAM of 2D and 3D woven textile structures, *Journal of the Textile Institute*, 87, Part I, No. 2, 356–370.

Milasius V, and Reklaitis V (1988). The principles of weave–coding, *Journal of the Textile Institute*, 79, 598–605.

Ping G, and Lixin D (1997). Study of the mathematical model for fabric weave design and its applications, *Journal of the Textile Institute*, 88, Part I, No. 3, 265–381.

Further Reading

Grosicki Z J (2004). *Watson's Textile Design and Color*, Woodhead Publishing, Ltd., Cambridge, UK, ISBN: 1 85573 995 X.

Appendix

For any two arbitrary matrices, $A = (a_{ij})$ and B, we have the direct product or Kronecker product $A \otimes B$ defined as

$$\begin{bmatrix} a_{11}B & a_{12}B & \cdots & a_{1n}B \\ \vdots & \vdots & \cdots & \vdots \\ a_{n1}B & a_{n2}B & \cdots & a_{mn}B \end{bmatrix}$$

It is to be noted that if A is m-by-n and B is p-by-r, then $A \otimes B$ is an mp-by-nr matrix. This multiplication is not commutative.

For example,

$$\begin{bmatrix} 1 & 2 \\ 3 & 1 \end{bmatrix} \otimes \begin{bmatrix} 0 & 3 \\ 2 & 1 \end{bmatrix} = \begin{bmatrix} 1\times0 & 1\times3 & 2\times0 & 2\times3 \\ 1\times2 & 1\times1 & 2\times2 & 2\times1 \\ 3\times0 & 3\times3 & 1\times0 & 1\times3 \\ 3\times2 & 3\times1 & 1\times2 & 1\times1 \end{bmatrix} = \begin{bmatrix} 0 & 3 & 0 & 6 \\ 2 & 1 & 4 & 2 \\ 0 & 9 & 0 & 3 \\ 6 & 3 & 2 & 1 \end{bmatrix}$$

7

Primary and Secondary Motions of a Weaving Loom

Warp and weft yarns are interlaced and converted to a woven fabric on a loom. A weaving loom can be classified in many ways, some of which are listed here:

- Driven by electrical power (power loom) or by human power (hand loom).
- A shuttle loom or a shuttleless loom.
- Flat loom or circular loom, depending on the arrangement of warp sheet in a plane or along the periphery of a cylinder, the product, accordingly, being either a flat sheet or a tube.
- Monophase or multiphase: One cycle of all basic motions of a loom being completed in a specific order within one loom cycle in the case of a monophase loom, and in a multiphase loom, basic motions of succeeding loom cycles are executed simultaneously.

All discussions in this chapter relate to flat-bed, monophase, power-driven, plain-shuttle looms.

7.1 Basic Machine Elements

The prime mover is a motor, which transmits motion to the loom pulley by V-belts (Figure 7.1). This pulley is mounted on one end of a crankshaft. The crankshaft carries two cranks along its length. These cranks occupy the same plane, and while the shaft rotates, its cranks trace out the periphery of a circle. The radius of the crank and that of the corresponding circle are the same. On the opposite end of the crankshaft is mounted a gear wheel that meshes with another gear wheel of the same pitch but with double the number of teeth. This larger wheel is carried by the bottom shaft, which, therefore, rotates at half the speed of the crankshaft. The bottom shaft may carry the shedding tappets (cams) or may drive the tappet shaft over a pair of gear wheels. Tappets are eccentric discs, which displace antifriction bowls that

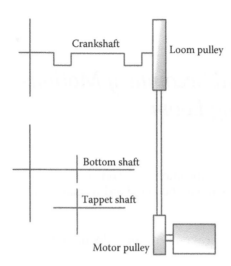

FIGURE 7.1
Transmission of motion on a plain loom.

are pressed against their active surfaces. Thus, a rotational motion of tappet is translated into a to-and-fro or up-and-down motion of the follower bowl. Such bowls are either linked to or are directly mounted on levers. Hence, the corresponding levers execute a specific time-displacement profile governed by the contour of the corresponding tappet.

In the sectional view of a loom (Figure 7.2), it is observed that the pair of shedding tappets (cams) operates a pair of bowls mounted on two treadle levers. The treadle levers are fulcrummed at one end while their free ends are linked to vertical heald shafts. The shedding tappets are positioned at a phase difference of 180°. Thus, the two tappets press the corresponding treadle levers down in a sequential manner such that when one heald occupies the uppermost position, the other would be in the lowermost. In this process, a split, known as the shed, is created in the warp sheet. A weft is propelled through this shed during the weaving process. The net displacement of a heald and therefore of the corresponding warp sheet controlled by the heald can be easily worked out as depicted in Figure 7.3.

The bottom shaft, as shown in Figure 7.3, also carries two more tappets, namely the picking tappets, at its two ends. These tappets are also positioned 180° apart and are employed for picking a shuttle once from the left-hand box of the loom and then from the right-hand box. A shuttle carries a pirn containing weft yarn. It is therefore apparent that in one complete rotation of the bottom shaft, two cycles of shedding and picking take place. During this period, the crankshaft too rotates twice. Hence, one rotation of the crankshaft accounts for one loom cycle. Indeed, the loom rpm represents the number of crank rotations per minute, which, in turn, stands for the number of weaving cycles per minute.

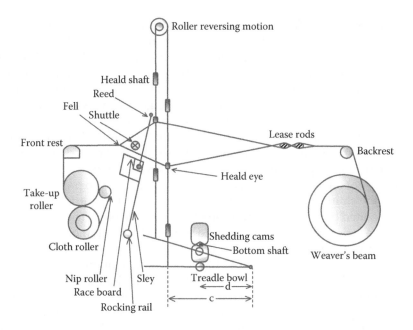

FIGURE 7.2
Sectional view of a plain loom.

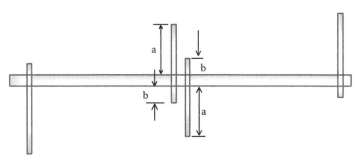

Net movement of Tr.bowl = (a - b)
Net movement of H.shaft = (c - d)(a - b)

FIGURE 7.3
Tappet shaft.

Positioned in front of the heald frames is a reed through which warp yarns pass and finally merge into the fell of the cloth. The reed is mounted on a sley, which is a horizontal beam supported near its two extremities by two upright frames, known as sley swords. The two sley swords are mounted on a rocking shaft, which, in turn, is supported at its two ends by bearings mounted on the loom frame. The rocking shaft acts as a fulcrum for the sley swords and supports the entire sley system. In this way, the entire load of the sley is passed onto the loom frame. The two sley swords are linked to two cranks by two crank arms. As the cranks execute a rotary motion, the ends of the crank arms linked to the cranks also rotate, but their other ends that are linked to the sley swords execute a to-and-fro motion. This results in an oscillatory motion of the sley. The rotational motion of the cranks is thus converted to an oscillatory motion of the sley by a four-bar linkage system. As the reed is carried on a sley, it also executes an oscillatory motion with the sley. This motion of the reed is responsible for smooth passage of the shuttle through the shed and for pushing the weft, left behind by the shuttle in the shed, to the fell of the cloth (beating-up motion) for creation of the woven fabric.

The sley not only supports and oscillates the reed, but also provides a pathway for the shuttle. While flying through the shed, the shuttle glides along the lap created between the bottom shed line and the reed. The bottom shed line rests a little on or just above the race board, which is a smooth plate fixed to the top of the sley. Hence, a lap is effectively created between the race board and the reed. As the shuttle has a certain height and width, the triangle enclosed between the shed lines, the fell of the cloth, and the reed must be appropriately large to accommodate the shuttle. The area of this triangle actually keeps on changing continuously during a loom cycle because the two shed lines execute vertical movements and the reed an oscillatory one. Evidently, the motions of the warp lines (shedding), of the reed (beating-up), and of the shuttle (picking) must be synchronized for a smooth functioning of the loom.

Besides the important machine elements described, Figure 7.2 also shows the material flow. The warp sheet is unwound from a weaver's beam by a let-off motion and is taken through a system of drop wires (not shown in the figure) and lease rods. Each drop wire rests on a warp yarn and, in the event of a thread break or a thread falling slack, signals the warp stop motion to bring the loom to a halt. The warp sheet is then taken through a pair of lease rods, through respective healds, and through the reed before it merges into the cloth fell. The woven cloth is pulled across the front rest and down into the cloth take-up zone and is subsequently wound properly onto a cloth roll.

The motions of various loom elements can be broadly grouped into three classes, namely the primary, secondary, and auxiliary motions:

- The primary motions are shedding, picking, and beating up.
- The secondary motions are let off and take up.
- The important auxiliary motions are warp stop, weft stop, and warp-protecting motions.

All motions are synchronized with respect to the angular position of the crank. For this purpose, the circle traced out by the crank during one loom cycle is employed in a figurative sense. This circle is divided into four segments, each covering 90° of displacement of the crank from its basic zero position. At its zero position, the crank is closest to the loom operator stand-ing on the side of the cloth roll. This position is conventionally termed as the front center. While moving away from front center, the crank normally passes through bottom center (90° position), back center (180° position), and top center (270° position) before returning to the front center position in one complete loom cycle. These positions are illustrated in Figure 7.4. As the crank rotates with a certain angular velocity of ω rad/s or at a speed of *n* rpm, the time period for one complete loom cycle is given by $2\pi/\omega$ seconds, which is equivalent to $60/n$ seconds. This time is consumed by the crank in tracing out one circle, and therefore, the circle can be treated both as a time scale and a crank displacement scale. Such a scale employed for synchronizing different motions of a loom is termed the timing diagram of corresponding motions.

The figure additionally shows a hatched segment of 120° duration span-ning the 30° to 150° position of the crank. A heald expected to occupy the bottom shed line reaches its location at 30° crank displacement. So too does the heald expected to be in the top shed line. They remain in their respective positions during this time of 120°, that is, they dwell in their positions. Hence, this segment is indicative of the period allotted to the dwell of shed. After the 150° crank position, healds start moving in opposite directions such that, at the top center, both are level and all warp yarns occupy the same plane. Subsequently, the healds cross over and reach their extreme ends at the 30° position of the crank in the subsequent cycle. Hence, a heald moves during a period of 240° for changing position from one shed line to the other.

As the crank moves through its centers, the reed is oscillated by a four-bar linkage system (Figure 7.5). A reed is closest to the fell of the cloth at around front center and most distant from it around back center. Therefore, the area of the triangle enclosed between the fell of the cloth, reed, and the two shed lines is equal to zero at the front center and keeps on growing as the reed

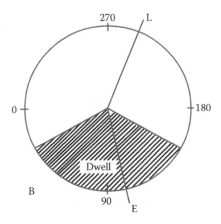

FIGURE 7.4
Timing diagram.

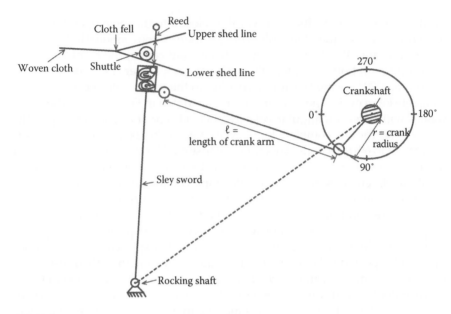

FIGURE 7.5
Four-bar linkage sley drive.

moves toward back center. However, between the 150° position and back center, the shed lines start moving toward each other. During this period, therefore, the reed and heald movements have an opposite effect on the area of the triangle. A shuttle can enter the shed only when this area is sufficiently large and must leave the shed before this area becomes critically small. The entry point of the shuttle in a shed is indicated by the line E (approximately 105° to 110°) in Figure 7.4, and its exit point (approximately 220°) is shown by the line L. It is, however, very odd that although the shed is fully open at 30°, picking cannot effectively take place for want of required space in the triangle. It is also somewhat surprising that the shuttle continues to remain in the shed for approximately 70° duration even after the shed lines have started moving closer.

Part of this paradox is resolved if the time displacement equation of the sley is considered. This can be stated as

$$S = r[1 - \cos\theta + (r/2\ \ell)\sin^2\theta] \qquad (7.1)$$

where
   S = displacement of the reed away from the front center position
   r = crank radius
   θ = angular displacement of the crank from its front center
   ℓ = the length of the crank arm

The time displacement function $S = f(\theta)$, normalized with respect to r has been plotted in Figure 7.6 for the period between front center and back center. The ratio (r/ℓ) in Equation 7.1 is a parameter of the loom, commonly termed eccentricity. In order that eccentricity e can be equal to zero, the crank radius has to be zero. Under this condition, the reed does not move at all. On the other hand, r is never so large as to be equal to ℓ as the space between the crank and the sley is taken up by healds. Increasing the value of r would mean increasing the sley sweep. This would not only slow down a loom, but also result in higher abrasion of the warp threads with the reed wire and reduction in effective space available for the healds between the reed and the crankshaft. However, a higher sley sweep permits the use of a larger shuttle, which is required when weaving thicker weft yarn. Similarly, reducing ℓ would effectively reduce the space for healds. Hence, the value of e lies between 0 and 1.

In Figure 7.6, two curves are shown for values of e equal to 0.2 and 0.5, respectively. Both lie above the SHM curve given by $[S/r] = 1 - \cos\theta$. The line drawn parallel to the x-axis within the plot refers to the minimum displacement of the reed from its zero position for allowing entry of the shuttle into the shed. It is observed that, for higher values of loom eccentricity, the shuttle can enter the shed earlier. As the reed would return to front center along a path that would be a mirror image to the plot, it can be inferred that, for higher values of loom eccentricity, the shuttle can also leave the shed later. Hence, higher values of loom eccentricity provide for a larger flight time to the shuttle.

The velocity and acceleration profiles of the reed for values of e equal to 0.2, 0.3, and 0.4 are plotted in Figure 7.7. The beating-up process starts logically

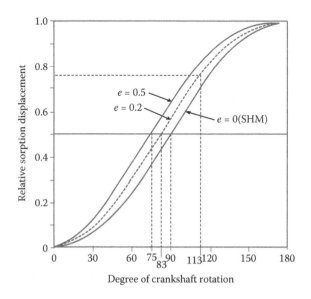

FIGURE 7.6
Eccentric motion of sley.

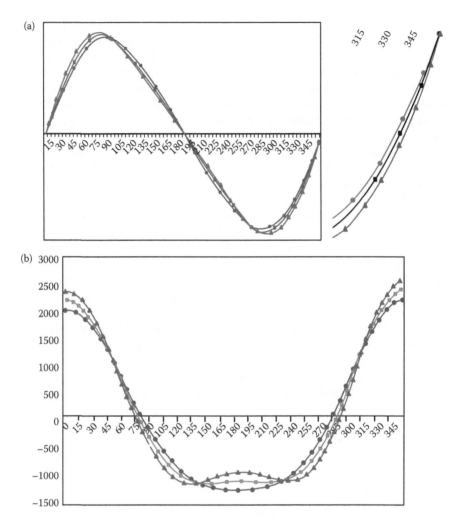

FIGURE 7.7
(a) Velocity profiles of the reed. (b) Acceleration profiles of the reed.

much earlier than the front center as, at front center, the weaving cycle is completed. Indeed, the higher the resistance to beating up, which is usually associated with heavier fabrics, the earlier the beating up process would start. In this process, energy of the sley is transferred to the weft thread so that the same can move to its destined location, overcoming resistance offered by the large number of interlacing warp threads. Considering the velocity profile of the sley in the blown-up segment from 315° onward, it is observed that at the same crank position, a higher loom eccentricity makes the sley move at a higher velocity toward the front center. This effect is reinforced when a higher resistance is offered by the weft as an earlier onset of the process would result in a still higher energy transfer. Hence, higher loom eccentricity is beneficial for weaving heavier fabrics.

7.2 Principles of Shedding

Shedding, one of the three primary motions in weaving, is aimed at splitting threads in the warp line into two groups, namely the top (or the upper) shed line and the bottom (or the lower) shed line. A simplistic representation of a shed geometry is shown in Figure 7.8. The various elements are listed here:

> **ABCA:** the front shed; **BCDB:** the rear shed; **ABD:** the top shed line; **ACD:** the bottom shed line
>
> **1:** breast beam; **2:** heald; **3:** lease rods; **4:** back rest
>
> α: shed angle; L_F: length of front shed; L_R: length of rear shed; h_T: lift of top shed line; h_B: lift of bottom shed line

The most important element in the shedding process is the heald. Warp threads of the top and bottom shed lines are controlled by the respective heald eyes. The nature of movement of the healds as well as their lift along with their positioning with respect to the fell of the cloth A and the point D

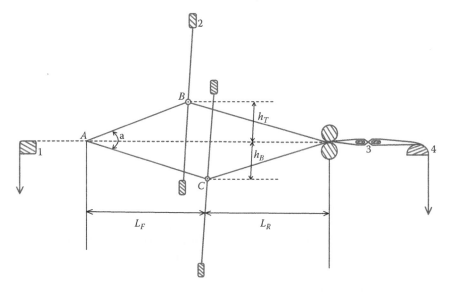

FIGURE 7.8
Geometry of a shed.

play decisive roles in the weaving process. Healds are moved by shedding systems, the simplest of which is controlled by shedding tappets.

7.2.1 Shedding Tappet

The principle of construction of a shedding tappet for a plain weave is illustrated in Figure 7.9. A plain weave is generated over two picks during which the bottom shaft rotates only once. Hence, the corresponding tappet is mounted on the bottom shaft. This tappet has to keep the corresponding treadle bowl in the lowermost position during 120° rotation of the crankshaft, which accounts for dwell of the corresponding heald at the bottom shed line. This dwell duration is equivalent to a bottom shaft rotation of 60° only. Subsequently, the bowl has to be allowed to move to the topmost position over a 240° displacement of the crankshaft, equivalent to a 120° displacement of the bottom shaft. The bowl dwells at the topmost position for a bottom shaft displacement of 60° and then is pushed down to the lowermost position again over a bottom shaft displacement of 120°. This sequence is tabulated in Table 7.1.

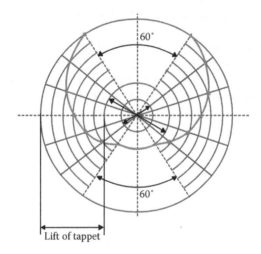

FIGURE 7.9
Construction of a tappet.

TABLE 7.1

Equivalence of Motions of Treadle Bowl, Crankshaft, and Bottom Shaft

Movement of Treadle Bowl	Crankshaft Displacement	Bottom Shaft Displacement
Dwell at bottom position	120°	60°
Change over from bottom to top position	240°	120°
Dwell at top position	120°	60°
Change over from top to bottom position	240°	120°

The innermost circle in Figure 7.9 represents the bottom shaft, and the next concentric circle has a radius equal to the distance between the center of the bottom shaft and the line of contact of the treadle bowl with the tappet at its closest approach to the bottom shaft, that is, when the bowl is at its topmost position. The outermost circle has a radius equal to the distance between the center of the bottom shaft and the line of contact of the treadle bowl with the tappet at its farthest point from the bottom shaft, that is, when the bowl is at its lowest position. The annular ring between these two circles is divided into four segments of 60°, 120°, 60°, and 120°, respectively, corresponding to the sequence listed in the table. The arcs of these two circles enclosed within each of the two 60° sectors represent the contour of the tappet during the periods of dwell. A complete profile of the tappet is generated by joining these two arcs by smooth curves. If linearity is assumed for the displacement function of the bowl, then the following procedure is adopted for joining the two arcs.

The surface of the tappet enclosed between the two 120° segments is first divided into a large number of concentric segments of equal thickness. A large number of radial lines are then drawn on the two 120° segments so that these lines are spaced apart at equal distance along the circular arcs, the number of radial lines being equal to the number of concentric arcs. The points of intersection of these radial lines with the arcs provide coordinates of the tappet profile during the changeover period. The fundamental principle governing this approach is based on the manner of displacement of a point along the diameter of a circle. If, instead of linearity, some other displacement functions, such as SHM or cycloidal or trapezoidal (see Figure 9.11 later in the chapter), is chosen, then corresponding changes have to be adopted for working out the coordinates of the tappet profile along the changeover period, keeping the same guideline in view.

For weaves other than plain and repeating on a larger number of picks, the corresponding tappets are mounted on a counter shaft or tappet shaft that is geared to the bottom shaft. The tappet shaft rotates once during one complete weave repeat. Thus, for a weave repeating on four picks, the counter shaft would rotate once during four revolutions of the crankshaft. The table relating treadle bowl displacement has then to be worked out, keeping the relationship between angular displacements of the counter shaft and crankshaft in mind. Such a table for a 7/1 weave is listed in Table 8.1.

7.2.2 Shed Geometry

7.2.2.1 Shed Angle

The magnitude of shed angle α has functional importance. Tensions in the top shed line and in the bottom shed line can be resolved into vertical and horizontal components. The sum total of the horizontal components would be balanced by the cloth tension. The vertical components have a shearing effect between yarns of the top and bottom shed lines. If the shearing effect is not large enough, yarns of the two groups may not separate properly and result in an unclean front shed. This would impair the picking process. As the magnitude of vertical components grows with shed angle, a minimum value of the latter must be maintained for ensuring a clean front shed. On the other hand, too large a value of α would put excessive strain on warp yarns, leading to breakages. Commercially, an angle of 20° has proved to be acceptable, whereby this angle can be smaller for smooth yarns with low warp density (Hollstein, 1978). On the other hand, rough and hairy yarns with high warp density would necessitate a higher front shed angle.

7.2.2.2 Cyclic Variation in Yarn Strain

On shed opening, an additional length of yarn sheet is called for to maintain warp tension at the same level. This cyclic demand can be met either through extension of the warp sheet or through release of an additional supply of warp length controlled by a compensating system. In practice, both factors play some role in satisfying this demand, whereby the type of material being woven determines their relative proportion. For very elastic yarn material having low modulus, most of the additional demand can be met by yarn extension itself. On the other hand, for a less elastic and brittle material with high modulus, most of the demand has to be met from an additional supply released by a compensating system. In both cases, the important criterion is the elastic extension of the yarn.

As each yarn segment has to undergo many shedding cycles before the same gets embedded into fabric, fatigue resistance of the material also plays an important role in this matter. Yarns of lower fatigue resistance need to have more support of the compensating system than yarns with higher resistance. The latter can absorb a greater amount of dynamic extension and still not suffer breakage.

7.2.2.3 Shed Envelope

Splitting of warp threads into two shed lines by different healds takes place in many parallel planes. If the lift of each heald is kept the same, then a situation depicted in Figure 7.10a results. Such an unclean shed would not permit smooth insertion of weft threads. A clean shed, on the other hand, would be formed if the lift of successive healds were progressively increased in the manner shown in Figure 7.10b. The disadvantage of such a shed formation is that the warp yarns controlled by different healds would need different quantities of additional thread supply during the opening of a shed. This raises the question of the desirable lift that different healds should follow so that strain in each yarn group remains nearly the same while a sufficiently clean front shed forms. An elliptical envelope of the fully open shed

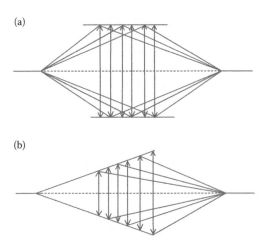

FIGURE 7.10
Linear envelopes of shed. (a) Equal displacement of each heald. (b) Progressively higher displacement of succeeding healds.

works out to be the optimum solution (Hollstein, 1978). If the extreme ends of the front and rear sheds are stationed, respectively, at the two foci of an ellipse, and the heald eyes at their top and bottom positions are adjusted to be always on the periphery of said ellipse, then yarns from each heald would be along two lines joining the two foci with a point on the periphery of the ellipse. Hence, as per rules governing the geometry of an ellipse, the extended length of threads from each heald would be the same, and therefore, requirement of additional yarn from each heald would also be the same although the healds are individually lifted to different extents.

7.2.2.4 Asymmetric Shed

Although the front shed needs to be suitably designed, keeping the constraints imposed by weft insertion in view, the rear shed can be so designed as to influence the additional cyclic requirement of yarn (Hollstein, 1978). If

$$\delta = \text{permissible yarn strain}$$

$$\ell_1 = \text{length of the front shed}$$

$$\ell_2 = \text{length of the rear shed}$$

$$h_1 = \text{lift of the heald in the upper shed,}$$

then

$$\delta = h_1^2/2 \, \ell_2 \, \ell l_1 = \text{constant}/\ell_2 \tag{7.2}$$

The length of the rear shed can therefore be used to control strain in warp yarns during shed formation. A longer rear shed results in a lower warp strain than a shorter one. With increasing strain, the stress in the warp goes up. Therefore, a longer rear shed is chosen when the warp tension in the open shed needs to be low. On the other hand, a shorter rear shed reduces the chances of adjacent warp threads clinging to each other and creating an unclean passage for the weft carrier. A longer rear shed would therefore be employed for smoother yarns, which do not tend to cling to each other. A substantial difference in length between the front and rear shed leads, however, to a large relative movement of warp yarn in the horizontal direction with respect to the vertical planes of shedding as the difference in additional yarn requirement between the front and rear sheds in the open shed is quite high. This leads to a greater rubbing between the heald eyes and the threads.

For plain weave and equal amount of lift of upper shed and lower shed lines, the vertical components of warp tension balance each other, and the fabric supports only horizontal force. When, however, the lifts of the upper and lower shed lines are unequal, an unbalanced vertical component acts on the fell of the fabric, displacing the same from the equilibrium position of the closed shed. This creates an angle between the fabric plane and the warp line. A similar situation would crop up when shedding the warp line for unbalanced weaves, such as 1/2 twill or 1/4 satin, even if lifts of the upper and lower shed lines are the same. In fact, weaving unbalanced weaves without deflecting the fell of the cloth can be accomplished by suitably choosing the lift of the upper and lower shed lines such that

$$[h_1/h_2] = [n_2/n_1]^{1/3} \tag{7.3}$$

In the expression, n_1 and n_2 represent the number of yarns in the upper and lower shed lines, respectively, at any instant (Hollstein, 1978).

Weaves for which the ratio of n_2/n_1 does not remain constant, unbalanced vertical forces would act both on the cloth fell as well as on the lease rods. These movements are undesirable for the weaving process and need to be controlled through suitable measures.

7.2.2.5 Staggered Shed

According to the timing of shed line movement, changeover of shed lines takes place during a 240° displacement of the crank in such a manner that all crossing healds are usually level at top center (270°). At this point, all heald wires and the yarns contained in them are forced into the same plane. If warp yarns are very closely packed, then a situation may arise in which abrasion among yarns as well as between the yarns and heald wires becomes critically high. To avoid such a possibility, two solutions can be adopted either in isolation or even together. One may opt for a larger number of healds than that required by the weave repeat. This would space out the heald wires

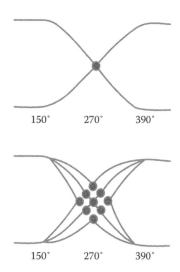

FIGURE 7.11
Staggered motion of shed lines.

along the plane of symmetry of the warp sheet, reducing therefore the wire density. One may also adopt a staggering of healds such that different healds follow different displacement lines during the changeover period. An illustration of this phenomenon is depicted in Figure 7.11, in which three healds move from the bottom to the top shed line while another three move in the opposite direction following different displacement profiles. As a result, instead of one crossing point for all yarns, a total of nine crossing points displaced from one another in time as well as in space arise.

7.2.2.6 Geometry of Warp Line

A warp line is illustrated in Figure 7.12. The parallelogram ABCD represents a symmetric shed with the warp line inclined to the horizontal plane. In such an event, warp yarns of the top and bottom shed lines undergo similar

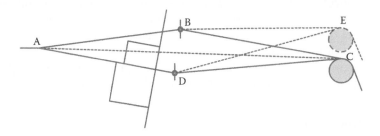

FIGURE 7.12
Differential tensioning of shed lines. (From Marks, R., and Robinson, A. T., *Principles of Weaving*, The Textile Institute, Manchester, UK, 1976.)

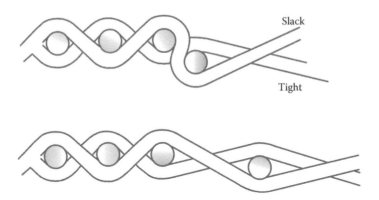

FIGURE 7.13
Effect of differential tensioning of shed lines on fabric cover and packing of yarns. (From Marks, R., and Robinson, A. T., *Principles of Weaving*, The Textile Institute, Manchester, UK, 1976.)

cyclic fluctuation in tension. If, however, the backrest is raised from C to E, then the top line ABE becomes shorter in length compared to the bottom line ADE. Hence, although the frequency in tension fluctuation in the two shed lines would be the same, their amplitudes would be different. Thus, at the extreme positions in the open shed, one set of yarn would always be slacker than the other. When such a strain difference between adjacent yarns can be absorbed by suitable means, the differential tensioning of two shed lines can be put to good use in packing more picks per unit of fabric length as well as in improving fabric cover. The effect of differential tensioning of warp lines on pick density is illustrated in Figure 7.13.

The pick number four from the left, inserted last, is forced to occupy a lower position than the fabric plane in view of higher tension in the lower shed line at the instant of beat up. Simultaneously, the slacker yarn line causes a slight upward movement of pick three. This results in a reduction in the horizontal gap between the adjacent picks. Further, the slacker yarns can move laterally much more easily than the tighter yarns during beat up in a crossed shed. Such a necessity arises owing to the crimping phenomenon of warp and weft yarn that develops during beat up. The resulting balance of forces decisively affects uniformity of thread spacing and, hence, the fabric cover.

The effect of beating in a crossed shed is also illustrated in Figure 7.13. A comparison of the top and bottom figures shows the restraining effect of a greater degree of yarn crossing in holding the pick being beaten up from springing back away from the cloth fell once the reed starts receding.

7.2.2.7 Types of Shed

Different types of shedding mechanisms yield different heald displacement profiles. This is illustrated in Figure 7.14 for one heald, which has to keep the yarns in the top line for two consecutive picks and bring them down and

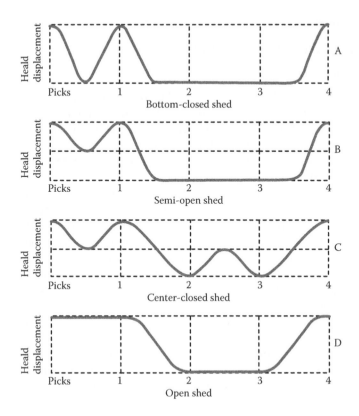

FIGURE 7.14

Types of shed. (From Marks, R., and Robinson, A. T., *Principles of Weaving*, The Textile Institute, Manchester, UK, 1976.)

keep them in the bottom line over the next two picks. The resultant curves plotted as a function of picks depict the motion of this heald.

The bottom (or the top) closed shed results in a great deal of wasted movement and, therefore, affects rate of production negatively. Such systems are encountered in a single lift dobby or Jacquard employed in hand looms. The semi-open shed, encountered in some double-lift Jacquards, involves reduced wasteful movement of yarns as compared to bottom or top closed sheds. Between successive picks, the yarns do not quite merge into a single layer. In a center-closed shed, encountered commonly in some form of tappet shedding, all yarns return to the central plane between successive picks. Here again, a considerable amount of wasted yarn movement is observed. The most economical yarn displacement is effected in open shedding. Cam-driven, double-lift dobbies generate this kind of shed. It must, however, be pointed out that in open as well as in semi-open shedding processes, an additional heald-leveling system is called for so that all yarns can be brought to the same level for the mending of a broken yarn.

7.3 Principle of Shuttle Picking

7.3.1 Mechanism

The shuttle is an oblong-shaped wooden body, hollow in the middle to accommodate a pirn and tapered at its ends with sharp metallic tips. Shuttle walls are thin, and the whole body is aerodynamically shaped for ensuring a smooth free flight. The shuttle for a plain loom is equipped with a tongue, which holds a pirn in position, and some shuttles have metallic flaps to hold the hollow weft spool securely. The shuttle for an automatic pirn-changing loom has a prong-shaped tongue designed to grip the head of a pirn between two spring-loaded clamps. Weft yarn contained in the pirn is drawn through thread guides and pulled out of a shuttle. When this free weft end is gripped securely at a fixed point and the shuttle made to move away, the weft yarn unwinds from the pirn. As a result, a moving shuttle leaves behind a trail of weft. The shuttle is propelled by a picking system and flies across the reed from one shuttle box to another. The reed and the shuttle boxes are mounted on a sley, which is supported by two sley swords.

The sectional view of a shuttle box is shown in Figure 7.15. The wooden sley (1) has a smooth top surface, namely the sley race (2), which supports and rubs against the bottom shed line, across which the shuttle moves. The shuttle (6), when it lands in a shuttle box, is guided along its walls by the box front (4) and the box back (3). An additional wooden strip (5) mounted on the inside of the box back limits vertical movement of the shuttle. Accordingly, the shuttle is constrained to travel along the sley from one box to the other. A spring-loaded element (7) named the "swell" is mounted on the box back for retarding the incoming shuttle to a halt. Each shuttle box is equipped with

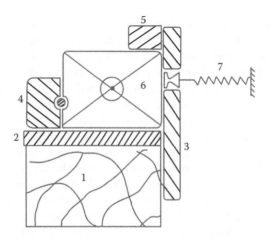

FIGURE 7.15
Sectional view of a shuttle box.

a picker, a solid rectangular parallelepiped slotted at the center and having recesses at each center of its two vertical surfaces. Each of these recesses matches the metallic tips of the shuttle. When the shuttle rests within a box, one of its tips remains firmly pressed against the recess of the corresponding picker. Hence, any movement of the picker toward the reed center would propel the shuttle in that direction.

A typical picking system is shown in Figure 7.16. The picker (1) is housed within a shuttle box, represented by two parallel discontinuous lines. The picking stick (2) passes through a slot in the picker and is fulcrummed at its end (3). A downward movement of a side lever (4) causes the band (5) to pull the picking stick laterally, thereby making the picker move from left to right, pushing, in this process, the shuttle into the warp shed. The downward motion of the side lever is realized by the smart blow of the antifriction bowl (6) on a shoe (7), which is mounted on the side lever. The side lever is fulcrummed at its end (8). The antifriction bowl is mounted in a slot of a disc (9), which, in turn, is mounted on the tappet shaft (10). The springs (11) and (12) ensure that the side lever as well as the picking stick return to their original positions after the antifriction bowl (6) has moved away from the shoe (7). The timing of a pick can be altered by adjusting the position of the bowl in the slot of the disc, and the intensity of the blow can be adjusted by moving the fulcrum point (8) vertically. A similar mechanism is mounted on the other side of the tappet shaft with a phase difference of π radians.

Owing to the nature of the picking mechanism, the picker would move along the arc of a circle if it is rigidly mounted on a picking stick. Such an eventuality would result in the loss of effective power transmitted to the

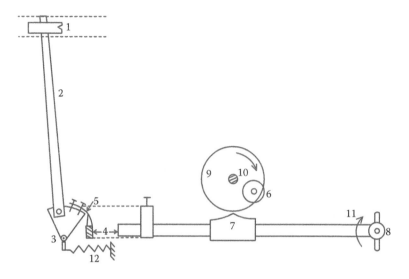

FIGURE 7.16
Side lever picking system.

shuttle. Hence, the picker is loosely mounted on a picking stick, and the wooden strip (5) mounted on the inside wall of the box back (Figure 7.15) ensures that the picker is prevented from moving upward.

Such a picking system results in a harsh pick suitable for propelling a heavy shuttle. However, the accompanying vibration would not permit a high loom speed. The harshness of the pick results out of the impactful nature of the blow by a bowl on the shoe. However, if the shoe and bowl were to remain always in contact with each other, then the energy transfer would not be so abrupt although the resultant picking would be gentler. In such a situation, loom speed could be raised, but the shuttle mass has to be correspondingly lower than in the case of a harsh pick.

The bowl and cone arrangement, found in cone under-pick motions, is illustrated in Figure 7.17. Here, the antifriction bowl is mounted at the end of a lever and always rolls along the surface of an eccentric cone. The cone (2) is mounted on the tappet shaft (1), and the antifriction bowl (3) is mounted on an L-lever fulcrummed at (4). The tail end of the L-lever is linked to the picking stick (6) through a band (5).

When a cone under-picking motion is employed on a loom equipped with pirn-changing motion, the wooden strip mounted on the inside of the box back has to be so thin that it can no longer restrict the vertical motion of the picker. Therefore, the picker is securely fastened to the picking stick, which, however, is not fulcrummed at its other end in the conventional manner. This end of a picking stick is attached to a shoe (7) that runs on a horizontal

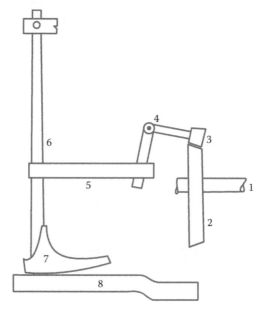

FIGURE 7.17
Cone under picking system.

plate (8). The lower surface of the shoe is curved in such a manner as to maintain a movement of the picker parallel to the base of the shuttle box. Such a system is known as a parallel pick. Because of the flexible nature of the system, the picking shoe has a tendency to lift off the surface of the plate, and therefore, the speed has to be kept at a relatively low value. This problem is overcome with a link-picking system.

7.3.2 Kinematics of a Picking System

If a loom is operated manually by turning the flywheel slowly, and the displacement of the shuttle noted as a function of crank position, then one finds that the shuttle moves a short distance toward the center of the reed and then comes to a halt. This displacement, when recorded graphically, yields the normal displacement curve, shown in Figure 7.18. The nature of this displacement is governed by the contour of the picking tappet. However, when the loom is operated by power, the nature of displacement of the shuttle as a function of the angular position of the crank is quite different. Indeed, there is a considerable lag, initially followed by an exponential rise as shown by the actual displacement profile.

The lag between the nominal and the actual displacement of the shuttle is caused by strain in the picking system, arising out of the impulsive nature of the input force and the inertia of the picking system. This strain results in a buildup of elastic energy within the system, which, on its release, accelerates the shuttle very quickly to its initial velocity of free flight. Indeed, the shuttle loses contact with the picker around the angular position of the crank at which the curves of nominal displacement and actual displacement intersect.

If M is the combined equivalent point mass of the shuttle and picker, λ is the stiffness of the system expressed in units of force/unit distance, s and x

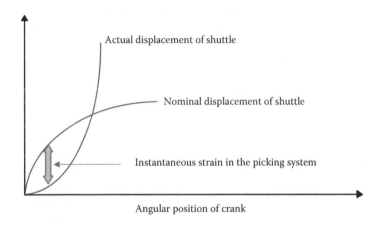

FIGURE 7.18
Profiles of shuttle displacement in shuttle box.

are the nominal and actual displacements of the shuttle at any instant, and a is the acceleration, then

$$Ma = \lambda(s - x)$$

If the ratio λ/M is written as n^2, then the equation of motion of the shuttle during the acceleration phase in the shuttle box can be expressed as

$$a = n^2(s - x) \tag{7.4}$$

The quantity n is the natural frequency of the system; the higher the values of n, the closer the profiles of the actual and normal displacements of the shuttle are.

If shuttle acceleration is to be maintained at a constant value A, then by integrating the expression (Equation 7.4), the shuttle displacement curves can be expressed by the following relationships.

$$x = (At^2/2) + Bt + C \tag{7.5}$$

and

$$s = (At^2/2) + Bt + C + (A/n^2) \tag{7.6}$$

The constants B and C can be eliminated if the boundary condition is imposed that at $t = 0$ both displacement and velocity are zero. This leads to expression of the nominal displacement profile:

$$s = (At^2/2) + (A/n^2) \tag{7.7}$$

This expression shows that if uniform shuttle acceleration is desired throughout the acceleration phase, then a finite nominal displacement at the beginning of the picking motion is necessary, a condition that can only be satisfied by a prestrained picking mechanism. Clearly, the under-pick motions described in the foregoing cannot therefore lead to uniform shuttle acceleration.

One can extend this exercise and impose other types of desirable natures of shuttle displacement, such as of constant nominal velocity, constant nominal acceleration, or of a sinusoidal actual acceleration, etc., and work out the corresponding profile of nominal displacement. After all, the design of a picking tappet—such as of the picking cone—would be governed by the desired profile of nominal displacement of the shuttle. A detailed account of this study is reported in an article authored by Catlow and Vincent (1951).

Nonetheless, one may simplify the issue and assume a uniform value of acceleration during the actual displacement of the shuttle so that some useful although approximate relationships can be derived. Based on this simplification, a velocity profile of the shuttle over one complete loom cycle has been suggested in Figure 7.19. Accordingly, the shuttle remains idle in the shuttle box during the time t_0, accelerates in the shuttle box during t_1, flies across the shed with a constant velocity v over t_2, then lands up in the opposite box and decelerates to a velocity of zero over the time t_3. Evidently, the time for one cycle T is a sum total of these four time intervals. If the loom rpm is n, the idle time t_0 is assumed to be exactly half of the total time T, t_1 is set equal to t_3, and s is the total distance traveled by the shuttle during one complete cycle (reed space plus twice the length of the shuttle box), then it is possible to derive the following expression (Hollstein, 1980):

$$n = (30va)/(v^2 + as) \tag{7.8}$$

A numerical analysis of this equation leads to the following observations:

- A combination of higher velocity and higher acceleration leads to a faster growth in loom rpm than that with a low acceleration and high velocity.
- A rise in total travel path of the shuttle lowers loom rpm.
- The distance traveled by the shuttle during the acceleration phase is considerably less than that traveled during retardation.

Further on, assuming that $s_1 = s_3 =$ the distance traveled by the shuttle in the two boxes and $s_2 =$ the reed space such that $s = s_1 + s_2 + s_3$, it is observed that a reduction in box length leads to a rise in loom rpm for the same flight velocity of the shuttle. Hence, from the point of view of productivity, a combination of short shuttle box, high acceleration, and high shuttle velocity is desirable.

A more realistic velocity profile of the shuttle shows a nonlinear rise with an initial lag followed by a rapid climb to the escape velocity. During the free flight, there is a steady drop to a lower velocity near the entry to the other box. Finally, within the shuttle box, a nonlinear drop to a state of rest

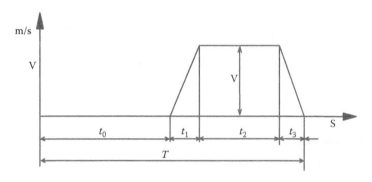

FIGURE 7.19
Simplified velocity profile of a shuttle.

happens owing to multiple impacts between the shuttle, the box front, and the swell. From the point of view of the weaving process, minimization of jerk on the shuttle and stoppage of the shuttle at a precise location within the box are critical. Axial jerk on a shuttle promotes sloughing off of coils of weft from the pirn, and variation in the stoppage point of the shuttle within a box results in variable picking force in the subsequent cycle.

Power consumed in picking can be derived from considerations of loom rpm (n), average velocity of free flight (v meters per second), angular displacement of the crank during free flight (θ degrees), the reed space (R meters), the length of shuttle (L meters), and the shuttle mass (m kilograms). The steps involved are as follows:

- Number of picks per second = $n/60$, which is equivalent to $6n$ degrees of crank displacement in one second.
- Time taken by the crank to get displaced by θ degrees is therefore $\theta/6n$ seconds.
- Distance covered by the shuttle in $\theta/6n$ seconds is approximately $R + L$ meters.
- Hence, the average velocity of the shuttle in free flight is $v = [6n(R + L)]/\theta]$ meters per second.
- Assuming no loss in velocity subsequent to loss of contact with the picker, the kinetic energy of the shuttle at the instant of its escape from the picker is $[\frac{1}{2}mv^2]$.
- This amount of energy has to be supplied by the picking system during each pick.
- Each picking cycle consumes $60/n$ seconds.
- Hence, power consumed by the picking system in kW is $[\frac{1}{2}mv^2]$ $[n/60]10^{-3}$.

On substitution of the expression of v in the equation, the resultant form is

$$\text{Picking power } P \text{ in kW} = [\{3m(R + L)^2n^3\}/\theta^2]10^{-4} \tag{7.9}$$

The effect of different variables on picking power is clearly discernible from the expression.

7.4 Beating Up

Once a pick of weft has been laid in a shed across the entire width of warp, and the weft insertion device has been safely withdrawn from within the shed, the beating-up process comes into play. The reed, which controls spacing between warp yarns, is firmly held in position by a sley. On rocking the sley forward, the pick of weft, which is trapped in the space between the reed wires and the fell of the cloth, is pushed against resistance offered by frictional forces arising out of interlacements between warp and weft. Energy supplied in pushing the just-inserted pick of weft in conjunction with warp yarn density and the coefficient of friction between crossing yarns as well as the tension in yarns affect cloth texture significantly. The machine element playing the central role in the beating-up process is the reed.

7.4.1 Reed

A reed is constructed like a comb and can be made of metallic wires, cords, and wooden strips. It can also be fully metallic. The fully metallic reed is the strongest and most expensive and finds application in weaving heavy and/ or high-quality fabrics as well as in shuttleless looms. Reeds on shuttle looms exhibit a curved profile for securely guiding the flying shuttle.

Reed wires are spaced evenly across the entire reed width and held firmly at the two extreme edges either in baulks or in metallic bars. In the case of a pitch-baulk reed, each wire is held between a pair of semicircular wooden strips at each of its two ends. A cord dipped in pitch is wrapped around each pair of wooden strips so as to firmly grip the reed wires and keep the same in position. Such cords also space the neighboring reed wires uniformly. The entire assembly of reed wires, baulks, and cords is finally coated with an additional layer of pitch and covered with colored paper. Reed wires in a metallic reed are fastened to metallic bars firmly with additional wires, and the whole assembly is soldered by dipping it in a molten bath of tin and lead alloy.

The space between adjacent reed wires is known as the dent. Warp yarns are drawn individually or in groups of two or more through these dents. The quality and appearance of a woven fabric is influenced considerably by proper selection of the reed and the denting plan. The denting plan is so chosen as to distribute the total number of warp yarns evenly across the entire desired width of the resultant fabric. It depends on warp density, warp count, weave, drafting plan, desired appearance of fabric, and lastly, also on the reed count itself. The count of reed (R) in the Stockport system is expressed as the number of dents in 2 inches and can be expressed as

$$R = 2\,Z/i\,\ell \tag{7.10}$$

where

Z = total number of warp yarns

i = number of yarns in a dent

ℓ = the length of a pick of weft

It is clear that reed count depends on the denting plan. The rule of thumb for denting is to choose as low a number as possible so as to minimize reed marks. This calls for very fine dents and, therefore, very thin reed wires as well. On the other hand, dents should allow easy passage to the thickest part of a warp yarn that one expects to encounter. The ratio of the width of a dent to that of a wire should also be higher than unity. The following empirical equation (Hollstein, 1978) can be chosen as a first approximation for selection of the reed.

$$[50t/R] \geq [3\{4T/\pi\rho1000\}^{0.5}] \tag{7.11}$$

where

t = thickness of reed wire in mm

R = reed count in the Stockport system

T = count of warp thread in tex

ρ = density of yarn in g/cc

Thus, for a cotton yarn of density 0.66 and count 30 tex, drawn through a reed of count 40, the thickness of wire should be around 0.58 mm. This would effectively yield a dent width of 0.7 mm, a ratio of 1.2 with the wire thickness. A higher ratio and, therefore, a thinner wire can be chosen for high-quality steel wires.

7.4.2 Mechanics of the Beating-Up Process

7.4.2.1 Development of Crimp and Widthwise Contraction of Fabric

The schematic view of a straight segment of weft yarn located in front of the reed within a shed, created by separation of odd-numbered warp yarns occupying the bottom shed line and even-numbered yarns occupying the top shed line, is depicted in Figure 7.20. Individual wires of the reed would come in contact with the body of weft at multiple points and push the same toward the fell of the cloth. Each dent of reed houses two yarns, and it is

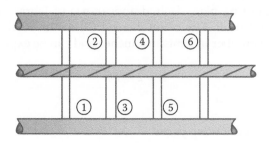

FIGURE 7.20
Warp and weft yarns in front of the reed.

observed that distance between the centers of neighboring yarns within a dent is less than that between the neighboring yarns across a reed wire. If this spatial distribution of warp yarns is maintained in the fabric, then each pair of warp yarns in the resultant fabric would appear to be separated by a split corresponding to the thickness of reed wire. Such a defect is termed as a reed mark.

As a reed moves forward to beat up the weft, warp yarns cross over and exchange places in respective shed lines. As a result, a situation depicted in Figure 7.21 develops. The straight segment of weft has been forced into a crimped form, requiring a total yarn length that is much higher than what was available in Figure 7.20. Part of this additional requirement would be met by yarn extension. During this extension, stress would develop in the arms of the weft yarn around each post of warp thread. The resultant force can be split up into a vertical and a horizontal component. The vertical component would force warp yarn into a crimped state, and the horizontal component would pull adjacent warp yarns closer. Thus, weft strain causes the warp sheet to be pulled inward toward the center, which leads to a narrowing of the warp sheet and, hence, to growth of a lateral force in warp yarns. Equilibrium is achieved when the cumulative lateral force generated in the warp sheet balances the weft stress.

The extra demand on weft length during the beating-up process is partly met by the extension of weft yarn and partly by the supply of an additional length accruing out of the difference between the inclined (2) and the straight length (1) of weft as depicted in Figure 7.22. In certain cases, as, for example, in shuttleless weaving systems, the inserted pick lies nearly parallel to the cloth fell in the open shed. Under such circumstances, the entire requirement of additional weft length upon beating up has to be met through yarn extension alone. Tension generated in the yarn can be assumed to be approximately proportional to the yarn extension. Weft tension in systems inserting picks parallel to the cloth fell would accordingly be high. This would have a significant effect on fabric properties. All variables remaining the same, the warp threads in the case of straight weft insertion would have to undergo a greater crimped path. Moreover, if breaking extension of the weft yarn does

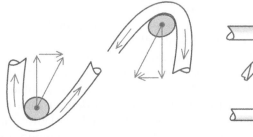
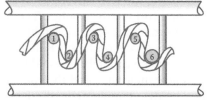

FIGURE 7.21
Development of yarn crimps.

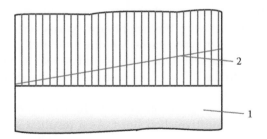

FIGURE 7.22
Inclined weft within shed.

not correspond to the strain generated during beating up, it may also break in places. In such an event, the loom would not stop, and the fabric would exhibit picks that have burst in places. A very similar problem would also be encountered even with a shuttle while beating up in an open shed. In such a situation, the weft yarn is first pushed to the fell and made straight before the shed lines cross over. This is similar to the situation prevailing with weft insertion in shuttleless looms and would result in very high weft strain.

The lateral force generated in the warp sheet during beating up leads to a widthwise contraction of the fabric near the reed. As a result, the selvedge yarns are pulled against the reed wires and are rubbed vigorously during the to-and-fro motion of the reed. This being undesirable, measures are taken to grip the fell of the fabric at the two selvedges or even over the entire width by means of temples so as to restrict widthwise contraction. The efficacy of a temple depends on the elastic nature of the weft yarn. For the same weft strain, yarns of higher primary creep deformation and, hence, of lower elastic deformation would load the temple to a lesser extent. As yarns with low twist multiple exhibit higher primary creep as well as an overall low modulus, weft yarns are normally made with lower twist multiple.

7.4.2.2 Stabilizing Fabric Width at the Reed

The manner in which the cloth on a loom is gripped and passed through temples is depicted in Figure 7.23. A pair of such systems is employed on a loom along the two selvedges. Temples can be of roller, ring, combination, disc, and full width varieties. The roller—especially when made of rings of rubbery material—and disc temples are, compared to the ring temple, relatively less severe on fabric. Roller temples can, however, have a pinned surface, in which case fabric would be pierced in places. Such a roller is suitable for coarse materials, and the rubber-ringed one is good for delicate materials. The disc temple grips the fabric only at the extreme edges. However, a perfect horizontal and concentric alignment of the discs is paramount to its functioning satisfactorily. The ring temple is very severe as the pins on the inclined rings not only pierce the fabric at several points but also pull it

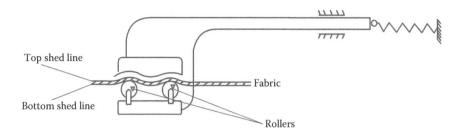

FIGURE 7.23
Passage of cloth through temple.

sideways. Such a temple is suitable for fabrics that are made of coarse materials and exhibit very high width-way contraction. The combination temple is a hybrid of roller and ring temples and can be used for fabrics of intermediate character. The full width temple is very gentle on the fabric and, because the nip is spread right along the entire fabric width, it is effective for a very wide range of fabrics. However, a length of fabric immediately beyond the fabric fell is hidden from view, creating operational problems.

7.4.2.3 Cloth Fell Displacement during Beating Up

In addition to causing considerable strain in the weft, the beat-up process also results in sharp peaks in warp tension. In fact, before a segment of warp can be embedded in the fabric, it has to undergo several cycles of such impactful loading. Hence, warp yarns have to be not only strong, but also must possess high fatigue resistance. Yarns spun with high twist multiple exhibit high strength as well as high fatigue resistance. Hence, warp yarns are made with high twist multiple.

The process of beat up causing a sharp rise in warp tension is demonstrated in Figure 7.24. The force F_A applied on the cloth fell causes a displacement ΔX (Greenwood, 1975) such that

$$F_A = (C_K + C_G)\Delta X \tag{7.12}$$

The constants C_K and C_G are stiffness of warp and of fabric, respectively. Thus, for a given beat-up force, the sum total of stiffness of warp and fabric influence the cloth fell displacement and, therefore, the magnitude of rise in warp tension. Such a movement of cloth fell toward the weaver during beating up and away from the weaver after the beat up takes place in every cycle and is termed "cloth fell displacement." This displacement is larger for heavier fabrics and smaller for lighter ones. Excessive cloth fell displacement is undesirable as, on one hand, the fabric might become slack upon beating up and create disturbed weaving conditions, and on the other, the translational motion of warp threads across heald eyes and reed wires may lead

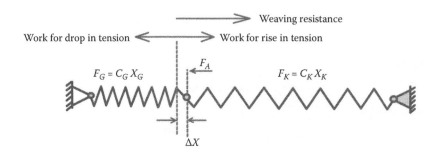

FIGURE 7.24
A model of warp and fabric in series connection.

to excessive abrasive wear. By adjusting the free warp length, the stiffness of the warp sheet (C_K) can be altered. This is an effective tool for controlling cloth fell displacement (ΔX) for the same value of beat-up force (F_A).

Beat-up force is needed to overcome frictional resistance to the sliding of weft yarn across warp yarns. The pick spacing adjacent to the cloth fell stabilizes only after a certain number of picks. During beat up, the pick of the weft is pushed in until the weave angle θ reaches a limiting value (Hollstein, 1980), given by the equation

$$\ell = \cos\theta\, e^{\mu\theta} \tag{7.13}$$

This limiting value of weave angle depends entirely upon the coefficient of friction μ between warp and weft yarns. Once this is achieved, the weave angle of the previous pick starts changing, gradually diminishing the instantaneous pick spacing ℓ. The number of picks that would be affected in a beat-up process would depend on the intensity or energy supplied in beat up. For very open fabric, this number would be zero whereas, for moderately heavy fabrics, the weave angle of two to four picks adjacent to the cloth fell would keep on changing. This zone at the cloth fell forms part of the weaving zone. Therefore, a stable pick spacing ℓ does not materialize in a fabric immediately on beating up but only after a certain number of weaving cycles.

7.5 Principles of Take Up

Taking up a woven fabric involves

- Pulling the fell of the cloth away from the weaving zone by a definite amount during each weaving cycle either continuously or intermittently, thus maintaining a constant pick spacing.

- Winding the cloth being formed on the machine onto a suitable cloth roller in such a manner that the required continuous fabric length can be rolled up at even tension without forming any crease. This system should be so designed that the wound cloth can be easily removed without disrupting the production process.

The manner of withdrawing the fabric length formed during a weaving cycle from the cloth fell can be classified in various ways, namely

1. Positive and negative
2. Direct and indirect
3. Intermittent and continuous

The principle of negative and direct take up is illustrated in Figure 7.25. The cloth roller on which cloth is wrapped up is driven directly by a take-up mechanism. The principle of the mechanism involves applying a certain amount of torque to the cloth roller continuously such that an opposite torque resulting from the fabric tension balances the same. As the fabric slackens during beat up, the magnitude of balancing torque goes down, allowing the torque applied by the take-up system to effect an instantaneous rotation of the cloth roller. This rotation would continue until the equilibrium of torque is established again. Obviously, then the amount of fabric taken up does not directly depend on the amount of fabric produced in each weaving cycle but purely on the torque balance. Such a take-up system is therefore negative in character. The driven roller, in this case, is the cloth roller. Thus, a direct drive is given to the cloth roller, and therefore, the system is characterized as direct.

Indirect take-up systems are illustrated in Figure 7.26. The drive from the take-up system flows to the take-up roller. The cloth roller is driven either

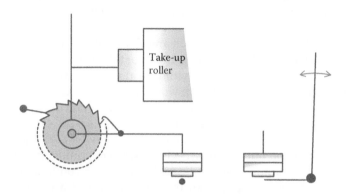

FIGURE 7.25
Negative and direct take up. (From Marks, R., and Robinson, A. T., *Principles of Weaving*, The Textile Institute, Manchester, UK, 1976.)

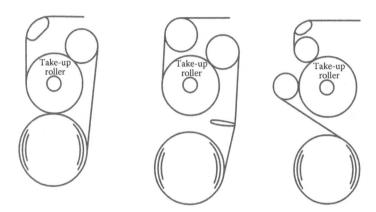

FIGURE 7.26
Indirect take-up systems.

by surface contact with the take-up roller or is loosely driven by a friction clutch system. If the cloth is wound directly on the take-up roller, then with the increasing roller diameter caused by successive layers of wound cloth, the angular rotation per pick has to be progressively reduced in a hyperbolic fashion. This calls for a feedback-control system, which is bypassed in an indirect system. If the fabric surface is sufficiently rough, then the cloth roller can be pressed against the take-up roller. However, a smooth or a delicate fabric surface would not permit such a solution. In such an event, the cloth roller is driven by a friction clutch system, permitting thereby a slippage in the drive as and when needed. The take-up roller is covered by a suitable material (emery paper or strip, pinned rubber strip, etc.) that allows a good grip between the roller surface and the fabric. The nip roller is also felt-covered. These would permit the fabric to remain firmly gripped even if the cloth is unwound from the cloth roller for doffing or for inspection. Such indirect driving systems are employed in conjunction with positive take-up systems.

In a positive take-up system, a train of wheels is usually employed to transmit a predetermined but constant angular rotation to the take-up roller in each weaving cycle. This angular rotation of take-up roller gets translated to a certain fixed amount of fabric withdrawal from the weaving zone. The final pick spacing in the relaxed cloth would be influenced by this amount of withdrawal as well as by the lengthwise cloth contraction during relaxation.

Each element in the take-up systems discussed so far executes a rotational motion. This motion is intermittent when a pawl-and-ratchet system is employed to drive the train of the wheels (positive intermittent). On the other hand, a positive continuous system is realized when a continuous rotational motion, sourced, for example, from the bottom shaft of a loom, is suitably scaled down by employing worm and worm wheels. A loom with a continuous take-up motion has to be equipped with a continuous let-off motion in order that tension variation in a warp sheet is maintained within acceptable limits.

7.6 Principles of Let Off

During each cycle of weaving, the warp sheet has to be released from weaver's beam in order that the woven fabric can be rolled up on a cloth roller without changing the location of the fell of the cloth. Release of warp can be caused by a rise in warp tension as a result of fabric take up or can be controlled by a system that drives the weaver's beam. The first type is termed negative, and the second one is a positive let-off motion. There is also an intermediate system that allows the weaver's beam to unwind by a definite amount in each cycle of the loom. This variety is referred to as semi-positive. In both positive and semi-positive motions, the position of the cloth fell as well as the magnitude of warp tension are maintained within a well-defined range during the entire weaving process without necessitating any manual intervention.

7.6.1 Negative Let-Off Motion

In a negative system of warp let off (Figure 7.27), tension in the warp is regulated by friction between a chain or rope of let-off motion (6) and the beam ruffle (4). This friction depends on pull applied to the chain or rope by the weighted lever (7), the coefficient of friction between the chain or rope and the ruffle and the arc of contact between the chain or rope and ruffle. Because tension on the slack side of chain is T_s, and the frictional force F between the chain and rope opposing rotation of the beam must balance the tension on the tight side T_t, the static condition of equilibrium is given by

$$F = T_t - T_s$$

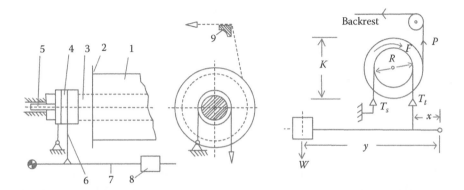

FIGURE 7.27
Principle of negative warp let off.

Taking moments about the lever fulcrum, it can be shown that

$$T_t x = Wy$$

Hence,

$$T_t = (Wy)/x$$

For a given weight W, the value of T_t changes with the distance y of W from the lever fulcrum. Moreover, under conditions of equilibrium,

$$T_t/T_s = e^{\mu\theta}$$

where

μ = coefficient of friction between the chain and ruffle surface

θ = angle of wrap between the chain and ruffle

The expression of T_t can therefore be stated as

$$T_t = Wy/x = T_s e^{\mu\theta}$$

In order that a length of warp can be released, warp tension P has to be greater than frictional resistance $F = T_t - T_s$ acting on the beam ruffle.

$$F = T_t - T_s = T_t - T_t e^{-\mu\theta} = T_t (1 - e^{-\mu\theta}) = Wy/x(1 - e^{-\mu\theta})$$

The weighting arrangement is applied at each end of a beam, and hence, under conditions of equilibrium for beam diameter K and ruffle diameter R,

$$PK \le 2FR$$

or

$$P \le (2FR)/K$$

Hence,

$$P \le \{2y/K\}\{W(1 - e^{-\mu\theta})(R/x)\} \tag{7.14}$$

Therefore, if warp tension P is to remain constant as the beam weaves down, the distance y must be proportionately decreased. In other words, the ratio y/K must be kept constant to maintain constant yarn tension. This is the reason why a weaver is seen constantly changing the position of weight at

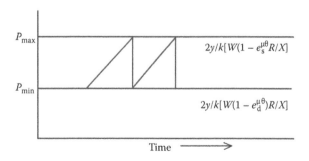

FIGURE 7.28
Cyclic tension variation in warp due to negative let off.

the back of looms equipped with negative let-off motion. The warp tension would, in the short term, vary between a maximum and a minimum value given respectively by the static (μ_s) and dynamic (μ_d) coefficients of friction between the chain and ruffle as depicted in Figure 7.28. This would cause the warp to be let off intermittently, resulting in uneven pick spacing in the resultant fabric.

Warp tension varies considerably during opening and closing of the shed; the variation also depends upon the kind of shedding and on whether a rope or chain is used at the beam ruffle. When the warp is slack, the tension T_s at the slack side of the rope or chain is equal to the tight side tension T_t.

7.6.2 Positive Let-Off Motion

Positive let-off motion is defined as a motion in which the weaver's beam is turned at a rate that maintains a constant length of warp sheet between cloth fell and weaver's beam, the means of applying warp tension being separate from the beam-driving mechanism.

In order to ensure steady warp tension, a constant length of warp sheet should be released during each loom cycle. In other words, the angular velocity of the rotation of the weaver's beam should be increased steadily as the beam weaves down.

In positive motions, the most common form of tension arrangement consists of a back rail, which is made to press against the warp sheet as shown in Figure 7.29. The force applied to the warp sheet by the back rail is represented by X. Obviously, if this force remains constant and if the warp tension depends on no other factors, then the warp tension must also remain constant.

A weight on lever F, pivoted at N, applies a downward force W at one end of the swing lever pivoted at B. At the other end of this lever is mounted the back rail, which is, therefore, made to press against the warp sheet with force X. This force can be changed by moving the weight to a different

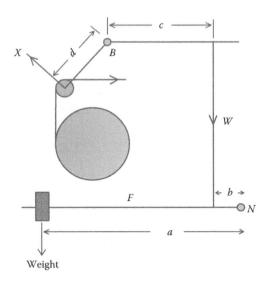

FIGURE 7.29
Principle of positive warp let off.

position on the weight lever *F*. For any given position of weight, the force *X* must have a fixed value. In fact, in the simple case illustrated, this force can be expressed as

$$X = \{W(a/b)(c/d)\} + m \tag{7.15}$$

where *m* is the force applied to the warp sheet by the weight of the lever and connecting rod. The swing lever is duplicated at the other end of the loom, the back rail being suspended between the two. These two levers are carried on studs *B*, which are bolted to the loom framing. These studs are adjustable in height in order to raise or lower the back rail as required. This adjustment enables the geometry of the rear shed to be altered, which is of considerable advantage in improving fabric cover and increasing pick density (see Section 7.2.2.6).

7.6.2.1 Variation in Warp Tension

It was stated earlier that a constant tension would be obtained with such a system if the tension depended only on force *X*. There are, however, three other sources of variation, namely

   i. The effect of continuous reduction in weaver's beam diameter
  ii. The effect of angle between the back rail arm and the swing lever
 iii. The effect of back rail mounting

7.6.2.1.1 *Effect of Weaver's Beam Diameter*

Because beam diameter gradually diminishes during the weaving process, the angle between the top and back warp sheet passing around the back rail also decreases gradually. This change in angle is responsible for a slow change in warp tension. Figures 7.30 and 7.31 illustrate why this tension change must occur with such a system.

The back rail mounting and the warp sheets are shown with full lines, and the dotted lines show forces acting at the back rail. Figure 7.30 represents a condition in which the beam is fairly full. A downward force W is applied to the end of the swing lever C, and this, in turn, applies a force in the X direction to the warp sheet at the back rail at right angles to the arm of lever C, which carries this rail. Tension t acts along the two warp sheets and pulls the back rail inward and downward toward the loom. These two tensions may be represented by a single force R_1. Because it is assumed that there are no frictional effects at the back rail, the tension t must have the same value in both sheets, and the direction R_1 is such that it bisects the angle between the two sheets. Because the system is in equilibrium, R_1 must be balanced by an equal and opposite force R_2, this being the result of the forces in the X and Y directions, which represent stresses in the arm of the swing lever, which is carrying the back rail.

When the beam is woven down to a degree, the angle between the warp sheets goes down. This causes the resultant forces R_1 and R_2 to swing counterclockwise while the force in X direction remains unchanged as W has

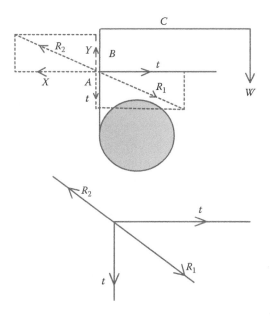

FIGURE 7.30
Forces acting on warp sheet.

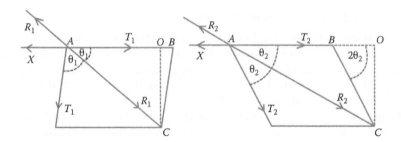

FIGURE 7.31
Tension variation in warp due to change in beam diameter.

remained the same. These changes in direction of R_1 and R_2 affect the warp tension as demonstrated in the following.

Let X represent the force exerted by the back rail, which has the same value in both Figures 7.30 and 7.31. Now, from the triangle BOC (Figure 7.31), we have

$$BO/BC = BO/T = \cos 2\theta.$$

That means

$$BO = T \cos 2\theta.$$

From the triangle AOC, we find that $AO/AC = AO/R = \cos\theta$.
So

$$AO = R \cos\theta, \text{ but } AO = AB + BO.$$

Hence,

$$AB + BO = R \cos\theta.$$

Because

$$AB = T \text{ and } BO = T \cos 2\theta, \text{ we obtain}$$

$$T + T \cos 2\theta = R \cos\theta$$

or

$$T(1 + \cos 2\theta) = R \cos\theta.$$

Hence,

$$T = R \cos\theta/(1 + \cos 2\theta) = R/(2\cos\theta).$$

Again,

$$X/R = \cos\theta.$$

Hence,

$$R = X/\cos\theta.$$

Therefore,

$$T = X/2\cos^2\theta \qquad (7.16)$$

Accordingly, if T_1 is warp tension when the beam is fairly full, and T_2 is the corresponding tension when the beam is nearly empty, the ratio T_2/T_1 can be treated as an index of fall in warp tension during the weaving process.

$$T_2/T_1 = X/(2\cos^2\theta_2) \times 2\cos^2\theta_1/X$$

$$= \cos^2\theta_1/\cos^2\theta_1$$

7.6.2.1.2 Effect of Angle Between Back Rail Arm and Swing Lever

Let the back rail be inclined to the swing lever such that the force X acts at an angle α to the warp sheet as shown in Figure 7.32. It is observed from the diagram that for equal value of T in both arms of the parallelogram

$$\angle AOB = \theta, \text{ and } \angle AOC = \alpha,$$

And clearly, therefore,

$$\angle BOC = (\theta - \alpha).$$

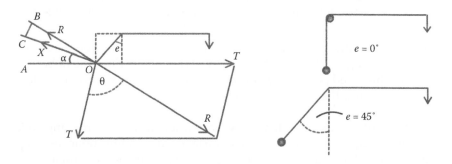

FIGURE 7.32
Effect of angle of inclination between arm of back rail and swing lever.

Again,

$$X/R = \cos(\theta - \alpha),$$

thus,

$$R = X/\cos(\theta - \alpha)$$

as

$$T = R/2\cos\theta.$$

One can state that

$$T = X/\{2\cos\theta \cos(\theta - \alpha)\}$$

when

$\alpha = 0°$ $\qquad\qquad\qquad T = X/(2\cos2\theta).$

When, however, $\alpha = 90°$,

$$T = X/(2\sin\theta \cos\theta) = X/\sin2\theta$$

The effect of the angle between the back rail arm and the swing lever is minimum when they are at right angles to each other, that is, when $\alpha = 0°$, a condition given by $e = 0°$.

7.6.2.1.3 Effects at the Back Rail

The effects can be summarized as follows:

    i. A fixed back rail will induce a frictional resistance to the sliding of the warp sheet because of which tension in the top sheet will be greater than in the sheet between the back rail and the weaver's beam. With the increase in angle of contact due to the weaving down of the beam, tension in the top sheet tends to rise further due to increasing wrap angle with the back rail.

    ii. With a rotating back rail, frictional resistance will be encountered at the bearings of the back rail. By employing a superior type of bearing, this could perhaps be minimized. Warp tension variation would also occur if the back rail is not perfectly concentric.

A fixed rail results in a steady rise in tension whereas a rotating roller may give a significant cyclic tension variation depending upon its shape and behavior of its bearings. The ideal condition would be a nonrotating rail together with some means of maintaining a constant wrap angle at the back rail while maintaining very low frictional resistance.

References

Catlow M, and Vincent J J (1951). The problem of uniform acceleration of the shuttle in power looms, *Journal of Textile Institute*, 42, T413–T487.

Greenwood K (1975). *Weaving: Control of Fabric Structure*, K. Merrow Publishing Co. Ltd., UK, ISBN: 0-900-54165-2.

Hollstein H (1978). *Fertigungstechnik Weberei*, Vol. 1, VEB Fachbuchverlag, Leipzig.

Hollstein H (1980). *Fertigungstechnik Weberei*, Vol. 2, VEB Fachbuchverlag, Leipzig.

Marks R, and Robinson A T (1976). *Principles of Weaving*, The Textile Institute, Manchester, UK, ISBN: 0-900-73925-8.

Further Reading

Ormerod A, and Sondhelm W S (1995). *Weaving: Technology and Operations*, The Textile Institute, Manchester, UK, ISBN: 1-870-812-76-X.

8

Developments in Shedding Motions

8.1 Limitations of Shedding Tappet

Two concentric shedding tappets for a 7/1 construction, similar in all respects except their size are shown in Figure 8.1. In view of a repeat size of eight picks, the shaft carrying the shedding tappet, namely the counter shaft, rotates once for every eight rotations of the crankshaft. The resultant distribution of crankshaft and counter shaft displacements for the entire repeat is listed in Table 8.1.

The two segments of each of the two tappets accounting for changeover periods, located to the right and to the left of the 15° dwell segment of the tappet nose, differ in respect of pressure angle θ. For the smaller tappet, the corresponding segments are steeper as compared to the larger one. This is explained with the help of a triangle drawn on the left-hand side of the figure.

The two sides of the equilateral triangle can be mentally equated to the two rays of line that originate at the common center of the two tappets and spread out at an angle of 30° across one of the two sides of the tappet noses. If the two inclines of the two noses are simplified as straight segments, then they can be represented in the equilateral triangle by the two lines OB and ED, respectively, for the larger and the smaller tappet. As the two tappets have the same lift, the equivalent distances OA and EC drawn on the representative triangles must also be equal in magnitude. It follows then from the simple trigonometric relationships shown in Figure 8.1 that pressure angle θ_2 must be larger than angle θ_1. Treadle bowls remain pressed against these segments along a line of contact while rolling along the tappet surfaces. A force P acts between the tappet and the bowl normal to the instantaneous line of contact between the two objects. Hence, corresponding to any particular angular position of the two tappets, the vector P for the larger tappet would be directed closer to the vertically downward direction than that for the smaller tappet. Hence, the corresponding vertically downward components F, which effectively account for the downward movement of the treadle bowls, would be larger for the bigger tappet. As a result, larger tappets are able to transmit energy more efficiently.

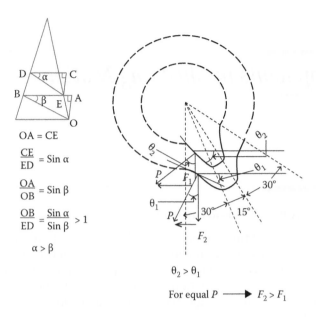

FIGURE 8.1
Effect of weave repeat and tappet size on energy transmission to treadle bowl. (From Marks, R. and Robinson, A. T., *Principles of Weaving*, The Textile Institute, Manchester, UK, 1976.)

TABLE 8.1

Extent of Counter Shaft and Crankshaft Motions over One Repeat of 7/1 Weave

Nature of Heald Motion	Degrees of Crankshaft Displacement	Degrees of Counter Shaft Displacement
Dwell at top	2280	285
Changeover to bottom shed	240	30
Dwell at bottom	120	15
Changeover to top shed	240	30

From the foregoing, one may draw the following conclusions:

1. Bigger tappets permit better utilization of energy.

    Therefore, when the magnitude of force required for moving the heald goes up, as is the case with weaving heavy fabrics or weaving very broad fabrics, a bigger shedding tappet is employed.

2. A bigger tappet is required for weaves with a larger number of picks per repeat.

    However, bigger tappets occupy greater space and have higher inertia. Hence, there is a practical limit to increasing tappet size. This calls for other forms of shedding devices for weaves repeating on a large number of picks per repeat.

8.2 Functional Principles of Dobby

A dobby is a more flexible shedding device than a shedding tappet because of the following features:

1. The program of displacement of each heald over a complete weave repeat is stored in a dobby in such a manner that the same can be changed at will. In a shedding tappet, the program is frozen forever; a change in program calls for a new shedding tappet but not a new dobby. In fact, shedding tappets are cast in a block for the entire group of healds. As an example, the shedding tappets for a plain weave come in pairs, and those for a 2/2 weave come in a block of four tappets. For a dobby, the program of displacement of one heald is totally independent of that of the other healds.

2. A dobby is an attachment on a loom and can be freely exchanged. Shedding tappets are more loom-specific. For example, the shedding tappet of a broader loom would be larger than that of a narrower loom.

8.2.1 Principle of Programming

A majority of weaves are realized by selective lifting of warp threads to one shed line, leaving the rest to occupy the other shed line. This selection is varied from pick to pick in a manner determined by the construction. In principle, then the shedding process involves generating and processing binary signals and transmitting the amplified signals in a suitable manner to healds for their upward and/or downward motions. Signal generation is executed through a suitable programming strategy.

For dobby shedding the programming involves

- Translating the targeted weave to a peg plan
- Storing this plan in a sequential manner for all picks in a suitable medium

Accordingly, a specific sequence of binary signals is transmitted by this peripheral system of dobby to its central drive every pick. In turn, the driving system transmits motion selectively to the healds under its control.

A typical example of the generation of a peg plan is illustrated in Figure 8.2. The weave in question is a regular 2/2/1/1 weave. The straight draft can assume two forms, one displaying positive slope and the other having a negative slope. Both these variants are depicted below the weave in the figure. The corresponding lifting plans are shown to the right-hand side of these drafting plans. It is important to remember that the rows of a drafting plan

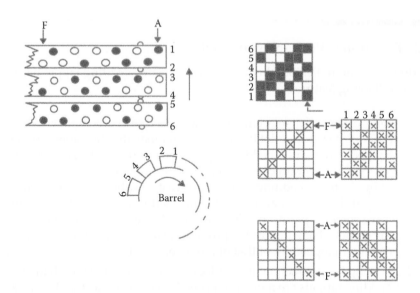

FIGURE 8.2
Principle of programming a dobby lattice. (From Marks, R. and Robinson, A. T., *Principles of Weaving*, The Textile Institute, Manchester, UK, 1976.)

represent healds, and the columns of a lifting plan represent the corresponding picks. In Figure 8.2, healds are numbered A to F, and the picks are numbered 1 to 6. A cross in the matrix of the lifting plan indicates lifting of a heald aligned against the adjacent row in the corresponding pick. Thus, the three crosses in the second column of the lifting plan indicate lifting of the healds A, B, and D in the second pick.

Translation of these drafting and lifting plans into a peg plan is illustrated in the left half of Figure 8.2. Three wooden lattices, each having a trapezoidal shape, are shown linked together into a chain. This chain is mounted on a cylindrical barrel having grooves matching the cross-section of the lattices. An angular displacement of the barrel causes a rotational movement of the lattice chain by a specific amount. Each lattice has two rows of holes. The distance between adjacent holes within and between rows is determined by the specifications of the dobby on which this lattice would be mounted. The two rows of holes on each lattice are also offset with respect to each other. In each of these holes, wooden pegs can be hammered.

In the illustration, a dark circle indicates a peg in the hole of the corresponding lattice, and a blank circle indicates a hole without a peg. The lattice on top of the figure has the two rows numbered 1 and 2, indicating that the pegging plan in this particular lattice corresponds to the picks 1 and 2 of the weave. The six holes in each row of this lattice starting from right to left have been numbered A to F. The hole earmarked for heald A for the first pick is located half a pitch to the right of the corresponding hole for the second pick. Effectively, a peg or its absence in a particular hole of a particular lattice carries a binary signal corresponding to the particular heald in the particular pick as, for example, the first row in the first lattice shows pegs corresponding to healds A, C, and F. On cross-checking with the drafting and lifting plans of the referred weave, it is found that these three healds are supposed to be lifted up in the first pick. Hence, a peg in the hole of a lattice would effectively be interpreted by the dobby as a command to lift the corresponding heald in the corresponding pick.

A pegged lattice chain mounted on a dobby barrel carries specific signals to the sensors of the dobby mechanism for the entire weave repeat or a multiple thereof. If a repeat is of 24 picks, then the chain would contain 12 lattices and, hence, information for one complete weave repeat. The leading edge of the first lattice would be linked to the trailing edge of the 12th lattice, giving rise to a continuous chain. However, a repeat of 23 picks would call for 23 lattices. The resulting continuous chain would carry signals for two consecutive repeats.

The presence of a peg on a lattice results in mechanical displacement of the corresponding sensor in the dobby. Such sensors (or feelers) are arranged in a row, pressing down on the surface of the dobby barrel and therefore on the lattice chain. As a barrel having, for example, six grooves on its surface rotates by 60°, a fresh lattice is brought to bear against the dobby feelers. A pair of neighboring feelers feels for the lifting plan of a particular heald over the next two consecutive picks. For example, the pair of sensors for heald A in Figure 8.2 would bear upon the extreme rightmost segment of the lattice covered by the first hole for the first and second picks. Evidently, the thickness of the sensors decides the lateral distance between these two holes. Similarly, the pitch of the sensors must be equal to half of that of the holes in each row of a lattice. Hence, thinner feelers would increase the capacity of a dobby for receiving and processing information for larger number of healds. Conversely, a thinner feeler would also mean smaller and therefore weaker pegs, which may break or simply get dislocated from their respective holes, resulting in the wrong signal to the respective heald.

The drawbacks of peg and lattice systems—large space occupied, tedious process of pegging, chances of breakage or dislocation of pegs, and heavy mass of the assembly—led to a changeover to punched paper systems for storage of the program. The binary signal can be stored in a matrix form on a sheet of paper by either punching a hole at suitable locations or leaving them blank. Such a sheet of paper, when joined end to end, results in a chain. This chain of paper can be positively mounted on a barrel and made to move across the sensors of the dobby in discrete steps. Every angular movement of the dobby barrel and, hence, of the chain brings a fresh selection to the dobby sensors for the next pair of picks. Evidently, the sensors for such a programming medium have to be different from that of the peg-lattice system.

A sheet of paper, made however strong and durable, wears out in the course of time owing to various mechanical and atmospheric strains. For example, the mechanical feelers of the dobby striking the paper surface repeatedly at the same locations or the high humidity and temperature of a loom shed would surely deteriorate a paper chain over a period of time. If, however, the binary signal of a hole or a blank is read optically or, even better, if the signal is stored in a digitized form to be subsequently used for controlling the flow of current to the solenoids, then reliability and operational speed can be improved considerably. The evolution of the dobby system to its present day form has been possible partly due to this transition of information storage from a bulky and clumsy pegged wooden lattice to data bits on discs.

8.2.2 Driving System

One system of classification is based on the number of picks of the lifting plan that is transmitted to the driving system per cycle of dobby. In a double-lift dobby, the information transmission is for two picks per cycle of dobby whereas in a single-lift dobby the transmission is for one pick only. The double-lift dobby has two sets of the principal sensing and driving elements as opposed to only one set in the single-lift ones. The two sets of the sensing system of a double-lift dobby receive lifting programs for the odd pick and the even pick separately and transmit them to the two driving sets, which operate sequentially. Thus, in a double-lift dobby, each set of driving elements and, therefore, the dobby itself operate at half the speed of loom. Consequently, a loom equipped with a double-lift dobby can run at twice the speed of its counterpart equipped with a single-lift dobby. The principal benefit of a double-lift dobby over a single-lift system, therefore, is the higher loom speed that can be achieved with the former.

The line diagram of a typical mechanical double-lift system, which is operated by pegged wooden lattices, is depicted in Figure 8.3. The tip of feeler F_1 senses a peg on the lattice and therefore moves up while that of the adjacent

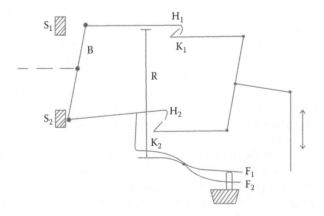

FIGURE 8.3
Working principle of a mechanical double-lift dobby. (From Marks, R. and Robinson, A. T., *Principles of Weaving*, The Textile Institute, Manchester, UK, 1976.)

feeler F_2 encounters none and remains in its running position. Centers of both these feelers are fulcrummed on a common stud, and hence, the tail of feeler F_1 moves down while that of feeler F_2 remains in its uppermost position. These movements of the tails of two neighboring feelers are transmitted by two upright needles to a pair of hooks, H_1 and H_2. As a result, the tip of hook H_1 drops down to a lower level while that of hook H_2 dwells in a raised position. The tail ends of these hooks are linked to two ends of the upright arm B, which, in turn, is linked to a particular heald by means of the flexible link shown by the dotted line. The two blocks S_1 and S_2 provide support to the two ends of arm B against the pull created by the healds and the other elements of the flexible link. The two hooks can be pulled to the right by two knives K_1 and K_2, respectively, as and when they get engaged, thereby pulling the corresponding heald to occupy one of the shed lines. The two knives execute a to-and-fro motion with a phase lag of 180°, being driven by a T-lever, which, in turn, is linked to a crank mounted on the bottom shaft of the loom. Thus, the knives complete one cycle of motion in two loom cycles.

Figure 8.4 depicts the manner in which the hooks get engaged with the knives and move the upright arm. The three figures in the top row show that, although the two knives oscillate sequentially, the tips of the hooks are not engaged, caused by the absence of a peg in the corresponding lattices. Hence, the respective heald is not moved from the shed line it has been occupying. On the other hand, both hooks in the two figures of the lower row get engaged sequentially by the respective knives, and therefore, the corresponding heald is made to move to the other shed line and stay there for the two picks. A slight gap has to be provided between the back of a lowered hook and the extreme position of the respective knife so that the knife in its forward journey can engage the hook and drag it forward. This would result in some idle movement of the knife. Therefore, no movement would be

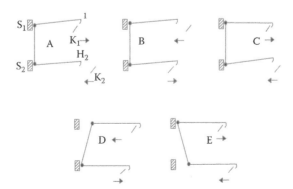

FIGURE 8.4
Principle of shed formation with a mechanical double-lift dobby. (From Marks, R. and Robinson, A. T., *Principles of Weaving*, The Textile Institute, Manchester, UK, 1976.)

transmitted to the corresponding end of the upright arm for a while. During this short period, the other knife in its return journey allows the engaged hook to slide back toward the rest position, causing a backward movement of the corresponding end of the upright arm. Therefore, even though the heald occupies the same shed line for two consecutive picks, it would have moved toward the central plane for a while. This results in a semi-open shed.

The driving system described in the foregoing receives power from the bottom shaft of a loom, and the nature of the linkage results in a near simple harmonic motion to the knives. Hence, the engaged hook and therefore the heald attached to it would also execute a nearly simple harmonic motion. Such a motion is devoid of any dwell of shed line. To compensate for this, either the loom speed has to be lowered so that effective time for passage of the shuttle is increased or the depth of the shed and/or the sweep of the sley have to be increased. Lowering of loom speed is detrimental to productivity while an increase in shed depth and/or sley sweep is detrimental to the warp and power consumption.

The aforesaid drawbacks are overcome in cam dobbies, in which the knives are linked to antifriction bowls as shown in Figure 8.5. These bowls are acted on by suitably designed matched cams, which operate with a suitable phase lag and are mounted on a common shaft. The common shaft rotates once in two picks, and thus each knife completes one cycle in the time taken by two loom cycles. The hooks, as and when they get engaged with these knives, exhibit a displacement profile corresponding to that of the cam profile.

The drawback of a pegged lattice chain is overcome in a paper dobby by punching the lifting plan on a sheet of paper that is then put around a barrel as shown in Figure 8.6. The sensors are a stack of fine needles arranged in two rows along the width of the paper. Each row senses the absence or

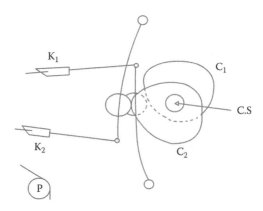

FIGURE 8.5
Functional principle of a cam dobby. (From Marks, R. and Robinson, A. T., *Principles of Weaving*, The Textile Institute, Manchester, UK, 1976.)

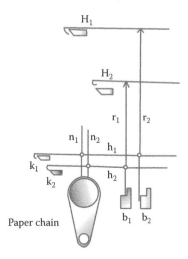

FIGURE 8.6
Functional principle of a paper dobby. (From Marks, R. and Robinson, A. T., *Principles of Weaving*, The Textile Institute, Manchester, UK, 1976.)

presence of holes corresponding to the particular pick. The difference in the deflection of needles, encountering a blank as opposed to a hole, is equal to the thickness of the paper. This weak mechanical signal needs to be amplified sufficiently so as to cause the tip of the corresponding hook H to move by a distance large enough to be either in the path of the knife or be totally clear of the knife. A method adopted for said amplification is shown in the figure. A duplicate set of knives and hooks, k and h, linked to the needles n, convert the downward deflection into a lateral deflection, which accounts for shifting the support r for the driving hooks H away from the lifting blocks b, which eventually permits hook H to become engaged with driving knife K and drag the link of the corresponding heald away from its idle position.

Evidently, the solution arrived at while aiming to overcome the drawbacks of a pegged lattice chain system is a fairly cumbersome driving system as a much larger number of moving elements is required as compared to the driving system associated with a pegged lattice chain system.

It can be noticed that the driving systems described in the foregoing are negative in character, imparting motion to the heald in one direction only. This means that another system must be in place for the return journey of the heald. Such heald-reversing motions have their own problems of inertia and synchronization with the dobby driving system, leading to a drop in speed of operation along with a loss of precision. Positive dobby systems impart drive to the heald in both directions and are therefore preferred for faster and more precise operations.

The working principle of a positive dobby is illustrated in Figure 8.7. The knives K are driven by a positive grooved cam. This an improvement over

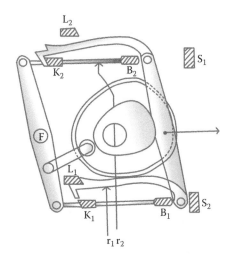

FIGURE 8.7
Working principle of a positive dobby. (From Marks, R. and Robinson, A. T., *Principles of Weaving*, The Textile Institute, Manchester, UK, 1976.)

the pair of negative matched cams employed in a cam dobby although, effectively, both result in a to-and-fro motion of the knives with the required amount of dwell.

The knives K are linked to pushers B in such a manner that while a knife engages the tip of a hook, the corresponding pusher engages the base of the same hook. This simple extension of the pulling action of the knife with the pushing action of the pusher results in a positive motion of the hook and therefore also of the heald operated by the hook. In its idle position, the hook remains lifted to a position clear of the path of the knife, held fast between the locking system L and the stopper S. Hence although a signal from the program may lead to a lowering of the needle supporting the hook, the hook would not come down until it is released by the lock. This extra measure permits a more precise coupling between the hook and the knife, which results in steadiness of the shed line during the weaving process.

In all the systems described in the foregoing, the prime mover is the reciprocating knife while the driven hook is selected by the program either to engage or to remain clear of the path of knife.

Reciprocating systems cannot be operated at a very high speed owing to the two acceleration and two deceleration phases associated with one complete cycle of such a motion. If, however, the inertia of the reciprocating systems is reduced considerably or the reciprocating system is removed altogether by a suitable design of the coupling between the upright arm B (Figure 8.3) and the source of the eccentric motion, then speed of the dobby can be increased significantly. The driving systems described in the foregoing are equipped either with a pair of matched cams or a positive groove cam for providing the eccentric motion. The flexible coupling between such eccentrics and the upright arm B is formed by the pair of knives and hooks and the associated linkages. Dispensing with these knives and hooks altogether and devising an on-off coupling system forms the basis of the rotary dobbies.

The functional principle of a rotary dobby is illustrated in Figure 8.8. The heald is linked to one arm of the L-lever, which is fulcrummed in the manner shown. A rocking motion of this lever about the fulcrum causes the coupled heald to move between two shed lines. The other end of the lever ends in a sleeve in which is mounted an annular ring. This ring can rotate freely within the sleeve housing supported by a suitable bearing. This ring is, in turn, mounted in an eccentric manner on the driving shaft. A flexible coupling between the driving shaft and the ring is responsible for transmission of rotational motion from the shaft to the ring, which, in turn, imparts an up-down motion to the sleeve of the L-lever. This is illustrated in Figure 8.9. As and when the annular ring gets coupled to the driving shaft, the ring rotates, resulting in a change in the value of h. One observes that the prime mover here is the driving shaft itself (as opposed to the knives in reciprocating dobbies), which is a rotational element. The

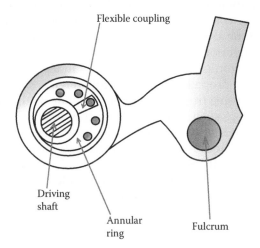

FIGURE 8.8
Functional principle of a rotary dobby.

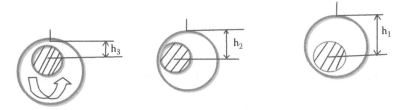

FIGURE 8.9
Conversion of rotary motion to translational motion.

FIGURE 8.10
View of an electronic dobby drive to healds, CCI Tech Hong Kong.

intermediate element between the prime mover and the hooks has been dispensed with totally, and the element equivalent to the hook, that is, the annular ring also rotates. The conceptually clever step of putting a sleeve around the ring for conversion of the rotational motion into a positive to-and-fro motion of the arm connected to the heald is at the root of the transformation into a superior system.

In a modern electronic dobby, each heald is independently operated by cylinders, which are propelled pneumatically (Figure 8.10). The signals for lifting motion that come from the computer operate on the valves that control the flow of air to such cylinders. With this solution, the complicated dobby systems have been simplified to the basic elements. With the help of a CAD system, it should be possible to create a weave and generate the lifting plan automatically. Such a plan, stored suitably in a digitized form and synchronized with the loom movement, could be made to operate the air-flow valves to the cylinder. Such a system permits an online control of cloth construction.

8.2.3 Limitations of Dobby

The dobby system is designed to move healds. Healds occupy space, and therefore, a limited number of the same can be accommodated in a loom between the rearmost position of the sley and the position of the lease rods ahead of drop pins of the warp stop motion. Moreover, the further the heald is from the sley, the greater has to be its net displacement during the shedding motion. This would increase variability in strain among warp yarns, not only leading to an increase in warp breakages, but also to nonuniformity of the fabric being formed. For all practical purposes, dobbies are known to have the capability of controlling not more than 30 healds. Hence, for any weave requiring more than 30 ends per repeat while employing a straight draft, a conceptually new form of shedding becomes necessary.

8.3 Functional Principles of Jacquard

8.3.1 Selection System

The total number of warp yarns on a loom can be broken up into groups, the size of which depends on the number of ends per repeat. Thus, in a warp sheet containing 15,000 ends weaving a construction that repeats on 100 ends, the total number of groups works out to be 150. If it were possible to accommodate 100 healds in a loom, then in such a case, each heald would have had to control the movement of 150 ends. If a straight draft is assumed, then, between the first and the second yarn in each heald, there would have been a gap of 100 yarns, and the resultant fabric would have exhibited 150 repeats of the construction along its width. If, on the other hand, the construction were to repeat on 15,000 yarns, then the number of healds required would have been 15,000, and each heald would have carried only one yarn. It is observed that with a rise in the number of ends per repeat, the number of healds increases, occupying more space along the depth of the loom while the gap between adjacent yarns in a heald rises, causing wastage of space. If, instead of being arranged only along one direction, the elements controlling warp yarns were arranged along two dimensions, as in the rows and columns of a matrix, then a better utilization of space would be possible.

Considering Figures 8.11 and 8.12, one observes that 20 heald eyes, each represented by the symbol X, that are arranged along the depth of a loom, as in a straight draft, occupy the space $H_1 \times H_2$ while the same number of elements, when arranged in 2×10 or 4×5 or 5×4 matrix format, occupy the same width H_2 but a different amount of loom depth, namely H_{12} or H_{13} or H_{14}. These lifting elements, namely the heald eyes, are linked to hooks in the dobby via corresponding healds. Hooks in a dobby are arranged along a horizontal plane

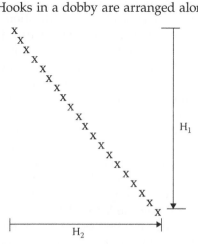

FIGURE 8.11
A spatial arrangement of heald eyes employing straight draft.

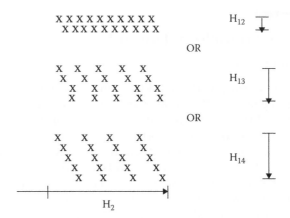

FIGURE 8.12
A spatial arrangement of heald eyes employing a matrix format.

along the loom depth. Hence, they can be linked to the heald and therefore to the corresponding heald eyes only in the manner depicted in Figure 8.11. On the other hand, if the hooks were arranged along parallel vertical planes in the shedding system, then their tail ends would occupy a two-dimensional plane along both the depth and width of a loom. The tail ends of the hooks could then be linked to the different lifting elements. However, as heald eyes along a row of any of these matrices must be able to move independently of each other, healds cannot be used in a shedding system in which hooks are stacked in a matrix format. Such heald eyes, which are not grouped in healds and which are operated directly by the shedding system, are termed mail eyes.

The resultant system, employing needles, hooks, and knives that are arranged—from the point of view of space utilization—in a more compact manner than in a dobby and which operate the yarns directly through mail eyes, is known, after the name of its inventor, as Jacquard. Indeed this term has become generic, implying a method of individual selection of yarns for the purpose of patterning, be it a woven, knitted, or braided fabric.

Just like mechanical dobbies, a mechanical Jacquard also employs needles for sensing holes or blanks punched on cards. As needles are arranged in a matrix format, the holes and blanks in a card also need to be arranged in the form of a matrix. A typical example of a punched card is shown in Figure 8.13. This card displays five large holes along a row as well as three columns, one in the center and two farthest from there, each column having four holes. These holes, indicated by line segments, are for lacing the cards into a chain form and guiding the laced chain positively from one pick to another. The rest of the holes and blanks are meant for the needles and are arranged in 16 rows (four rows have not been punched) and 32 columns. Hence, a total of 512 needles would be impinging the cards for sending signal to 512 hooks. These hooks are arranged in a 16 × 32 matrix controlling 16

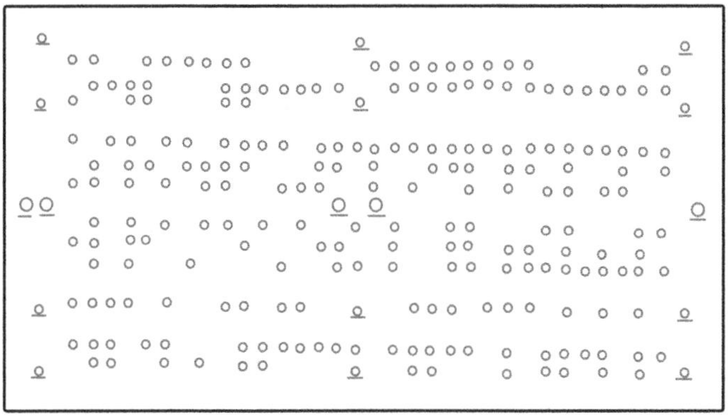

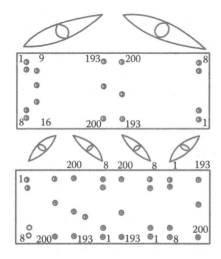

FIGURE 8.13
View of a punched card.

mail eyes along the depth of the loom and 32 mail eyes along its width. The resultant weave can have a repeat size of 512 ends.

Hooks are linked to harness cords, which, in turn, are fastened to mail eyes. By suitably linking the hooks to mail eyes, the repeat size can be either a fraction of 512 ends or a multiple thereof. This is demonstrated in Figures 8.14 and 8.15. In the upper part of Figure 8.14, a Jacquard of 8×50 hooks has been split up into two segments, and in the lower part, four segmentations have been resorted to on an 8×100 Jacquard. On the other hand, by drawing the harness cords through the comber board (Figure 8.15) according to principles of broken draft/divided draft (Figure 6.13), it is also possible to create repeats that are multiples of the number of hooks in Jacquard.

FIGURE 8.14
Variations in linking hook to mail eyes for varying repeat size and nature of repeat.

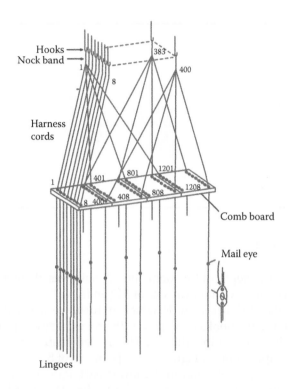

FIGURE 8.15

Multiplying repeat size through harness cord arrangement in comber board. (From Marks, R. and Robinson, A. T., *Principles of Weaving*, The Textile Institute, Manchester, UK, 1976.)

The basic principle of functioning of a single-lift Jacquard is illustrated in Figure 8.16. One set of knives acts on hooks, which are either not disturbed from their vertical positions by the needles or are pushed away from the path of the corresponding knives. To each hook is linked one needle, and the needles are looped around the hooks in the manner shown in Figure 8.17.

Figure 8.17 illustrates the hook arrangement in a 3 × 6 matrix. The first row of needles is linked to the hooks in the first column, and the second row of needles accounts for the second column of hooks. The corresponding arrangement of the needles must therefore be in a 6 × 3 matrix format.

Movement of a needle is dictated by the presence or absence of a hole in the card. The actual displacement of a needle facing a blank on the card is larger than the thickness of the card as a swinging cylinder carrying the chain of cards pushes the affected needle by a distance high enough to bend the corresponding stem of hook and thereby move the tip of the hook away from the knife. Eventually, this arrangement ensures that a hole in the card would account for the lifting of the corresponding warp yarn to the top shed line.

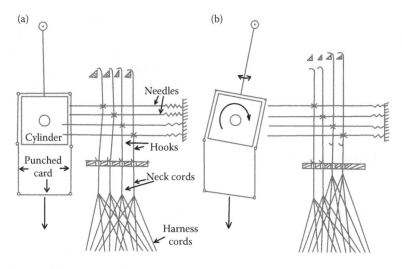

FIGURE 8.16
Principle of functioning of a single-lift Jacquard. (a) Cylinder striking needles. (b) Cylinder swinging away for fresh selection.

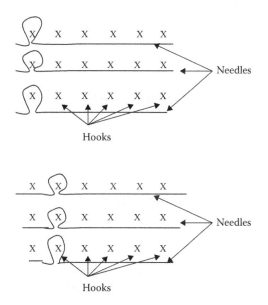

FIGURE 8.17
Relationship between hook matrix and needle matrix.

In a double-lift Jacquard, there is a pair of reciprocating knives, acting on a matrix of pairs of hooks that is acted on by two sets of needles. There is also a pair of cylinders, one carrying the chain pertaining to the odd picks, and the other carries a chain pertaining to the even-numbered picks. In one cycle of such a Jacquard, two picks are woven on the loom. Such systems also lead to semi-open or open shed, and the single-lift system can only yield a bottom-closed shed.

8.3.2 Limitations of Mechanical Jacquard

The limitations may be listed as follows:

1. The manner of storage of the program in punched cards results in a large inventory as well as a complex card-punching system.
2. Sensing the signals stored in punched cards with the help of a matrix of spring-loaded needles can lead to malfunctioning of the springs in a spring box and of buckling of the springs. The natural frequency of such a mass spring system also affects frequency of operation.
3. The operation of cards in terms of lacing them together into a chain and intermittently driving the chain with the help of a rocking cylinder means managing a long chain and moving it in either direction as per requirement. This implies a very high response time.
4. Providing reciprocating motion to the cylinder limits the speed of operation.
5. Imparting reciprocating motion to the set of knives limits the speed of operation.
6. Repeated bending of hooks away from knives might lead to fatigue-induced faults.

8.4 New Generation (Electronic) Jacquards

With the advent of digital signal processing systems, the yarn-lifting program and its storage as well as its transmission to corresponding hooks have been modified drastically in electronic Jacquards. Such systems dispense with cards, cylinders, and needles. Thus, at one stroke, many essential elements of a conventional Jacquard, such as the chain of cards, the card-punching system, the cylinders with their driving systems, and the entire needle–spring box assembly are eliminated. An electronic Jacquard exhibits reciprocating knives and hooks along with

- An actuator system that influences the location of hooks
- A pulley system that links the hooks with harness cords

The functional principle of a typical assembly is shown in Figure 8.18. The actuator is an electromagnet, which, upon getting energized, retains the adjacent hook, thus preventing it from either (1) moving down with the reciprocating knife (Type Bonas/Staubli) or (2) moving up with the reciprocating knife (Type Grosse/Schleicher). If the magnet is not energized, then hooks would remain engaged with the reciprocating knives, resulting in no effective motion to the pulleys and hence to the harness cords. As a result, the shed line remains stationary either on the top line or on the bottom line. An open shed is formed by such a mechanism.

In spite of considerable simplification of the Jacquard assembly due to the introduction of electronics, the reciprocating hooks and knives and the retraction springs that are attached to the mail eyes limit the speed

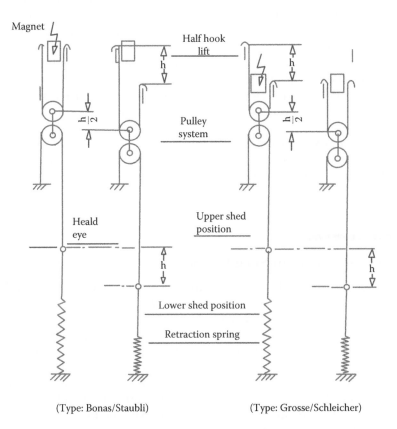

(Type: Bonas/Staubli)   (Type: Grosse/Schleicher)

FIGURE 8.18
Functional principle of electronic Jacquards.

of operation. However, very compact modular construction of the basic operating units—comprising the reciprocating systems and the double pulleys—result in a very large capacity of such Jacquards. Such systems may have in excess of 10,000 hooks.

8.5 New Concepts of Jacquard Shedding

Further advances in the concept of Jacquard shedding have led to the elimination of hooks, magnets, pulleys, pull-down springs, and the massive supporting frame. In one development, namely that of the UNISHED of Grosse, even the harness cords are eliminated as each mail eye is directly connected to a leaf spring, which, in turn, is controlled by an actuator. A prototype of such a system has been exhibited in operational condition at the rate of 800 cycles/min. It results in a positive shed formation. The other development from Staubli is of a system that employs stepper motors. Each motor moves a heald wire through a harness cord. This UNIVAL 100 machine not only does away with many conventional elements of Jacquard, but it permits the movement of each yarn by the desired amount. The manufacturers claim a capacity of 5120 to 20,480 stepper motors (Seyam, 2003). Such a system with 7920 motors, spread over a width of 2.4 m has been exhibited in operation at 1025 cycles/min.

Another approach to simplifying the system involves the use of specially designed heald wires, which can be directly moved up and down by actuators located at both ends of each wire.

8.6 Next-Generation Shedding Systems

Shedding of the warp line is primarily a two-position or, at times, a three-position location of individual warp yarns. The ultimate goal of a shedding system is to achieve independent positioning of all yarns across a warp sheet. With an increasing number of yarns in a warp sheet, the possible number of combinations of these yarns in their different locations, that is, the different types of shed, go up exponentially. Hence, a shedding system should not only be able to handle each yarn independently, but also must have a very large memory. Developments in hardware and software would aid in continuously extending the limits of these frontiers. The shedding system as such should get further simplified to a tag, say, a ring with a memory, through which a warp yarn passes.

References

Marks R, and Robinson A T (1976). *Principles of Weaving*, The Textile Institute, Manchester, UK, ISBN: 0-900-73925-8.

Seyam A M (2003). Weaving and weaving preparation at ITMA '03, *Journal of Textile and Apparel Technology and Management*, 3, 1–20.

Further Reading

Hollstein H (1980). *Fertigungstechnik Weberei*, Vol. 2, VEB Fachbuchverlag, Leipzig.

References

Further Reading

9

Developments in Weft Insertion Systems

9.1 Drawbacks of a Conventional System

In conventional looms, a shuttle housing a pirn of weft yarn is propelled to and fro across the warp shed by either an over-pick or an under-pick shuttle propulsion mechanism. Such a system suffers from the following drawbacks:

1. A large shuttle mass of about 450 g (for cotton weft yarn) is employed for transporting a pick of weft, which is about 0.2 g/m on average, leading to a considerable waste of energy.

2. Unguided free flight of the shuttle can lead to shuttle fly and abrasive damage to the reed and the shuttle as well as uncontrolled weft tension.

3. Shuttle checking within the confined space of a shuttle box results out of multiple impacts. Consequently, the exact location of the rest position of the shuttle within a box becomes indeterminate, affecting efficient transfer of picking energy during acceleration of the shuttle. Moreover, such impacts result in damage to the shuttle and pirn.

4. The mass of the shuttle with a full pirn and with a nearly empty pirn can differ by about 10%. As a result, the nature of shuttle flight varies in a periodic manner, which, in turn, affects the weft tension profile and therefore properties of the resultant fabric.

5. Noise emanating from weft insertion systems employing a conventional shuttle can be as high as 110 dB.

6. The shuttle is made of an assembly of various elements, which may come apart over a period of time due to multiple and severe impactful cyclic loads. Moreover, the major component of a shuttle is good wood, which is becoming scarcer by the day.

7. Limited space within a shuttle limits pirn size, which leads to frequent pirn changing. This is a source of additional workload and frequent defects. A plain loom stops running when a pirn is exhausted. Restarting the loom with a new pirn invariably causes a starting

mark in the fabric while imposing a periodic workload on the loom operator. On a pirn-changing loom, one needs an additional worker in the form of battery filler. Unintended pirn mixing is a frequently encountered problem during battery filling. Moreover, faulty working of the pirn-changing motion very commonly results in a lashing-in problem.

8. Although a shuttle is in a state of rest within a shuttle box, the weft yarn that it carries in the pirn extends from the instantaneous unwinding point on the pirn body, passing through the shuttle eye, and ending up at the fabric selvedge. It is observed from Figure 9.1 that, due to the asymmetric location of the shuttle eye on the shuttle body, the length of this thread segment is higher when the shuttle is in the right-hand box than when it is at the other end. Thus, as the shuttle is accelerated from its state of rest in a shuttle box, the weft yarn initially falls slack and remains in such a condition until the shuttle has traveled a fair distance into the shed to have completely taken up this slack. Evidently, the distance that the shuttle has to travel before the weft yarn can become taut again is higher in alternate picks than in the intervening ones. During the subsequent displacement of the shuttle, the weft unwinding rate depends to a large extent on the flight trajectory of the shuttle as one end of the weft remains firmly anchored to the stationary fabric selvedge and the three-dimensional coordinates of the shuttle eye undergo a

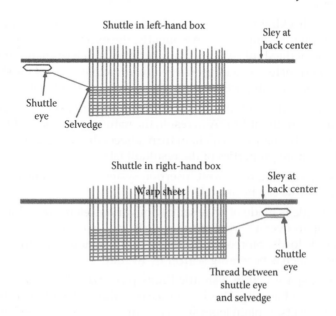

FIGURE 9.1
Effect of asymmetric location of shuttle eye.

continuous change. Over and above, the effect of the fourth dimension, that is, time, also plays a role as the flight velocity varies with the continuously changing combined mass of the shuttle and the depleting pirn. As yarn tension depends on its unwinding rate, it is evident that weft tension even during this phase of shuttle flight would keep on varying from one instant to other. Superimposed on this tension variation would be the effect of the variation of unwinding point on the pirn itself. Finally, as and when the shuttle enters the deceleration phase, multiple impacts cause further uncontrolled and indeterminate changes in tension in the inserted pick of weft, which, during this phase, still lies untrapped and free within the warp shed. Thus, at the end of a picking cycle with a conventional shuttle, a curvilinear segment of yarn lies stretched across the warp shed in an undefined and nonuniform state of tension. During the subsequent phase of shed change and the beating up process, a heterogeneous crimp distribution between warp and weft yarns across the fabric width would be the logical consequence. This effect viewed over a series of consecutive picks leads to very nonhomogeneous woven fabric, as crimps in warp and weft yarns and their homogeneity primarily govern properties of woven fabrics.

Recalling that power consumed by the shuttle during its acceleration phase is proportional to the product of its mass and (loom rpm)^3 (see kinematics of a picking system in Section 7.3.2), it is evident that a lowering of shuttle mass can be translated into higher loom speed and hence higher loom productivity. Moreover, smooth passage of a large shuttle through the confined space within a warp shed and over a large reed width demands a high sweep as well as a high eccentricity of sley, factors that adversely affect loom speed. Hence, a reduction in mass and dimensions of the shuttle provides an efficient route to raising loom productivity. The mass of a conventional shuttle can be reduced by reducing its dimensions and by opting for a lighter and stronger material. The first option would aggravate problems related to pirn changing, and the second solution has not found any commercial acceptance. An elegant solution to this vexing issue is found in the unconventional, that is, the so-called shuttleless weft insertion systems.

9.2 Basic Principle of the Unconventional System

A conventional shuttle performs three functions: It stores a certain length of weft within its hollow in the form of a pirn, transports this package across a warp shed, and permits the desired length of pick to be unwound smoothly from the pirn stationed within its hollow. This multitasking demand on a conventional shuttle is simplified in the unconventional systems by delinking the function of weft transport from the ones of weft storage and weft unwinding. The equivalent of a shuttle in an unconventional system simply functions as carrier of the tip of a weft yarn through the warp shed, unwinding the same in the process from a cone while the cone itself remains suitably stationed on one side of the loom. Additional devices ensure a controlled weft unwinding from respective cones. The implications of such a conceptual change are listed in the following:

1. The weft carrier can be drastically reduced in mass and size, leading to reduced consumption of energy per pick. Alternately, supplying the same power to the carrier as to the one with higher mass enables the lighter carrier to travel a longer distance. This translates into a higher reed width of the loom for equivalent picking power.

   A reduction in carrier size automatically leads to a reduction in shed depth and sweep of sley. Reduction in shed depth suppresses the amplitude of dynamic load on warp yarns and permits a higher frequency of shed change, and a reduction in sweep of sley promotes a higher frequency of sley oscillation. Higher frequencies of shed change and sley oscillation mean a higher loom rpm.

   It can hence be concluded that, as a result of reduction in mass and size of the carrier, a loom employing an unconventional principle of weft insertion can be run at a higher speed with a wider reed. This effectively translates into a higher weft insertion rate and, hence, higher productivity.

2. Being stationed outside the oscillating sley, the weft supply package remains visible and accessible while weaving continues. Such a scenario permits intervention of suitable control systems for switching weft supply from an exhausted to a full cone. This literally guarantees inexhaustible weft supply, contributing to higher loom efficiency. Moreover, intervention of suitable control systems operating from within the weft supply zone ensures controlled tension in inserted picks.

3. The unwinding of weft from a pirn stored within the hollow of a conventional shuttle during a picking cycle is governed by the location of the shuttle at any instant of time. As a result, tension in an inserted pick of weft varies from one segment to the other in an uncontrolled manner. The unconventional weft insertion systems permit a continuous and effective control of weft tension, guaranteeing, in the process, a superior fabric quality.

From the foregoing, it is observed that a weaving loom equipped with an unconventional weft insertion system is more productive and more efficient while producing qualitatively superior fabric compared to its counterpart equipped with a shuttle propulsion system. Such looms are generally referred to as shuttleless looms and constitute a different genre of looms altogether.

It is worth noting, at this point, that multitasking is usually associated with higher forms of systems. However, this example of switching over from a multitasking shuttle to a monofunctional weft carrier reveals that a step in exactly the opposite direction can also lead to a quantum jump in performance.

9.3 Functional Principles of Shuttleless Weft Insertion Systems

Commercially successful shuttleless weft insertion systems can be categorized with respect to the nature of the carrier and its flight through the warp shed. A carrier may either be solid or fluid, and its flight may be positively controlled, partially guided, or absolutely free. Viewed purely against these two criteria, a system involving a positively controlled fluid carrier should theoretically provide the best solution. However, such a system is yet to be invented. Similarly, the worst combination of a freely flying solid carrier, akin to a conventional shuttle, is not encountered in shuttleless systems. The four other combinations are outlined in the following.

9.3.1 Partially Guided Solid Carrier

A flat, oblong-shaped ballistic element (Figure 9.2), having a mass between 20 and 60 g and designed to fly with minimum air resistance while gripping the tip of the weft in a clamp located near its tail forms the carrier. This element, commonly referred to as a projectile or gripper, is accelerated from its state of rest to a finite velocity within a very short distance in the designated chamber and then, upon entering the reed space, is constrained to follow a specific path created by a set of rakes in front of the reed within the warp

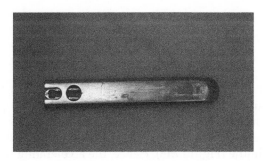

FIGURE 9.2
View of a gripper.

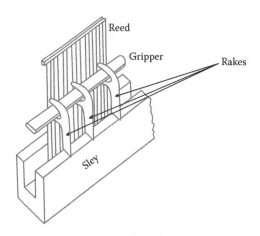

FIGURE 9.3
Guidance of gripper in warp shed.

shed (Figure 9.3). These rakes preclude the possibility of a projectile flying out of the warp shed. However, any obstruction in the path of the projectile, created by an unclean shed, would result in warp breaks and therefore a reduction in loom efficiency and deterioration in fabric quality. Hence, a high-quality weaver's beam is a prerequisite to successful projectile weaving. Indeed, this aspect is equally critical for all shuttleless weaving systems.

The heart of a projectile propulsion system is a torsion rod (Figure 9.4). This element is subjected to a specific angular distortion in every loom cycle. If torsional rigidity of the rod is K_t Nm, and it is twisted through ϕ radians, then the torque M_t generated is given by

$$M_t = K_t \phi$$

$$= (\pi d^4 G)\phi/32\ell \tag{9.1}$$

where

    d = diameter of torsion rod (m)
    ℓ = length of torsion rod (m)
    G = shear modulus of torsion rod (N/m^2)

This torsional energy is stored in the rod and released later to projectile instantaneously. On the other hand, in a conventional picking system employing a shuttle, picking energy is generated gradually through elastic deformation of elements of the picking system and transmitted simultaneously to the shuttle (Figure 7.18). Hence, the accelerating force acting on a projectile is impulsive in nature, and the corresponding vector acting on a conventional shuttle has the nature of a ramp. This fundamental difference

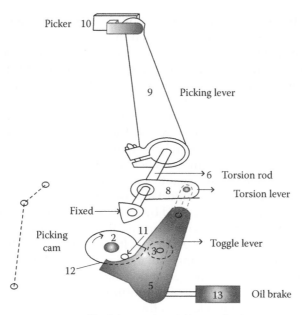

The Sulzer torsion-picking mechanism

FIGURE 9.4

Projectile picking system. (From Marks, R. and Robinson, A. T., *Principles of Weaving*, The Textile Institute, Manchester, UK, 1976.)

in generation and transmission of picking energy in conjunction with a much lower mass is instrumental in the much higher value of average acceleration and, consequently, much higher initial velocity in flight of up to 50 m/s for a projectile as compared to a maximum of 15 m/s for a conventional shuttle.

A typical layout of elements that enables time-specific generation, storage, and release of torsional energy to projectile is shown in Figure 9.4. The toggle lever, linked to an oil brake at one of its ends and to the torsion lever via an intermediate link at its other end functions as the switch. In its state of rest, the three link joints are not collinear and resemble an elbow as depicted by the line diagram on the left. At a given instant of time, the nose of a picking cam starts pushing an antifriction bowl mounted on the toggle lever in a clockwise direction. This displacement forces the joint between the toggle lever and the intermediate link to travel clockwise along the arc of a circle, pushing, in the process, the joint between the intermediate link and torsion lever in an upward direction. This upward displacement of one end of the torsion lever results in a twisting of the torsion rod to which the other end of the torsion lever is rigidly fastened. Incidentally, one end of the torsion rod is rigidly linked to the machine frame, and its other end is splined to the picking lever. The potential energy in the twisted rod keeps on building up until, at the end of the stroke of the picking cam, the torsion lever, connecting the link and the toggle lever assume a configuration such that the link joints become collinear. Such a state is depicted in the main body of the diagram. As the nose of the picking cam moves away from the antifriction bowl, the downward force acting through the three link joints maintains the system in a state of equilibrium. Evidently, the system would collapse in the event of the minutest play in any of the links.

As long as the three link joints remain perfectly collinear, the energy in the torsion rod remains trapped and stored. The moment this condition is disturbed, the collinear link joints collapse to the elbow configuration, forcing a sharp clockwise rotation of the joint between the picking lever and the torsion rod. This rotation is translated to a linear motion of the picker shoe and hence to the projectile. The all-important disturbance to the equilibrium state of the collinear link joints is created by a collision between a small antifriction bowl mounted on the picking cam and the sharp upturned nose on the extreme left of the toggle lever extension. It is observed that the design and location of the nose of the picking cam vis-à-vis that of the antifriction bowl on the toggle lever governs the generation and storage of picking energy in the torsion rod, and the location of the antifriction bowl on the picking cam governs the instantaneous release of this energy.

The sequence of functioning of various machine elements during the period of forced clockwise displacement of the toggle lever from its state of rest may be viewed overall as the suitable alignment of an L-shaped three-way rotating valve (Figure 9.5) that allows a requisite amount of power for every pick to flow from the source to a reservoir (Figure 9.5a), and attainment of a self-locking state of the mechanical system, that is, energy stored in the reservoir remaining stored over a period of time is equivalent to the valve transiting from Figure 9.5a to b. The collision of the antifriction bowl on the

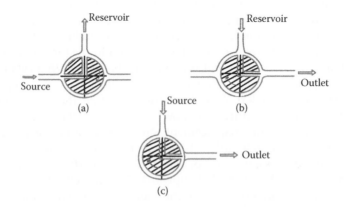

FIGURE 9.5
Model of energy transmission of picking systems. (a) Power flow from source to reservoir, (b) power flow from reservoir to outlet, and (c) power flow from source to outlet.

picking cam with the upturned nose on the toggle lever can be viewed as energy from the reservoir flowing to the outlet (Figure 9.5b) namely to the projectile. Further rotation of the valve in transiting from Figure 9.5b to 9.5a depicts the state of rest of the propulsion system. Hence, the entire system may be considered equivalent to that of an L-shaped three-way valve rotating with the uniform angular velocity and suitably connected to a power source, a reservoir of energy, and an outlet. A conventional shuttle propulsion system can, in this sense, be represented by a much simpler two-way L-shaped rotating valve (Figure 9.5c), in which, at the given instant of time, power flows directly from the power source to the outlet.

In order to form an idea about the magnitude of critical variables involved in a gripper propulsion system, two typical sets of combination of width (m), rpm (min⁻¹), and weft insertion rate (m/min) of modern gripper looms, namely 3.9, 333, and 1300 and 5.4, 275, and 1500, can be considered in conjunction with relevant loom timing details. In such looms, a gripper is accelerated to its free flight velocity over a period of 8° movement of the loom shaft. For a machine running at 333 rpm, an 8° displacement consumes 0.004 s, and the corresponding value for a machine running at 275 rpm is 0.0048 s. Over a period of about 150°, the corresponding grippers cross the entire reed width at average flight velocities of 52 and 60 m/s for the 3.9 and 5.4 m looms, respectively. Assuming a uniform acceleration of gripper from its state of rest, values of 13,000 and 12,500 m/s² are expected to be encountered by grippers of the two looms of 3.9 and 5.4 m width, respectively. However, if acceleration is assumed to rise and fall as a linear function of time, then maximum values twice as large, namely 26,000 and 25,000 m/s², would be encountered. The corresponding average accelerating force for the lightest gripper of 20 g works out to be 260 and 250 N, respectively. An interesting observation ensuing out of these numerical values is that, in spite of differences in critical values of commercial interest, such as reed width and weft insertion rate, the propulsion systems of the two looms can be operated with the same elements incorporating only very minor alterations in settings.

Simple expressions that guide calculations carried out in the foregoing paragraph are stated in the following:

Let

m = mass of shuttle (kg)

n = loom rpm (min⁻¹)

R = reed width (m)

W = weft insertion rate (m/min⁻¹)

θ = angular displacement of loom shaft during acceleration of picking element (degrees)

ϕ = angular displacement of loom shaft during flight of picking element across reed (degrees)

Then

(Rθ/6W) s = duration of acceleration of picking element

(Rφ/6W) s = duration of flight of picking element across reed

(6W/φ) m/s = average velocity of flight of picking element

$(36W^2/R\theta\phi)$ m/s^2 = magnitude of uniform acceleration

$(72W^2/R\theta\phi)$ m/s^2 = maximum value of acceleration rising and falling linearly at the same rate

$(36mW^2/R\theta\phi)$ N = average accelerating force on picking element

For a conventional shuttle loom, the following values can be considered for estimating equivalent magnitudes of relevant variables:

$m = 0.5$ (kg)

$n = 200$ (min^{-1})

$R = 1.5$ (m)

$W = 300$ (m/min^1)

θ = 15 (degrees)

φ = 120 (degrees)

Then

0.0125 s = duration of acceleration of picking element

0.1 s = duration of flight of picking element across reed

15 m/s = average velocity of flight of picking element

1200 m/s^2 = magnitude of uniform acceleration

2400 m/s^2 = maximum value of acceleration rising and falling linearly at the same rate

600 N = average accelerating force on picking element

Comparing the estimated values for the shuttle with that of a 20-g gripper, one infers that a 25 times lighter gripper aided by a superior propulsion system consumes about one third the energy in being accelerated to a three to four times higher velocity at one third the time required by a conventional shuttle. The numerical values chosen in this illustration indicate that the accelerating distance in both instances would be around 5 cm (see Appendix A.1). Incidentally, even if the acceleration profile of gripper is changed to that of variant B or variant C, the net displacement would be 6.2 and 5.4 cm, respectively.

Typical velocity and acceleration profiles of a picker shoe along with a velocity profile of a gripper are illustrated in Figure 9.6. This diagram refers to a relatively old model of a projectile loom. The corresponding approximate values of rpm, reed width, and weft insertion rate can be worked out

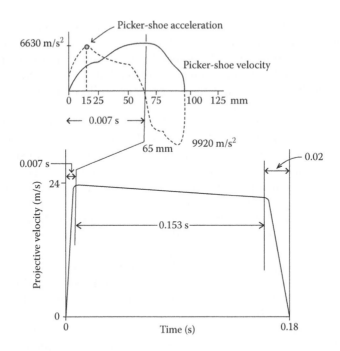

FIGURE 9.6
Velocity and acceleration profiles of projectile picking elements. (From Marks, R. and Robinson, A. T., *Principles of Weaving*, The Textile Institute, Manchester, UK, 1976.)

from the diagram as 163, 3.7 m, and 600 m/min, respectively (see Appendix A.2). Application of the formula for estimation of maximum acceleration that rises and falls at the same rate yields a value of 6868 m/s². This is quite close to the value of 6630 m/s² shown in the diagram.

Far from being uniform, the picker-shoe acceleration profile shows three distinct phases. It has a finite value at the start itself, resulting out of the instantaneous release of stored energy in the torsion rod. The subsequent phase may be approximated to a linear rise to the maximum value (a_{max}) followed by a linear drop to the value of zero before entering the deceleration phase. Incidentally, this profile is not a function of time but of picker-shoe displacement.

One may tend to equate the instants at which the picker-shoe and gripper attain their maximum velocities, as has been shown in the diagram, but only under the condition that these two elements follow same velocity profiles. Indeed, as per the catapult theory proposed by Catlow and Vincent (1951), this may not be the case, and gripper might as well leave the picker-shoe much earlier. Considering the possibility of the velocity profile of the picker shoe and of the gripper following an identical pattern until they come apart after 0.007 s of the acceleration phase, the distance covered by the gripper in the given time period works out to be somewhere between 4.3 (Variant A in Appendix A.1) to 5.3 mm (Variant B in Appendix A.2) whereas the picker shoe is shown to attain its highest velocity at 65 mm displacement. Hence, this diagram indeed reinforces the proposition that the two elements follow different velocity profiles, and they disengage well before the picker shoe achieves its highest velocity.

A 40-g gripper flying away at the velocity of 24 m/s—values commensurate with the system described in Figure 9.6—possesses kinetic energy of about 12 N m, and a typical corresponding propulsion system stores about 67 N m energy in the torsion rod for effecting the propulsion (see Appendix A.3). This means that approximately 17% of picking energy is effectively used in such a system, and the remaining 83% is wasted at each pick through noise and heat. Most of this heat is generated in an oil brake to which one end of the toggle lever is connected (Figure 9.4) through a suitable plunger. During the anticlockwise rotation of the toggle lever, caused by the collapse of the collinear link joints, the picker shoe enters the acceleration phase while the plunger linking the toggle lever to the oil brake starts a simultaneous inward movement. The subsequent rise in accelerating force is thus quickly damped, and a drop in acceleration sets in followed by a strong deceleration, bringing the system quickly back to its initial condition. In the absence of a suitably strong damper, the system would have gone into uncontrolled vibration, leading to instability. Hence, a certain amount of energy wastage is inevitable in such mass-spring-damper systems. With a twofold rise in frequency of picking—from 2.72 Hz in the older machines to about 5.5 Hz in the more modern ones—as well as a sharp rise in the cost of energy, it becomes a challenge to a machine designer to reduce energy wastage while ensuring the stability of the gripper propulsion system.

Unwinding yarn from the supply package at a very high speed with minimum variation in tension during the propulsion of a pick is critical to efficient functioning of this system. Yarn unwinding tension is directly proportional to the square of yarn withdrawal speed. It is, moreover, affected by ballooning that occurs during overhead unwinding. Yarn tension in a balloon is directly proportional to the square of the balloon height and inversely proportional to the square of the unwinding radius. As a result, a short-term tension variation due to a continuous change in balloon height and a long-term tension variation due to a gradual reduction in package diameter get imposed on the basic yarn tension. In order to eliminate the deleterious effect of ballooning, recourse is taken to the storage feeder, the principle of action of which is illustrated in Figure 9.7.

A storage feeder, also referred to commonly as an accumulator, unwinds yarn at a constant speed from a supply package with its own drive and stores the same in parallel coils on a suitable rotating drum. In Figure 9.7, yarn from a cone is shown being passed through a suitable guiding system to the winding drum. An inclined ring keeps pushing yarn coils toward the nose of the winding drum as soon as the coils are wound, and when a sufficient

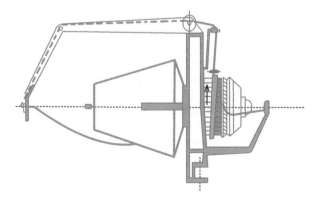

FIGURE 9.7
Storage feeder.

length of yarn has been wound on the drum, it switches off the motor in the accumulator thus stopping the yarn unwinding from the package. The free end of yarn near the drum edge passes through a suitable tensioning device of the accumulator to the projectile feeder of the picking system, which, in turn, transfers the same to the projectile. Hence, during its flight, a projectile unwinds yarn from the accumulator drum and not from the supply package, doing away thereby with the associated tension variations. As soon as yarn starts getting unwound from the drum, a sensor in the accumulator switches it on, and yarn from the supply package starts getting wound on the drum afresh. Unwinding yarn from the supply package and hence winding coils around one end of the drum while allowing the unwinding of yarn coils from its other end during projectile flight can occur simultaneously.

As and when the projectile crosses the warp shed and reaches the braking zone located on the opposite side of the picking assembly, the projectile feeder moves close to the corresponding selvedge and grips the pick of weft firmly while a pair of weft-end grippers, one near each selvedge, also grips the same pick very close to the corresponding selvedges. Subsequently, a weft cutter moves in between the projectile feeder and the weft-end gripper and cuts the yarn. As a result, a length of pick, part of which is held at a predetermined tension across the warp shed by weft-end grippers while a part hangs out on each side of selvedge, gets separated from the supply package conveniently. Subsequently, the warp shed crosses over, the projectile feeder withdraws to its rest position while keeping the weft tip firmly gripped, and tucking systems fold protruding pieces of weft yarn at the two selvedges back into the next shed. Resulting selvedges are therefore made up of double picks. Hence, construction of selvedge in terms of yarn count and yarn spacing needs to account for effectively thicker picks than that of ground fabric in order that eventual warp yarn consumption for each pick remains the same for ground fabric and selvedges.

During flight of the projectile, the loom sley dwells at its back center. Hence, the body of an inserted pick of weft remains parallel to the fell of the cloth. At the same time, a weft tension–control system ensures reasonably constant tension in the inserted pick of weft. This predetermined constant tension and also an exact alignment of the pick is maintained by weft-end grippers until the shed lines cross over. Such control on the pick of weft is maintained in each pick, resulting in very uniform crimp distribution and therefore very homogenous fabric.

A projectile loom operates with a large number of projectiles. While one projectile is in flight, many others are on their return journey to the starting position adjacent to the picking shoe. On being released by the brake shoe, projectiles drop onto a conveyor belt, which transports them to the picking side. A thumb rule governs the total number of projectiles required on a loom as a function of reed width.

9.3.2 Fully Guided Solid Carrier

In such a system of weft propulsion, the carrier is rigidly linked to the propulsion system at each instant of time, bringing into the equation effects of inertia and natural frequency. Evidently, such a system can operate at high speed only when its natural frequency is high and mass is low.

The essential elements of such a system comprise a carrier, a source of oscillating motion, and a link between the two. The carrier and a certain portion of the link move in and out of the warp shed quite in the manner of a rapier being thrust in and out. This apparent similarity with the action of a rapier has led to the coining of the term "rapier system of weft insertion," and as an extension of the same logic, a loom equipped with a rapier system of weft propulsion is conventionally termed a rapier loom. The three elements of a rapier system may be equated to a mass-spring system oscillating to and fro at a certain frequency. Spring stiffness of a rapier system is primarily governed by the buckling rigidity of the link.

A rapier system is characterized by the nature of the carrier, namely Gabler or Dewas, and by the nature of the link, namely rigid or flexible. Accordingly, four different rapier systems come into question, and each combination enjoys its own domain in the woven fabric formation process.

The Dewas type of carrier, named after its inventor R. Dewas, which brings weft from the supply package is equipped with a spring-loaded tip (Figure 9.8a), and the corresponding element to which the weft is transferred has a spring-loaded clamp (Figure 9.8b). In such a system, the tip of the pick of weft is gripped by the rapier head and dragged to the center of the warp shed where a tip-to-tip transfer (Figure 9.8c) takes place from the right-hand element to the one on the left. Subsequently, both heads retract to their origin. In this way, a complete pick of weft is inserted in the warp shed by a Dewas type of rapier.

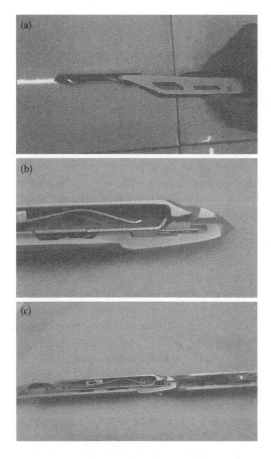

FIGURE 9.8
Rapier heads on Dewas system. (a) Giver rapier head, (b) receiver rapier head, and (c) rapier heads during yarn tip transfer.

In the case of a Gabler system, named after its inventor J. Gabler, the corresponding head that brings the weft from the supply package exhibits a fork-like opening, which can fasten onto and drag a yarn loop from the supply package to the center of the warp shed and transfer the loop into the head of the hook like a receiving head. During retraction of the receiving head, the loop of weft gets unfolded over the corresponding part of the warp shed as a result of which, eventually, a single pick of weft gets inserted.

Both types of rapier exhibit one commonality, that is, the handing over of weft yarn from one head to the other at the center of the warp shed. Subsequently, both heads move back in opposite directions. This directional change in the movement of the rapier heads necessitates a slowing down of rapiers midway through their insertion phase so that, at the reed center, their respective velocities approach the value zero. Transfer of weft yarns at the reed center thus imposes a severe restriction on the velocity profile of rapiers and, consequently, on the weft insertion rate.

The rapier heads of the Dewas system grip a pick of weft at one point only, somewhat in the manner of a gripper shuttle. They inflict no further physical damage to the rest of the inserted pick. On the other hand, rapier heads of the Gabler system subject the weft yarn to vigorous abrasive strain, both during insertion of the loop into the shed center as well as during unfolding of the loop over the second half of the warp shed. This strain is caused by the relative motion between the rapier head and the yarn, which results in the sliding of yarn across the inside wall of the respective rapier heads. Evidently, the Gabler system is suited for rough and robust yarns and the Dewas system for relatively delicate yarns. The issue of a fail-proof point grip both during insertion of weft into the warp shed as well as during its transfer from one rapier head to the other at the center of the warp shed has an element of uncertainty that puts a further restriction on the types of yarns suitable for the Dewas system. Factors such as yarn thickness, its compressibility, and the coefficient of yarn-metal friction govern the reliability of tip-to-tip transfer. To overcome these uncertainties, there is a novel solution of transferring from one head to the other not the gripped yarn but a sub-element that grips the yarn throughout the insertion process.

A rapier head is moved in and out of the warp shed by a link, which, in turn, gets its motion from an oscillating element. The link may be rigid, in the form of a rod, or it may be flexible, in the form of a belt. Accordingly, a rapier loom is termed either rigid rapier or flexible rapier. A schematic view of one cycle of weft insertion on a rigid rapier machine employing Dewas heads is illustrated in Figure 9.9. The diagram at the top shows two rapier heads on two sides of a reed and a magazine creel of weft yarn cones located on one side. Each rapier head is mounted on a rigid rod, which is equipped with a rack of teeth meshed into an oscillating pinion. During certain durations of the loom cycle, the pinions oscillate simultaneously in opposite directions, thus moving the rods and therefore the respective rapier heads in and out of the warp shed. The throw of rods is adjusted to ensure that rapier heads cross each other at the warp shed center by a specific amount in order that a smooth tip-to-tip transfer can take place. Design of rapier heads ensure that such an overlap does not cause a head-to-head collision.

The rapier head on the creel side picks up yarn from a designated cone and is carried toward the center of the warp shed by the rod. Simultaneously, the matching rapier head from the opposite side moves in as shown in the central diagram of Figure 9.9. The picking cycle is completed when both rapiers retreat to their original positions, outside the two selvedges. At this instant, the rapier head opposite to the creel holds the weft tip while the other end of the weft is held firmly by feeding elements near the magazine creel.

In their fully retracted state, the rods of a rigid rapier machine may occupy a space nearly equal to the loom width. For saving this wasted space, rigid

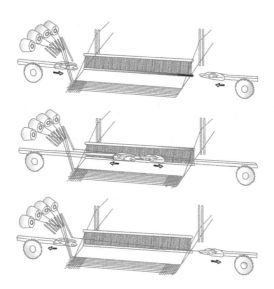

FIGURE 9.9
Weft insertion on Dewas system.

rods are designed in the form of telescopic elements. Alternately, rigid rods may be replaced altogether by flexible belts. Such belts are wound on spindles and housed in drums. As the spindle rotates alternately in opposite directions, the belt is unwound and rewound, and its tip enters into and withdraws from the warp shed. Rapier heads are fastened to the tips of such belts, and as the belts move to and fro, so do the rapier heads. Evidently, a flexible belt can buckle while accelerating toward the center of the warp shed if the inertial resistance of the equivalent mass of the moving system at the rapier head is greater than the buckling resistance of the belt. Properties of the material of the belt as well as its dimensional characteristics play decisive role in this respect. Let

E = Young's modulus of belt material

I = polar moment of inertia of belt

w = belt width

t = belt thickness

ℓ = belt length

m = equivalent mass at rapier head

a = acceleration of rapier

In order that the belt does not buckle,

$$[EI\pi^2/4\ell^2] > ma \tag{9.2}$$

In Equation 9.2, the product *EI* represents flexural rigidity of the belt while its polar moment of inertia *I* is given by

$$I = wt^3/12.$$

Thus, a wider, thicker, and shorter belt made from a material of higher modulus would be able to support a higher accelerating force on the rapier head. One can also infer that a rapier head of lower mass and a belt material of lower density are desirable for a flexible rapier system. Clearly, the demands on quality of the belt material and rapier mass become more demanding as the width of the machine goes up.

The other aspect in which weft propulsion by flexible rapier differs from that of a rigid rapier is in respect to the requirement of vertical support needed by the head during its movement across the warp shed. If the machine width is on the higher side, then the belt would sag after a certain length is inserted within the shed, causing the rapier head to slide along the bottom line of the warp shed at least for part of its journey. Hence, ensuring that the bottom shed line is nearly flush with the sley race is more critical for the flexible rapier than for the rigid one.

The width of the rapier loom can be increased substantially while operating at a reasonably high weft insertion rate by resorting to a rigid rapier system. An innovative solution to offset the accompanying wastage of floor space involves installation of one propulsion system between two looms in such a way that retraction of the rapier head from within the warp shed of one loom is associated with the insertion of another rapier head, mounted on the opposite end of the rod, into the warp shed of the adjoining loom. The pair of looms operates with a certain phase difference such that when the warp shed of one loom is in the process of opening, the shed starts closing in the adjoining loom. In this solution, the rigid rod carries two rapier heads, one at each end, and each rapier head moves the entire width of the corresponding machine. Consequently, there is no transfer of yarn from one rapier head to another at the warp shed center, leading to an improved velocity profile of the rapier during weft insertion. Hence, such machines can operate at a much higher effective rapier velocity than on other rapier looms on which a pair of rapiers operates. However, this very feature of sequential rapier insertion in two adjoining looms imposes the condition that when one loom stops for some reason, the other too must become idle. Moreover, any loom movement associated with attending to the stoppage of one loom, such as a broken pick, for example, becomes a complex issue of ensuring that the weaving condition of the paired loom does not get disturbed. Incidentally, such a system of coupled weaving on two adjacent looms is termed bi-phase weaving.

In principle then, a rapier loom might involve a pair of rapiers operating from two ends of fabric selvedge with their associated complex profile of velocities or may even operate with one rapier only that traverses the entire fabric width with a velocity profile that is more congenial than that of a double rapier system insofar as the weft insertion phase is concerned. However, such a rapier has to retract the entire shed width during which fabric production enters an idle phase. Effectively, therefore there is not much to choose between the two in terms of weft insertion rate. Additionally, such rapier systems, when employed on reasonably broad looms, need to be of the rigid type with associated space wastage. However, if, instead of inserting a single pick of weft, a single rapier inserts a loop of yarn across the entire fabric width, then the idle production phase as well as space wastage can be more than compensated. Evidently, not many cloth constructions permit such an option, limiting thus the commercial viability of single rapier systems.

Thus, although, theoretically, there could be a very large variety of commercial rapier looms in terms of combination of rapier heads (Gabler or Dewas), type of link (rigid or flexible), number of rapiers (single or double), and location of rapier (between two looms or either on one or both sides of one loom), one encounters, in practice, a limited number with proven practical utility.

A superior weft-handling regimen is the main reason for the commercial success of rapier systems. There are two dimensions to this issue: a successful rapier system enables mixing a wide range of weft yarns through a simple weft-selection device, and at the same time, it subjects different types of weft yarns to their appropriate levels of strain.

The weft creel shown in Figure 9.9 holds four yarn packages. On a modern commercial rapier machine, as many as 12 packages can be creeled at the weft-selection system. Yarn from each package is drawn through a finger whose motion and location are controlled by the weft-selection system. During a picking cycle, a selected finger is brought into the path of the rapier head, which grips and inserts the corresponding yarn into the warp shed. Viewed purely from this angle, the gripper and rapier systems are quite alike. However, the motion of the rapier can be designed to follow a certain specific function, depending on the nature of the weft being handled. This aspect is illustrated in Figure 9.10.

Four different plots are shown in Figure 9.10. There are two plots of rapier speed as a function of angular displacement of the main shaft of the machine and two more of corresponding weft tension. The two sets belong to two different functions of rapier motion, one designated as P (previous) and the other as D (developed).

A typical profile of weft speed in a Dewas system shows two symmetrical humps side by side. The first hump represents the velocity profile of the rapier that brings the weft into the warp shed, and the adjoining hump represents that of a rapier that takes over weft at the center of the shed. The instant at which weft is picked up by the rapier is indicated in the figure by a sharp rise in weft tension. In the "P" trace, this occurs about halfway to the peak of the first hump, while in the "D" trace, it occurs near the base of the first hump. Although both the "P" and "D" traces of rapier speed exhibit

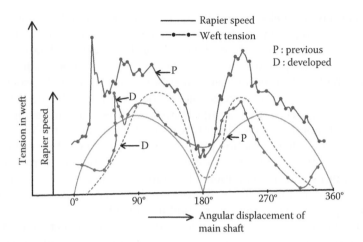

FIGURE 9.10
Displacement profile of rapier and tension in weft.

certain similarities of periodicity, their amplitudes and profiles are markedly different. The "D" profile shows a greater sluggishness at the start and at the end, whereby velocities increase more rapidly to a much higher peak value as compared to the "P" trace. Incidentally, the areas under the "D" and "P" curves are the same, indicating that equal distance is covered in the given time. The corresponding weft tension profiles clearly demonstrate the superiority of the "D" profile. Expressed in simple terms, it can be stated that a rapier that moves more sluggishly at the beginning and picks up weft during this phase does not cause a great deal of stress to the yarn even though it may accelerate subsequently to a much higher peak velocity if the nature of acceleration is such that no serious jerk is encountered. Filing weft yarns being usually weak in nature can be most appropriately handled by a rapier system because of this one singular aspect of controlled rapier propulsion.

The ability to select and impart the most appropriate velocity profile to weft characterizes the essence of the rapier propulsion system. By way of illustration, the displacement, velocity, and acceleration profiles of three typical functions, namely sinusoidal, cycloidal, and modified trapezoidal functions, are shown in Figure 9.11. Their intrinsic differences are least noticeable in the displacement function, more noticeable in the velocity function, and strikingly demonstrated in the acceleration function. Considering the velocity function, it is observed that the cycloidal function exhibits the highest sluggishness at the beginning and at the end of a cycle. However, this same function exhibits the highest peak acceleration among the three. The modified trapezoidal function, on the other hand, exhibits an intermediate sluggishness, and its acceleration profile indicates a certain period of uniformity and a moderate peak value. The acceleration function of sinusoidal function exhibits its peak

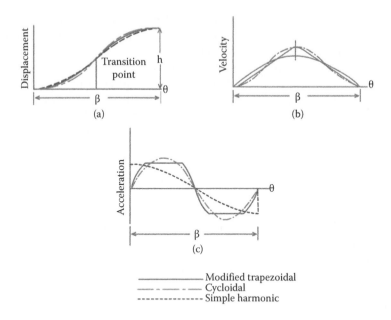

FIGURE 9.11
(a) Displacement, (b) velocity, and (c) acceleration profiles of some standard functions.

value at zero velocity, a very damaging proposition for picking up delicate weft yarns. Tumer and Dawson (1988) investigated implications of employing these and other functions, such as double harmonic, parabolic, or fifth-order polynomials for shaping displacement profile of rapiers.

9.3.3 Basic Concepts of Fluid Carrier

As compared to solid carriers, fluid carriers, such as water and air, offer considerably lower inertia as the density of pure water is 1 g/cc at 4°C and that of dry air at sea level is 0.001275 g/cc at 0°C. Just as potential energy can be stored in the torsion rod of a gripper loom, so can energy be stored in fluid by compressing the same. Releasing some of this energy at a suitable instant has proved to be an elegant way of propelling a pick of weft. The released energy propels the weft along with a certain amount of fluid. However, the density of fluid as well as of the weft yarn being very low and the net mass of a pick of weft being in milligrams, both can attain very high velocity within a short time, which is well beyond the scope of any solid carrier. Hence, modern looms based on fluid propulsion systems offer productivity twice as high as those attained by solid carriers. Indeed shedding and beating up systems constitute the limiting factors affecting productivity of fluid propulsion–based machines.

Of the two fluids, air offers the more lucrative and practical solution as it is available everywhere in plenty and its density is a thousand times lower than that of water. No wonder then that technological improvements on air-jet looms have taken place in leaps and bounds, making it one of the most widely employed systems for woven fabric production.

Energy is stored in fluid through compression. Let fluid pressure within a compressor be denoted by p_1. Let this fluid be allowed to flow through a Venturi tube (Figure 9.12). Let fluid velocity over the wider section of tube be v_1 and that through the narrower section be v_2. Applying the concept of mass balance, one infers that v_2 must be larger than v_1. Assuming laminar fluid flow, Bernoulli's equation states

$$(p_1/\rho) + \tfrac{1}{2}\, v_1^2 = (p_2/\rho) + \tfrac{1}{2}\, v_2^2 \qquad (9.3)$$

In Equation 9.3, p_2 is fluid pressure in the narrower section, and ρ is fluid density, assumed to be constant. Evidently, p_2 would be lower than p_1 as v_2 is larger than v_1. Expressed in simple terms, the equation states that when fluid is forced to pass through a constriction, its velocity increases accompanied by a drop in pressure. In other words, a Venturi tube converts potential energy of compressed fluid to kinetic energy. The nozzle of a jet propulsion system is designed following the principles of a Venturi tube.

A typical air-jet nozzle along with its accessories is depicted in Figure 9.13. A pick of weft (1) is threaded through the nozzle opening (2) and weft channel (3). Compressed fluid (4) is supplied through the distribution circuit (5) and control elements (6) to the annular reservoir (7) of the nozzle from which the fluid passes into highly constricted flow-stabilizing ducts (8) and nozzle throat (9). In this zone, the fluid gains speed and subsequently impinges upon weft yarn in the cylindrical jet chamber (10) after passing through the annular duct (11). Weft yarn emerges from the nozzle exit aperture (12) dragged by the jet of fluid into the surrounding atmosphere. The equilateral triangle around the weft yarn indicates a zone of laminar flow, which extends up to approximately eight times the diameter of the nozzle exit. The surrounding air then starts disturbing this flow into the zone of turbulence. In the turbulent zone, the velocity of yarn along its axial path drops while components of velocity across its axis start growing. However, as the body of yarn in the laminar zone keeps on moving at a higher velocity along the yarn axis, the yarn body, as a whole, cannot remain straight and, after a while, would start

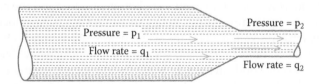

FIGURE 9.12
Flow through a Venturi tube.

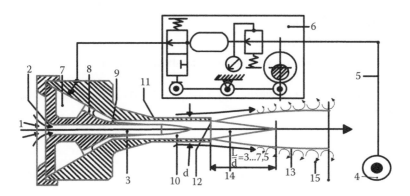

1. Weft thread
2. Weft thread entry
3. Weft channel
4. Compressor
5. Distribution circuit
6. Control components
7. Annual reservoir
8. Flow-stabilizing ducts
9. Throat of the nozzle
10. Cylindrical jet chamber
11. Annular duct

12. Nozzle exit aperture
13. Air stream
14. Core of the air stream
15. Boundary turbulence
    ------> Depends on external contour of the
       tube of nozzle exit.
An air flow and its velocity depends upon
    1) pressure differences
    2) air temperature
    3) changes in nozzle cross-section
    4) nozzle duct surface properties

FIGURE 9.13
View of an air jet nozzle.

buckling. It is apparent that a single nozzle by itself cannot throw a pick of weft to any appreciable distance.

A good nozzle design is aimed at maximizing gain in fluid velocity and, consequently, a high drop in pressure such that the condition of subatomic pressure prevails in a cylindrical jet chamber (10). As a result, a suction force develops between the weft yarn entry point of the nozzle and cylindrical jet chamber. Consequently, when a pick of weft is brought close to the weft thread entry point of the nozzle, it gets easily sucked into the weft channel.

Just as a solid carrier drags a pick of weft from one end of selvedge to the other, a jet of fluid also creates a drag force on the yarn body. The drag force is given by

$$F_\ell = \tfrac{1}{2}\, C_{d,\ell}\, \rho \pi d\ell\, (v_{a,\ell} - v_{y,\ell})^2 \tag{9.4}$$

where

    F_ℓ = longitudinal drag force along yarn axis

    $C_{d,\ell}$ = drag coefficient

    $v_{a,\ell}$ = velocity of fluid along yarn axis

    $v_{y,\ell}$ = velocity of yarn along its axis

ρ = density of fluid

d = diameter of yarn

ℓ = length of yarn on which the fluid stream is acting

The drag coefficient is governed by yarn and fluid characteristics and may be expressed as

$$C_{d,\ell} = \alpha Re^{-\beta}$$

where α and β are constants with values less than one, and Re is the Reynolds number.

$Re = uD\rho/\mu$

u = velocity of fluid

ρ = density of fluid

D = diameter of fluid stream

μ = viscosity of fluid

It is inferred from relationships stated in the foregoing that, for the same Reynolds number, the drag force is higher for

1. Greater difference between fluid and yarn velocity
2. Higher surface area of yarn
3. Higher density of fluid

9.3.4 Guided Fluid Carrier

After emerging from the nozzle, a jet of air tends to disperse rapidly into the surrounding atmosphere. Such an eventuality can be delayed by attaching a suitable tube to the nozzle exit (Mohammed and Salama, 1986). Of considerable importance is the area of the throat in the nozzle. The actual contour of the throat through which air flows are derived by subtracting the space occupied by the weft channel from that of the annular duct (Figure 9.13). This contour can be varied by adjusting the location of the weft channel tip within the annular duct. Such a throat behaves as a converging-diverging element, and its design should ensure subatmospheric pressure at the channel tip so that air flows from the weft thread entry point though the channel toward its tip, sucking in, as a result, the tip of weft. Consequently then, just beyond the weft channel tip, a mixing of air takes place. Part of this air comes from the compressor through the annular duct, and the other part comes from the surrounding atmosphere through the weft channel. Thus, two distinctly different zones of airflow evolve in the nozzle up to the tip of the weft channel.

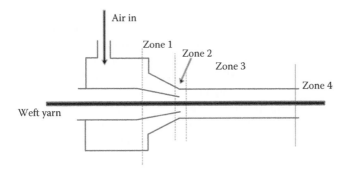

FIGURE 9.14
Air flow zones. (From Vangeluwe, L., *Textile Progress*, 29, 4, 1999, CRC Press.)

In the first zone (Figure 9.14), the potential energy of compressed air is converted into kinetic energy under a subsonic isentropic flow condition, that is, where adiabatic, compressible, and absence of frictional loss conditions prevail. In the second zone, a mixing of air takes place in the presence of the weft tip. As the weft tip moves beyond the channel tip, it is subjected to air drag force, and hence, it accelerates. The tubular tract in the rest of the nozzle and the air tube represents the acceleration zone for weft yarn. In the third zone, friction between the airstream and the inner wall of the tube plays an important role, and the flow can be categorized as Fanno-flow. It has been found that frictional effects rise with an increase in air tube length and a decrease in its diameter. A rise in Reynolds number, on the other hand, suppresses frictional effects. On exiting the air tube, the velocity of air remains constant for a while, and subsequently, it starts falling as turbulence grows. The velocity of the air at the tube exit grows with a rise in the supply of air pressure until a choking condition develops in the third zone within the nozzle–air tube system that is characterized by expansion and contraction waves. In general, a longer air tube needs a higher air supply pressure for achieving the same air velocity at the tube exit. Turbulence at the nozzle exit can be suppressed by opting for a longer air tube of smaller diameter.

Choosing then the most proper length and diameter of air tube in conjunction with the type of yarn—a thicker yarn requires an air tube of larger diameter—and supply air pressure in addition to designing the nozzle throat area suitably, constitute critical design aspects of a nozzle–air tube system. Such a system is mounted on the loom sley so that the tip of the air tube can be brought very close to the shed opening. Evidently, the air tube can have only a finite length beyond which the stream of air jet has to plunge into the surrounding atmosphere. Air velocity can reach the value of Mach 1, that is, approximately 340 m/s, in the nozzle throat and at tube exit under favorable design conditions.

Theoretical as well as experimental investigations on acceleration of the pick of weft by the main nozzle has led to provision of an additional tandem nozzle that helps in overcoming unwinding resistance from the yarn feeder and leads to more favorable conditions of air flow at the throat area of the main nozzle. A typical setup of four tandem and main nozzles is illustrated in Figures 9.15 and 9.16. Tandem nozzles are mounted on a stationary frame near the yarn feeders, and the main nozzles are mounted on the sley. Air tubes of the main nozzles converge toward the mouth of the profile reed channel. Yarn moving through the weft channel of the main nozzle is already accelerated to a degree by a tandem nozzle, and therefore, its second stage of acceleration in the main nozzle at the mixing zone near the nozzle throat is smoother.

Typical profiles of air jet velocity and the velocity of the pick of weft being dragged along after exiting the air tube tip of a nozzle are depicted in

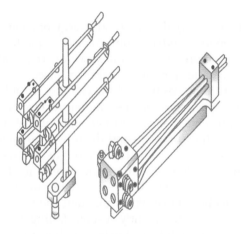

FIGURE 9.15
View of four tandem and main nozzles, Somet, ITEMA.

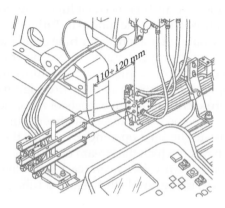

FIGURE 9.16
Setup of tandem and main nozzles on machine, Somet, ITEMA.

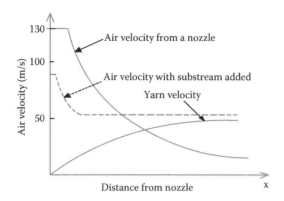

FIGURE 9.17

Velocity profiles of air jet and pick of weft. (From Uno, M., *Journal of Textile Machinery Society of Japan*, 18, 2, 37–43, 1972.)

Figure 9.17 (Uno, 1972). The corresponding profiles are drawn in continuous lines. It is observed that air velocity remains constant over a finite distance beyond which it starts rapidly falling owing to the onset of turbulence. Weft velocity at the tube exit has a finite value although it is much lower than that of air, and this velocity differential creates drag force on the yarn, which, therefore, keeps on accelerating. Even over part of the following turbulence zone, air velocity remains higher than that of the weft, and therefore, the weft velocity keeps on increasing until it equals the axial velocity of the turbulent jet. Beyond this point, the jet velocity falls below that of the weft, which then can move over only a very small distance owing to inertial forces. Some additional forces of suction created near the opposite selvedge can drag the pick for some more distance. A single nozzle–air tube system aided by a suction system has been commercially operated with a reed width of approximately 130 cm only. A modern air jet loom can, on the other hand, carry a pick of weft over a distance of 530 cm and more. This has been possible owing to development of two crucial guidance systems, namely the profile reed in conjunction with a relay jet system. Figure 9.18 depicts a side view of the two systems mounted on the sley.

A profile reed is made up of dents exhibiting prominent profiles. A front view of these dents reveals another dimension of the profile (Figure 9.19; Vangeluwe, 1999). When a large number of such dents are lined up in a reed, a clear channel is formed by the corresponding profiled dents. Weft is propelled along this channel by a stream of air jet. This channel supports and guides the air jet and plays the role of an extended air tube with a large number of very thin discontinuities.

The deleterious effect of discontinuities in the profile reed channel is compensated by additional streams of jet, which are blown into the profile reed channel by a series of nozzles that operate sequentially and hence are termed relay jets. These nozzles pierce through the bottom shed line when the sley

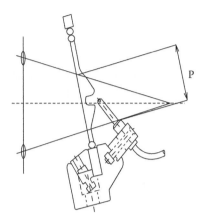

FIGURE 9.18
Side view of relay jet and profile reed wire, Somet, ITEMA.

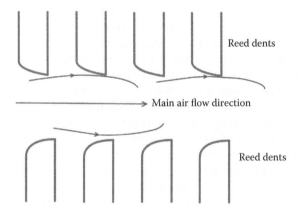

FIGURE 9.19
Front view of profile reed wires. (From Vangeluwe, L., *Textile Progress*, 29, 4, 1999, CRC Press.)

moves away from the cloth fell and blow jet streams as and when a pick of weft approaches the respective zones while traveling through the profile reed channel. Evidently, these jet streams do not blow along the axis of the pick being inserted. Hence, only a component of energy spent by each nozzle can be effectively utilized in boosting axial air velocity. A front view of the array of relay jets mounted on a sley is shown in Figure 9.20. It is observed that the relay jets are operated in groups by valves. When a valve opens, the nozzles controlled by it start blowing jets simultaneously. The instant of valve opening and the duration of its opening decide the operation of nozzles under its control. Sequential opening and closing of these valves result in the sequential blowing of jets through the respective nozzles. In the referred diagram, relay jets over the greater part of the sley are seen to be operated in groups of four, and the last two jets near the right-hand end are operated

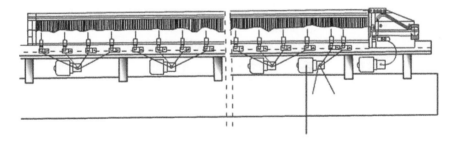

FIGURE 9.20
Front view of an array of relay jets mounted on an air jet machine, Somet, ITEMA.

by one. These last two nozzles blow on the end of the weft so as to keep it straight until shed lines cross and trap the same. Similarly, the diagram shows a valve operating the suction nozzle located directly in the path of the weft. Barring these nozzles with special functions, the rest of the nozzles are operated in groups of four. Such a setup results in a considerable waste of energy, and it can be demonstrated that one valve for each nozzle should be the ultimate solution insofar as energy saving is concerned. However, the necessary electronics may turn out to be too demanding.

The beneficial effect of relay jets or sub-streams in conjunction with a profile reed is revealed by the air jet velocity profile shown with discontinuous lines in Figure 9.17. Even though it exits the air tube at a much lower velocity, the jet slows down over a certain distance and then maintains its velocity at a slightly higher level than that of the weft over the rest of the journey. The velocity profile of weft measured across a reed width of nearly 2 m (Luenenschloss and Wahhoud, 1984), as shown in Figure 9.21, displays an initial acceleration phase followed by a steady movement for most of the remaining distance. Near the end of the picking period, the effects of application of the brake on the weft (deceleration) followed by the effect of suction (acceleration) are also clearly visible.

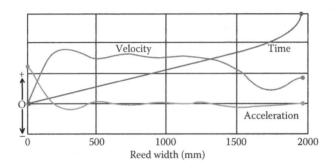

FIGURE 9.21
Velocity and acceleration profiles of weft on an air jet loom.

Conventional   Tapered
sub-nozzle   sub-nozzle

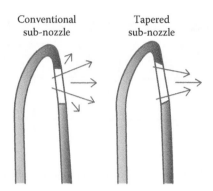

FIGURE 9.22
Diverging and converging jets from relay nozzle.

A relay nozzle is equipped with an opening, the design and location of which affect the effectiveness of the ensuing stream of air in boosting weft speed. Figure 9.22 reveals one dimension of this issue. Evidently, a converging jet would be more effective than a diverging one. A relay nozzle can also have different types of openings, such as a shower, star, rectangle, etc. (Figure 9.23). Effectiveness of such jets would be higher if the nozzles could get closer to the weft channel in the profile reed before they blow. In this process, however, they not only rub repeatedly against the same warp yarns in the bottom shed line, but also force them away laterally, thus subjecting them to repeated additional abrasive and tensile strains. Hence, dimension, profile, and surface properties of such nozzles are also of vital importance.

In terms of kilowatt hours of energy spent per square meter of fabric produced, a modern air jet loom consumes the highest energy among all weaving systems. Most of this energy is consumed by the air compressor. Interestingly enough, power actually consumed in propelling a pick of weft by a jet of air is 0.0004 times the power consumed in compressing the air (Krause and Kissling, 1980). Hence, machine makers currently devote more attention to lowering air consumption than increasing machine speed. In this respect, optimization of the blowing sequence of relay jets becomes a matter of central importance. This aspect is highlighted in Figure 9.24.

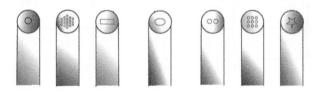

FIGURE 9.23
Different types of relay nozzle openings.

Injection air timing of sub-valve

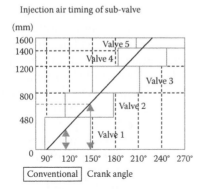 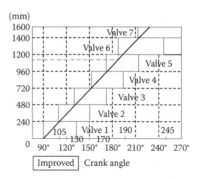

FIGURE 9.24
Hypothetical blowing sequence of relay nozzles.

Two hypothetical blowing sequences of 20 relay nozzles, spaced 80 mm apart over a distance of 1.6 m, are illustrated in Figure 9.24. The tip of a weft pick, whose locus is described by a straight line, enters reed space at approximately a 90° position of the loom crank and reaches the other end at approximately 225°.

In the conventional blowing sequence of relay nozzles, depicted on the left-hand side, a set of five valves operates the nozzles whereby the first 15 nozzles are operated in groups of five by three valves, the next three nozzles are controlled by the fourth valve, and the last two nozzles are controlled by the fifth valve. Each of the first four valves remains open for approximately 60°, and the last valve operates for approximately 70°. If it is assumed that an amount of δ cc of air is blown by each jet for every degree of crank rotation during which it remains open, then it is found that a total of 1100 δ cc of air is consumed by the 20 relay nozzles for insertion of one pick of weft. One can also work out from this diagram that, between the 120° and 225° positions of crank angle, a length of the leading end of weft varying between 200 mm and 600 mm remains under the influence of blowing jets. Hence, it can be inferred that, with the given blowing sequence, drag force on the weft caused by the impinging air jets would vary throughout weft insertion across the reed. Beyond 225° crank position, a length of 400 mm of the leading end of weft continues to be blown at by five jets over a period of 25°, which subsequently drops to 160 mm of leading end of weft blown at by two jets for another 20° for the purpose of keeping the pick of weft in a stretched condition until the crossing shed lines become level and trap the weft securely. Interesting information that this diagram also reveals relates to jets blowing in the weft channel much before the actual arrival of the tip of weft, thus resulting in creation of subatmospheric pressure over a stretch of path ahead of approaching weft. A perusal of the diagram reveals that this path length keeps on varying between 400 mm to start with, going down to 200 mm, and then jumping to 600 mm and so on. As a result, over the period between 120° and 180° positions of crank angle, a total length of 800 mm of path length in the weft channel remains under the combined influence of impinging air jets and subatmospheric pressure, which is half the length of the entire reed width. This value falls subsequently to 640 mm to jump again briefly back to 800 mm around 210° and then drops to 400 mm until 250°.

The blowing sequence shown in the adjacent diagram reveals a modified setup in which the first 18 relay nozzles are controlled in groups of three by six valves. The last pair of nozzles is controlled by a separate valve. A similar exercise with this modified sequence as that carried out with a conventional blowing sequence reveals that an amount of 875 δ cc of air would be consumed by the relay nozzles during insertion of one pick, resulting in a net saving of 20%. The range of leading weft length that remains under the influence of the impinging jet varies between 240 and 480 mm, ensuring a reduced variation in drag force during weft insertion. Moreover, a total length of 480 mm of path length in the weft channel remains under the combined influence of impinging air jets and subatmospheric pressure during the greater part of weft movement across the reed.

A logical extension of the process of optimizing air consumption would be directed at increasing the number of valves to a value nearly equal to the number of nozzles while devising a blowing sequence that guarantees smooth transport of a pick of weft across the reed.

During its journey across the reed, one part of the pick near the tip remains under the influence of impinging jets while another part trails behind. The length of this trailing part grows as a function of time, and the length of the leading part keeps on varying depending on the nature of the blowing sequence. The inertial resistance of the ever-growing trailing part can become an important issue for wide looms. A typical timing sequence and duration of blow of the main nozzle and relay nozzles, illustrated in Figure 9.25, reveals

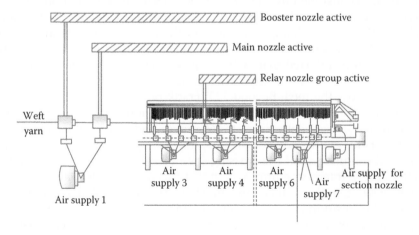

FIGURE 9.25
Typical blowing sequence of jets.

how this issue is negotiated in practice. It is observed that the main nozzle blows for a period long enough to permit a pick of weft to travel nearly half of the reed width. It stops blowing simultaneously with the third group of relay nozzles, much after the first and second valves have shut down and the fourth group of nozzles has been already blowing for half of its allotted time. Hence, it can be stated that, for the most part of weft insertion, the relay jets apply pulling and dragging forces to the weft while the main and booster nozzles keep on pushing the rear portion of inserted yarn.

The amount of air consumed per pick and the pressure at which such air needs to be supplied to nozzles depend on the type of weft yarn. Yarns may be made of continuous filaments or staple fibers. Continuous filament yarns may be flat or textured, and staple fibers may be spun into yarns by an array of techniques, such as ring spinning, open end spinning, air jet spinning, etc. The basic fibers or filaments may vary in thickness and surface characteristics. And, finally, the resultant yarn may be fine or coarse with varying degrees of hairiness. All these factors affect the magnitude of drag coefficient and, therefore, energy consumed per pick. Modern air jet looms are equipped with intelligent systems for automatic regulation of air pressure and duration of blowing in the main as well as the relay nozzles so as to enable weft mixing of up to eight different types of yarn.

9.3.5 Completely Unguided Fluid Carrier

Principles governing weft insertion by air jet also apply to water jets. However, water jet looms can handle only hydrophobic yarns and operate only with one nozzle (no booster and relay nozzles) although there may be more than one nozzle on a water jet loom for the purpose of weft selection. As a result, the maximum possible fabric width and range of fabrics that can be woven on such looms are limited. Pure water is, moreover, becoming scarcer by the day, and therefore, a production system based on water would only be resorted to if alternatives did not exist. Commercially, therefore, water jet looms are not popular and are no more displayed in international textile machine exhibitions of late. However, in terms of rate of power consumption expressed as kilowatt hours per kilogram of fabric produced, this system is the most efficient of all commercial weaving systems.

Although conversion of potential into kinetic energy with the help of a nozzle is common for both air and water jets, a much higher kinematic viscosity of water results in highly altered flow conditions near the downstream segment of the nozzle tip as well as over the entire length of the free jet stream, entrapping a length of weft. Leading up to the nozzle tip, a strong drag prevails between the inner wall of the nozzle and boundary layers of water. Therefore, a strong velocity gradient normal to the flow direction, in equilibrium with the pressure differential, affects viscous drag on weft. On emerging from the nozzle tip, the boundary layer of the jet stream becomes free of drag, and hence, the jet starts accelerating. This phenomenon is quite

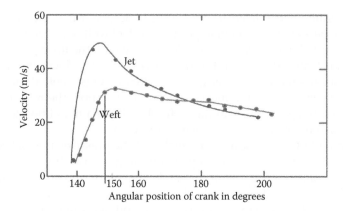

FIGURE 9.26
Velocity profile of water jet and inserted pick. (From Dawson, R. M. and Moseley, J. F., *Journal of Textile Institute*, 65, 639–658, 1974.)

opposite to the near laminar flow near the down stream as well as the up stream of the nozzle tip of the air jet. This striking difference is observed in Figure 9.26, which depicts the profiles of water jet velocity and that of the weft dragged by it as a function of time.

Experimental and theoretical studies on the configuration of water jets as well as on the velocities of jet and weft yarn after nozzle exit by Dawson and Moseley (1974) reveal that the leading end of a jet stream emerges with a low value of velocity from the nozzle, and then over about a 10° displacement of the crankshaft, it keeps on speeding up very rapidly at the initial stages and more slowly later on until the peak value is attained. Subsequently, its velocity starts falling gradually. The corresponding weft yarn follows the pattern of the jet head and reaches its peak velocity at around the same time as the jet head. The subsequent drop in its velocity is much more gradual and indeed, beyond a certain point of time, the weft velocity exceeds that of the jet head. This would tend to suggest that weft yarn beyond this critical time keeps traveling owing to its inertia overcoming drag forces at the yarn supply zone and also of the jet stream. A closer study of the constituents of the jet stream reveals this source of contradiction.

A water jet stream beyond the nozzle tip is not a continuous laminar stream of water (Kimura and Iemoto, 1978). More accurately, it can be described as a collection of water particles, each with its own velocity vector. Over a very short distance close to the nozzle exit, the particles continue to remain in contact with each other and move primarily along the jet axis. The jet appears somewhat transparent and smooth. Because of shear forces between layers of water particles as well as effects of surface tension and drag between particles of boundary layers and surrounding air, they subsequently start moving radially, imparting a wavy appearance to the jet stream. The jet becomes opaque. As a logical culmination of this process, the stream finally breaks up into individual droplets of water, somewhat akin to a spray. Among the various factors contributing to this transition from transparent to opaque and finally to the spray flow of the water jet, velocity profiles of particles emanating from the nozzle tip play an important role.

The displacement trace of individual water particles coming out of the nozzle along with that of the leading end of the jet head as a function of the angular position of the crank as reported by Dawson and Moseley (1974) is reproduced in Figure 9.27. The envelope on displacement profiles of individual water jet particles given by the discontinuous line represents the displacement profile of the water jet as a whole. Incidentally, the velocity trace of each particle would be given by the slope of its displacement trace, and hence, each particle exhibits a drop in velocity after emanating from the nozzle exit A perusal of the individual traces reveals that, initially over a period of about 10° or thereabouts, water particles keep emerging from the nozzle at progressively higher velocities than the preceding ones. This results in a push that particles lying ahead receive from the ones following them, causing the jet head to accelerate. Referring back to Figure 9.26, it is observed that net velocity of the jet head keeps on increasing and reaches a peak value at approximately 10° after exiting the nozzle. As a net result of this push from the rear, the jet head assumes the form of a fist instead of that of an arrow during its acceleration phase. Subsequently, over the next 30° to 40°, the velocities of emerging particles keep on falling at a low rate after which the jet head slows down and gradually breaks up into individual particles. Beyond 180°, the particles emerging from the nozzle exhibit progressively lower initial velocities. Hence, as far as the inserted pick of weft is concerned, it keeps on getting pushed from the rear as well as getting dragged around its tip over

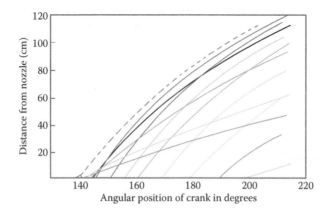

FIGURE 9.27

Displacement profile of water jet particles. (From Dawson, R. M. and Moseley, J. F., *Journal of Textile Institute*, 65, 639–658, 1974.)

a period of 40° after which there would be a sharp drop in both pulling and pushing forces forcing the weft to travel a distance on its own inertia.

As opposed to an air jet propulsion system, the cam-operated water jetting mechanism acts for a very short duration imparting a burst of energy to weft yarn only once and that too over a very short period of time. However, owing to the higher density of water, the quantum of energy is much higher, enabling the weft to accelerate much faster to a fairly high velocity within a very short time. As a result, the weft travels very fast in a water jet loom but buckles after traveling a certain distance in the absence of any relay and suction nozzles. Hence, although a water jet loom can run at a high WIR of 3000 m/min, the maximum reed width of a modern water jet loom hovers around 2 m only.

References

Catlow M, and Vincent J J (1951). The problem of uniform acceleration of the shuttle in power looms, *Journal of Textile Institute*, 42, T413–T487.

Dawson R M, and Moseley J F (1974). Some observations of weft insertion by water jet, *Journal of Textile Institute*, 65, 639–658.

Kimura S, and Iemoto Y (1978). Weft insertion by water jet; Part 1: Flow velocity and configuration of the jet in the vicinity of the nozzle exit, *Textile Research Journal*, 48, 604–609.

Krause H W, and Kissling U (1980). Experimental and theoretical analysis of weft insertion by air-jet, *Melliand Textilberichte*, 61, 780.

Luenenschloss J, and Wahhoud A (1984). Investigation into the behavior of yarns in picking with air jet systems, *Melliand Textilberichte*, 65, 242.

Marks R, and Robinson A T (1976). *Principles of Weaving*, The Textile Institute, Manchester, UK.

Mohammed M H, and Salama M (1986). Mechanics of a single nozzle air-jet filling insertion system, *Textile Research Journal*, 56, 683–690.

Tumer S T, and Dawson R M (1988). Filling insertion by rapier: A kinematic model, *Textile Research Journal*, 58, 726–734.

Uno M (1972). A study on air-jet loom with substreams added, *Journal of Textile Machinery Society of Japan*, 18, No. 2, 37–43.

Vangeluwe L (1999). Air-jet weft insertion, *Textile Progress*, 29, No. 4, 9–34.

Further Readings

Ormerod A (1983). *Modern Preparation and Weaving Machinery*, Butterworths and Co. Ltd., UK, ISBN: 0-408-01212-9.

Talavasek O, and Svaty V (1981). *Shuttleless Weaving*, Elsevier Scientific Publishing Company, Amsterdam, The Netherlands, ISBN: 0-444-4181-0.

Appendix

A.1 Distance Traveled by Picking Element during Linearly Rising and Falling Time Dependent Function of Acceleration

A.1.1 Variant A

Let the time-dependent acceleration function over time period T be described by

$a = mt$ for $0 < t \leq T/2$

$a_{max} = mT/2$

$a = a_{max} - mt$ for $T/2 < t \leq T$

A graphical representation of this function is illustrated in Figure 9A.1.

Distance traveled by picking element during time period $0 < t \leq T/2$ can be estimated by integrating the acceleration function twice with respect to time. Accordingly,

$$S_{at\ t\ =\ T/2} = mt^3/6$$

Substitution of

$m = 2a_{max}/T$

yields

$S_{at\ t\ =\ T/2} = a_{max}T^2/24.$

Distance traveled by the picking element during time period $T/2 < t \leq T$ can be similarly estimated, yielding

$$S_{during\ T/2\ <\ t\ \leq\ T} = a_{max}T^2/12.$$

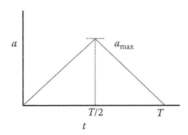

FIGURE 9A.1
Hypothetical acceleration profiles of gripper/shuttle.

Hence, total distance traveled by picking element over time period T works out to be

$$s = a_{max}T^2/8 = a_{average}T^2/4.$$

It may be noted that although the area under the acceleration curve in this time-dependent variant is the same as that under uniform acceleration, distance traveled by the picking element is half of that resulting out of uniform acceleration ($s = a_{average}T^2/2$).

A.1.2 Variant B

Let the time-dependent acceleration function over time period T be described by

$a = 0.5a_{max}$ at $t = 0$

$a = mt$ for $0 < t \leq T/3$

$a = a_{max}$ at $t = T/3$

$a = a_{max} - mt$ for $T/3 < t \leq T$

$m = 3a_{max}/2T$

This acceleration function exhibits a discrete value, equal to half of its maximum value of a_{max} at the start and climbs to its maximum value in one third of the total acceleration time. Subsequently, it diminishes at the same rate of climbing to its final value of zero. This function is illustrated in Figure 9A.2. Upon equating the area under the respective curves, it can be easily worked out that

$a_{max} = (12/7)a_{average}$

$s = (10/54)a_{max}T^2 = 0.317a_{average}T^2$

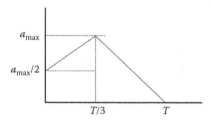

FIGURE 9A.2
Hypothetical acceleration profiles of gripper/shuttle.

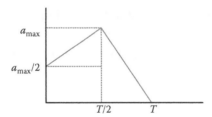

FIGURE 9A.3
Hypothetical acceleration profiles of gripper/shuttle.

A.1.3 Variant C

Let the time-dependent acceleration function over time period T be described by

$a = 0.5a_{max}$ at $t = 0$

$a = mt$ for $0 < t \leq T/2$: $m = a_{max}/T$

$a = a_{max}$ at $t = T/2$

$a = a_{max} - mt$ for $T/2 < t \leq T$: $m = 2a_{max}/T$

This acceleration function exhibits a discrete value, equal to half of its maximum value of a_{max} at the start and climbs to its maximum value in half of the total acceleration time. Subsequently, it diminishes at twice the rate of climbing to its final value of zero. This function is illustrated in Figure 9A.3. Upon equating the area under the curves, it can be easily worked out that

$a_{max} = (8/5)a_{average}$

$s = (1/6)a_{max} T^2 = 0.27a_{average}T^2$

A.2 Calculation of rpm, Reed Width, and Weft Insertion Rate from Figure 9.6

Time to travel reed width is 0.153 s. Allowing 150° angular displacement of the loom shaft for a gripper flight across the reed at approximately 24 m/s, the loom rpm, reed width, and weft insertion rate work out as follows:

Reed width = 24 (m/s). 0.153 s = 3.672 m

Loom rpm = $(150.60)/(360.0.153) = 163.4$ min^{-1}

Weft insertion rate = 3.672 (m). 163.4 (min^{-1}) = 600 m/min

A.3 Calculation of Energy Stored in the Torsion Rod of a Gripper Propulsion System

Torque M_t generated in a torsion rod when it is twisted through ϕ radians is given by

$$M_t = (\pi d^4 G)\phi/32\ell$$

where

d = diameter of torsion rod (m)

ℓ = length of torsion rod (m)

G = Shear modulus of torsion rod (N/m²)

For

$d = 0.015$ m

$\ell = 0.72$ m

$\phi = 0.49$ radians (28 degrees)

$G = 8163 \times 10^7$ N/m²

The value of M_t works out to be 272.2 N m. Assuming linearity between M_t and ϕ, the area under M_t versus ϕ plot yields the work done and hence the energy stored in the torsion rod. Accordingly, energy stored in the rod works out to be {0.5. 272.2. 0.49} N m = 66.7 N m.

10

Features of Modern Shuttleless Weaving Systems

10.1 Machine Drive and Power Consumption

Modern shuttleless looms come equipped with a number of microprocessor-controlled motors. The main motor supplies power for primary motions, and servomotors drive secondary motions. Other functions, such as automatic broken pick repair, control of weft tension, control of weft feeding, etc., are carried out by suitably controlled separate motors.

Remarkable developments in power electronics have contributed to an enormous freedom of the primary motions such that a loom can be started from rest with a very high torque for avoiding start-up marks, or the speed of a loom can be slowed down or increased during normal running as per the requirement of the weft yarn being inserted, or the shedding motion can be run at a very slow speed in the reverse direction for automatic broken pick repair while picking and beating up elements remain idle. An inverter-controlled AC-driven motor forms the basis of this improvement.

An electric motor works on the principle of attraction and repulsion between two sets of magnets, namely stator and rotor. Polarity of the permanent magnet (the rotor in an AC-induction motor) remains constant while that of the other alternates at a certain frequency. Consequently, when the polarity of the two sets of magnets is opposite in nature, a force of attraction prevails between the two, and in the event of similar polarity, the opposite happens. This phenomenon is responsible for a continuous relative motion between the two. The mechanical arrangement in a motor transforms this relative motion to a continuous rotational motion of the rotor. The frequency of power supplied to an AC motor dictates the rate at which polarity of the electromagnet changes. When the frequency is high, the rate of change in polarity is also high, and hence, the rotor rotates at a high frequency. The reverse would be the case with a reduced frequency. Hence, the speed of an AC motor can be varied by varying the frequency of power supplied to it. Similarly, if the supply voltage is suitably varied, then intake current and, therefore, the strength of the magnetic field in a motor can be higher or lower. The resultant output torque can accordingly be raised or lowered as per requirements of the weaving machine. Both functions of varying the magnitude and frequency of input voltage to an asynchronous AC motor are carried out by an inverter.

An inverter is a solid-state power electronics system, which converts the line AC supply first into DC by a rectifier. The output of the rectifier is smoothed out by a DC link, and the filtered DC is finally converted back to AC by active switching elements of the inverter. Both output voltage and frequency can be independently controlled by an inverter for satisfying requirements at any desired moment. However, higher current intake associated with higher input voltage to an AC motor can be sustained only over a limited period of time as excessive heating can damage the motor. For a shuttleless loom, such a requirement occurs only for a limited duration, such as during the starting of a loom from a standstill condition so that a starting mark can be avoided (Figure 10.1).

The power required by a loom is indicated by the rating of diverse motors (in kilowatts) employed for operating the machine. However, actual consumption of power may differ from the estimated value as demands on individual motors are cyclic in nature and sometimes even intermittent. Typical examples are the cyclic power demands of the sley (Figure 10.2) or intermittent demands on servomotors operating let-off and take-up motions. Ignoring this aspect in view of insufficient information from reliable literature and opting for the simpler route of a simple summation of ratings of individual motors, one can arrive at representative values, which can be employed for working out economics of woven fabric production by various shuttleless systems.

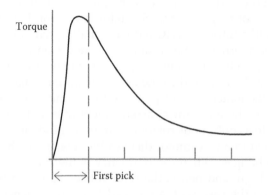

FIGURE 10.1
Instantaneous torque requirement.

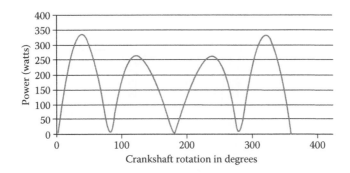

FIGURE 10.2
Fluctuation of power demand within a loom cycle.

10.2 Drive to Sley

Considering that the instantaneous power consumed by the sley is a product of its equivalent mass, concentrated at its radius of gyration; its instantaneous velocity, and instantaneous acceleration, a typical plot of power required by a crank-driven sley of a 1.9-m-wide plain loom of eccentricity 0.18 over one cycle is shown in Figure 10.2. This plot is based on the assumption that the angular velocity of the crankshaft remains constant throughout a loom cycle. The many cyclic peak power demands on the prime mover, a 1-hp motor, causes the loom to slow down periodically in order that other demands made, such as by shedding, picking, take-up motions, etc., can be satisfied.

A reduction in the mass of the sley, which constitutes one bar in the four-bar linkage system of the sley drive is one obvious step toward reducing the load on the prime mover. Leaving aside the fixed ground link of the four-bar linkage sley-driving system as well as the sley itself, the other two moving links also contribute to the overall inertia of the system and hence to power demand. Moreover, with such a drive, increasing the value of the crank radius for achieving higher values of loom eccentricity for permitting greater flight time to the weft carrier would lead to an increase in sley sweep. This not only slows down a loom, but also results in higher abrasion of the warp yarns with a reed wire and reduction in effective space available for healds between the reed and the crankshaft. Peak demands on power within a loom cycle also go up with higher eccentricity. Higher loom eccentricity is incidentally a desirable feature with shuttleless weft insertion systems.

It is evident, therefore, that shuttleless weft insertion systems not only require a modified sley design, but also a modified sley drive. Illustrations of typical sley-driving systems are provided in Figures 10.3 and 10.4.

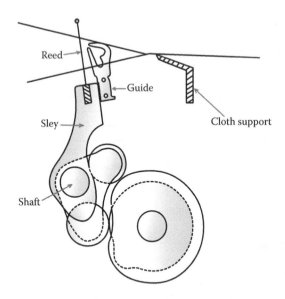

FIGURE 10.3
Sley drive by conjugate cams, Somet, ITEMA.

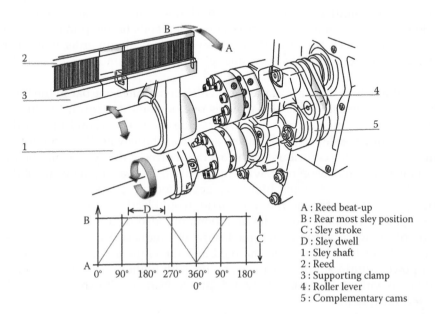

A : Reed beat-up
B : Rear most sley position
C : Sley stroke
D : Sley dwell
1 : Sley shaft
2 : Reed
3 : Supporting clamp
4 : Roller lever
5 : Complementary cams

FIGURE 10.4
Cam and follower drive to sley shaft, Somet, ITEMA.

A massive reed cap and shuttle boxes are dispensed with in the sley of shuttleless looms. The reed is gripped securely at its lower edge by the fully metallic sley, which, in turn, is made of a light and strong alloy.

The sley depicted in Figure 10.3 is an inverted Y-shaped lever fulcrummed around a rocking shaft. The two arms of the lever carry antifriction bowls, which are engaged by a pair of conjugate cams. The cams and antifriction bowls are encased in an oil bath and designed to remain always in contact. Such a system ensures the desired eccentricity of the sley as well as an exact displacement profile. Figure 10.4 depicts a modified design of another sley drive, which also functions with a pair of complementary cams and followers. The displacement profile of the sley shown in the inset underlines the advantage of such a drive. An improved four-bar linkage mechanism in conjunction with a modified sley may, however, prove effective enough in situations demanding very low eccentricity and dwell.

10.3 Fabric Selvedge

Selvedges of fabrics produced on shuttleless weaving machines are markedly different from those produced on weaving machines employing shuttles. Insertion of single picks by a weft carrier followed by the action of weft cutters at both selvedges leaves a small length of free weft hanging at the fabric selvedges. Unless these threads are secured properly, the fabric may start disintegrating along its edges during subsequent wet processing. Tucked selvedges (gripper and air jet), leno selvedges (air jet and rapier), and fused selvedges (water jet) are commonly encountered on fabrics produced on shuttleless looms.

A tucked selvedge can be created on a gripper machine with the help of tucking needles (Figure 10.5), which enter the warp shed from under the

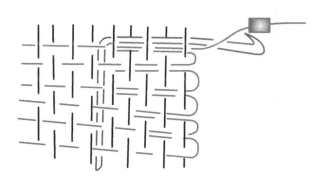

FIGURE 10.5
Tucked selvedge with tucking needle.

bottom warp line during the retreating motion of the sley after beat up and pull in the free segments of the pick into the subsequent shed. On an air jet loom (Figure 10.6), additional blowers A and B located near fabric selvedges perform this task. Tucked selvedges are fairly firm but contain double picks.

Leno selvedges (Figure 10.7) are generated by extra warp yarns supplied from spools that are mounted on disks, which are given rotational drive. These discs are, in turn, suitably mounted atop healds in such a way that yarns can land directly at the respective selvedges. When the disks are rotated every pick through 360°, a double-locked leno results, which grips relatively

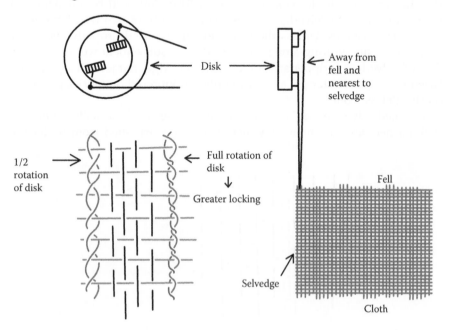

FIGURE 10.6
Tucked selvedge with air blower, Somet, ITEMA.

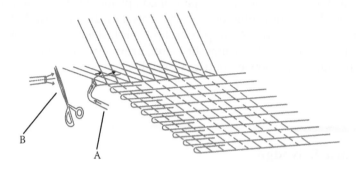

FIGURE 10.7
Leno selvedge.

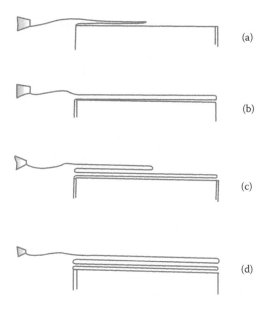

FIGURE 10.8

Woven selvedge by Gabler system. (a) Loop of 1st pick in shed center. (b) 1st pick inserted as double pick. (c) Loop of 2nd pick in shed center. (d) 2nd pick inserted as double pick.

smooth weft yarns fairly tightly, and a disk rotation through 180° can be sufficient for usual spun yarns.

Fused selvedges are created by employing glowing hot filaments located conveniently near selvedges. Such selvedges are found in fabrics made of thermoplastic yarns.

Whichever method may be adopted for securing the selvedge yarns on shuttleless looms, mechanical properties at these two extremities are bound to be inferior to those of the main body of fabric. This is caused by discontinuity of weft yarns. Incidentally, under the condition that double picks are acceptable, it is possible to ensure continuity of the weft yarn on a Gabler type of rapier machine. This is illustrated in Figure 10.8. For all its disadvantages, the shuttle-weaving machine proves its superiority in this one very important aspect, which comes into reckoning in technical textiles and more so in 3D woven fabrics. Modern machine manufacturers are switching back to this older method of fabric formation for creating this new-generation fabric!

10.4 Productivity and Fabric Quality

The act of replacing a wooden shuttle with a much smaller, lighter, and much harder steel gripper led to a substantial reduction in manpower and a radical improvement in woven fabric quality. One can state with a degree of confidence that this one development fueled the subsequent spate of developments in the woven fabric manufacturing process.

Focusing attention solely on the criterion of productivity of a loom, namely the weft insertion rate (WIR), it is observed that a steady but linear rise took place during the first 60 years of the 20th century through the advent of the projectile (gripper) loom. Since then, the rise has been rapid although, somewhat oddly, the growth in the WIR of the projectile system itself has been sluggish over the past 40-odd years. The projectile loom is now a very important type of shuttleless system for the production of an array of technical textiles that demands large width and versatility in terms of converting a wide range of materials into a broad palette of products. Nonetheless, the economics of projectile weaving, which was established during the '60s and '70s, fueled the development of alternate forms of weft insertion techniques. As a result, the technology of air jet, water jet, and rapier weft insertion systems improved by leaps and bounds. For example, a modern air-jet loom, which is equipped with a tandem nozzle, booster nozzle, relay nozzles, suction nozzle, and profile reed, can operate at a WIR value of around 3000 m/min, almost twice that of the most modern projectile loom. Evidently, limits are imposed on the acceleration and deceleration processes by the inertia of the projectile, which has a mass varying between 20 and 60 g. For the jets, however, the deceleration phase does not come into question at all, and limitation on the acceleration phase due to inertia of the fluid is also negligible. Similarly, the sley and the healds can be operated without any dwell in water jet looms, and therefore, the corresponding looms can run faster as the space requirement of the fluid stream within a shed is nominal. For the same reason, the amplitude of sley oscillation is also much lower than that of the shuttle or gripper loom. Thus, very serious bottlenecks, which restrict the development of both shuttle and gripper looms, are elegantly overcome by changing over from a solid propulsion to the fluid propulsion system. Indeed, a WIR of around 3000 m/min has been achieved by both forms of jet systems.

Restriction on the width and WIR is imposed upon rapier systems by the buckling rigidity of the band, the space occupied by the insertion systems on the two sides of a loom, and the inertia of moving elements. Similarly, the tip transfer from one rapier head to another in the center of the shed is a unique phenomenon that must be subject to a risk of failure. In spite of such heavy odds, the WIR achieved currently with this system is on par with that of a projectile, and the width of 4 m with a flexible rapier is a testimony to the development and application of new composite materials. The rapier happens to be the only positive weft insertion system, implying that the motion of weft throughout its journey can be controlled very precisely. This facility permits insertion of picks of diverse nature during the course of weaving. A typical example of a rapier-woven ladies' dress material exhibits a sequence of picks comprising textured PES of 78 denier, cotton yarn of 102 Ne, lurex of 69 Nm, chenille of 40 Nm, and Boucle silk of 3.8 Nm. Insertion of such a wide range of count and material is possible owing to the versatility of the weft insertion system along with some associated developments in machine-driving systems. As a result, this system has created a niche for itself in the apparel and household textiles sector. In addition to the versatility inherent in the rapier system, the ability of the modern loom to operate at different speeds and at different rates of take up and let off while different types of weft are inserted has contributed significantly to the scope of this system. This ability owes much to the growth of the application of electronics in modern looms.

10.5 Application of Electronics

The transformation of the air-jet loom from a low-WIR, single-nozzle, and narrow-width machine into the modern version of more than 5 m width and 3000 m/min WIR has been possible primarily due to the application of electronics. Indeed, the major problem of dragging the tip of the weft over a large distance at a speed larger than that of the trailing segments could only be solved by relay jets. Managing the sequence of timing and duration of blasting of neighboring groups of relay jets demands a precision and alacrity that can only be associated with microprocessor control. For that matter, the precisely controlled braking of weft after its tip has reached the end near the suction nozzle of an air-jet loom or the braking and positioning of the projectile upon its arrival on the receiver side is possible only because of microprocessor control.

Microprocessor control in a modern loom is, however, not restricted to the weft insertion system alone. Over and above, a modern shuttleless loom can be distinguished from the conventional automatic shuttle looms broadly in two respects, namely the transmission of motion and the high level of automation.

The classic prime mover from which power used to flow to various moving elements of the traditional loom has given way to a large number of servomotors. As a result, each unit is free to move in any direction and by any amount independent of the others. For example, in a modern loom, the normal sequence of movement of healds during the weaving process can be reversed during the pick-finding process without having to move the sley at all. The let-off or take-up motions can also remain totally idle during this process of repair. This is not possible in a conventional shuttle loom. Moreover, the speed of the healds during the reversed motion is kept much lower than its normal speed during the weaving process. All these are a result of replacement of the rigid mechanical links between the prime mover and the rest of the moving units as found in the traditional looms by flexible links between a central processing unit (CPU) and various servomotors responsible for the movement of individual units. These flexible links are the independent signals that are continuously sent to various servomotors by the CPU. The centralized power of the prime mover is replaced by multiple centers of power, which are coupled directly onto the units to be moved. The technical knowledge of the weaver, which is enshrined in the database and the software, gets translated into signals emanating from the CPU, which, in effect, are commands to the various units to function in a manner most appropriate to the type of work being carried out. The location of the back rest that would give the best cover of the fabric, the tension in the warp that would be most appropriate to the material being woven, the angular displacement of weaver's beam required at a particular stage of weaving, the proper time-displacement profile of individual healds, etc., and similar other settings and motions are stored in the CPU. The continuous commands flowing to the multiple centers of power are based on this knowledge. Any update on this knowledge can, in principle, be incorporated in such a system without needing a substantial change in hardware. The characteristic feature of this system is therefore the flexible space and supremacy accorded to this knowledge base. It is eminently possible to continuously improve upon solutions and offer better options to the customer without having to carry out major changes in the machine itself.

The other dimension acquired by a modern loom owing to incorporation of electronics is the quality of automation in domains that could not be taken under purview otherwise. The process of automation in looms was initiated with automatic weft replenishment, automatic stop motions, and automatic let-off systems. These systems are, in essence, open-looped in nature and rigid. Once the settings are in place, the operations would be repeated faithfully but without any possibility of variation that may be desirable at different stages of fabric production. Similarly, if the settings have not been exact or suffer some accidental alteration during the production process, no automatic rectification of the same is possible. With electronic controls, it is possible to build closed-loop systems, which may even be intelligent. As a result, complex operations that require an evaluation of the effect of an exercise undertaken followed by a course correction can be attempted with such systems.

An illustrative example is the automatic repair of broken picks. A considerable amount of damage to a fabric used to result out of the operator's exercises in locating the shed at which a pick had broken, removing the broken segment from within the shed, and then inserting a new pick and restarting the loom without causing a starting mark. This entire exercise is carried out in a modern loom automatically, thanks to the microprocessor control and some ingenuity of the machine manufacturer. The electronic let-off and electronic take-up systems permit a programmable variation in yarn tension and free length of warp during the course of weaving not only for accommodating different weft yarns at different phases of the production process, but also for varying the fabric structure through different warp and weft crimp distributions. Similarly, during the insertion of extra weft yarns the take-up and let-off systems may be brought to a complete halt automatically. Indeed, the surveillance systems associated with such automation permits a fairly high degree of reliability. The quality of automation achieved in modern weaving machines permits one even the fancy of looking ahead to an intelligent loom that can analyze a given fabric sample and weave the fabric when the warp and weft required by it has been supplied.

10.6 Application of Composite Materials

The course of innovations in looms has been strongly influenced by developments and application of composite materials. A modern loom running at nearly 1000 rpm would require the healds carrying a large number of warp yarns under a fairly high tension to reciprocate at a frequency of nearly 16 Hz. The peak acceleration and deceleration involved in such motions exceed 100 G, assuming the very conservative figure of 10 cm heald displacement. The enormous amount of power that would be required to maintain a number of 2.5–3 m wide healds in such a motion as well as the dynamic strains that the healds would be subjected to in this process clearly constitute serious hurdles. A part of this problem is negotiated by the positive cam or dobby drive to healds as well as the lateral guidance given to the healds in clearly defined channels, thus eliminating any unnecessary movement. Additionally, the light and strong fiber reinforced composite materials, which have replaced aluminum alloys, also play an important role in achieving such high frequencies. For example, carbon fiber–reinforced composite materials are very light (density of ~1.5 g/cc as compared to 2.7 g/cc of aluminum alloy and 7.8 g/cc of steel) and extremely stiff in tension (tensile strength of 800 MPa as opposed to 193 MPa of Al-alloy and 1100 MPa of steel). Heald frames made of such materials are capable of withstanding very high strain while offering less inertia for attainment of the very high frequency.

10.7 QSC and Automation in Drawing In

Application of electronics in weaving looms has resulted in qualitative improvements insofar as product ranges, product quality, and productivity are concerned. It has also enabled a certain improvement in flexibility as the changeover from one type of woven construction or woven design to another does not require a great deal of time with a CAD system. However, this facility cannot be fully exploited if the change of warp remains as time-consuming as ever. A change of warp involves, to begin with, removal of the reed, healds, and set of drop pins along with the partly finished weaver's beam in an orderly manner and then storing the whole set on a suitable stand until it is required again. This has to be followed up by bringing in a new set of a weaver's beam along with the drop pins, the healds, and the reed. Finally, all these have to be put in place on the loom and linked to their respective driving elements after which the new warp has to be tied up to a piece of fabric gripped by the take-up system in such a manner that weaving may be restarted. The entire process would usually take many hours, depending on the beam width and the number of healds in question. Speeding up the production process of woven fabric demanded a solution for removal of this bottleneck. This issue was addressed jointly by the manufacturers of looms and of accessories, leading to the commercialization of the concept of "quick style change" (QSC). Under this concept, the rear segment of the loom, designated as SCM (style change module) in Figure 10.9, is detachable from the rest of the machine. It is this segment that supports the weaver's beam. Once the reed is removed from the sley and the links joining the shedding system to the healds are disconnected, the SCM carrying the weaver's beam and accessories can be moved away from the main body of the loom by a suitable mechanized handling system and replaced by a new one. The two facilities, that is, the mechanized handling system and the detachable rear segment of the loom, permit quick and orderly removal of the weaver's beam along with healds, drop pins, and the reed followed by replacement with a new set. Hence, these two facilities are central to the first stage of the operation. The time-consuming tying-in of the new warp is taken care of by welding the free edge of the warp sheet that extends beyond the reed to a plastic sheet in the drawing-in section itself. Once on the loom, this sheet is simply tucked into the take-up system, and weaving can be started without any delay. A QSC loom as such does not involve any additional electronics although its developmental need was partly fueled by the flexibility that a modern loom has achieved owing to the large-scale application of electronics. Undoubtedly,

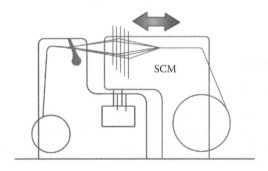

FIGURE 10.9
Principle of a QSC loom.

the rapidly changing style variation dictated by an ever-demanding market has also played a major role in this respect.

As a logical sequel to the mechanized replenishment system of warp on a loom, the entire process of drawing in of warp yarns through drop pins, heald eyes, and dents of a reed has undergone major changes. Here too microcontroller systems have taken over this manual and slow job. It is claimed that ends from the warp beam are being automatically drawn through drop pins, healds, and reed at the rate of up to 140 cycles/min.

10.8 Preparatory Process

The advances in the conversion process of a warp sheet and weft yarns into a woven fabric, outlined briefly in the foregoing, have had considerable repercussion on the preparatory processes. The warp sheet has become progressively broader, frequency and amplitude of dynamic strain on warp yarn have become higher, and tolerance on cleanliness of the shed has become narrower. Moreover, as very high-quality fabrics can be produced on these looms—meaning thereby a very narrow spread of the critical variables, such as width, areal density, thread spacing and crimp, and a very low count on defects, such as missing ends, starting marks, reed marks, etc., as well as a very high capacity of designing through warp and weft selection systems—the demands made from the downstream processes have been getting more and more exacting. Quantification of appearance, feel, and comfort for apparel fabrics or of permeability, modulus, and electrical conductivity for technical textiles demands that woven fabrics be engineered and produced to specifications. The performance of a product would obviously be better when the tolerance is narrower. The consequential demand pull on the quality of warp and weft fed to the loom has resulted in some remarkable innovations in the preparatory systems.

10.8.1 Modern Cone Winder

The most notable effect has been on the transformation of the cone-winding machine. The quality of package produced on such a machine can be specified in terms of conicity, height, diameters at the nose and base, hardness, weight, and length of yarn contained in a package. It can also be used to specify the level of acceptable objectionable faults in yarn or in that of the package geometry and build. Incorporation of an electronic clearer with the ability to detect foreign matters backed up by a suitable splicer in all modern winding machines enables cleaning and joining of yarn to be carried out with a very high degree of precision. However, innovative attempts at winding the cleaned yarn onto a shell at a constant tension through the package build have resulted in features that would have been impossible but for the wide application of electronics.

Each winding drum on a modern machine is driven directly by a servo-motor whose direction and speed of rotation at any instant of time are comman-deered by a central system based on the inputs that it receives continuously from various elements within a spindle unit. Thus, a thread break would cause the motor to stop at once, avoiding, in the process, unnecessary abra-sion between surfaces of the package and drum. Consequently, the package drum contact is not broken so that when the splicer comes into operation, the drum rotates slowly in a direction opposite to that of its normal mode, thus permitting the free end of the yarn to be located on the package surface, sucked in, and pulled out by the respective arm of the splicer. After yarn splicing is accomplished, the drum accelerates slowly until proper winding speed is reached. Similarly, the angular speed of the drum changes contin-uously as a function of the yarn unwinding point on the supply package. This is meant to cancel out unwinding tension fluctuation caused by the bal-looning effect. All of these speed variations of one drum are independent of those of other winding drums on the same machine.

The tensioner is driven by a miniature servomotor so that the tension it adds to the yarn can be varied continuously in order that the final tension in the yarn near the yarn trap is maintained within specified limits. The loca-tion of the balloon breaker is also continuously shifted so that as the unwind-ing point on the supply bobbin gradually shifts toward its base, the average balloon height remains constant. This measure results in minimizing long-term tension fluctuation in the yarn being unwound.

The cradle itself is controlled by servo systems so that the conicity of the package and its hardness are maintained within well-defined boundaries. In fact, the cradle setting can even be changed in a preprogrammed man-ner without interrupting the winding process so that the effective point of contact between the conical package and the cylindrical drum can be shifted away from the package base toward its nose and brought back conveniently to the original point over a certain period of time. Such a measure is adopted to alter the package rpm at diameters conducive to the formation of ribbons. Provision of yarn length and package diameter measuring units in conjunc-tion with a cradle pressure–regulating system on a modern winding machine results in packages of very precise dimensions.

The development of step precision winding systems must count among one of the major innovations in cone winding. It is well known that, in precision winding systems, embodied by the spindle-driven winders, yarn coils can be laid very accurately on the package surface on account of the number of coils per traverse remaining constant throughout the package build. This facili-tates the winding of very dense and ribbon-free packages. However, a con-stant change in coil angle occurs during this process that tends to destabilize the package and alter the package density. The attributes of a random winder, embodied by the grooved drum winding machine, are, however, exactly the opposite. The coil angle remains constant throughout the package build, but ribbon formation is a serious issue. A periodic disturbance to the effective package rpm is the accepted solution, which, however, affects the production. The hybrid step precision winder combines the positive attributes of the two by reducing yarn traverse speed without disturbing the drum rpm in a programmed manner over specific intervals of package diameter so that the number of coils per traverse remains constant during these intervals. Hence, over these intervals of diameter, the build of a package resembles that of a precision-wound one. At the transition points between two successive intervals, traverse speed is reversed to its original value. These transition points occur at specific package diameters, which are equidistant from those at which ribbon formation is most severe. Subsequent to the resetting of yarn traverse to its original value, it is reduced again in a programmed manner for keeping the coils per traverse during this diameter interval constant. However, this interval corresponds to a diameter value that is higher than that of the previous one. Therefore, the value of coils per traverse in this interval is lower than in the earlier one. Thus, there is an overall hyperbolic decrease in coils per traverse during the entire package build that takes place discontinuously in many finite steps, avoiding those values at which ribboning may occur. Hence, no measure of anti-ribboning is necessary while fairly large, stable, and dense packages can be produced on such a system.

In spite of rapid strides made, the grooved drum cone winding machine suffers a major flaw because of the grooved drum itself. A large short-term tension fluctuation is imposed on the yarn due to the forced variation in yarn path length and geometry between the yarn trap and the instantaneous winding-on point of the package. Moreover, abrasion between certain portions of the inside surface of the groove and the yarn is detrimental to surface properties of yarn. Thus, a part of the good work done by the advanced features of a modern winding machine is, to some extent, undone by this final element. A major development in this particular aspect can be expected in near future.

10.8.2 Modern Warper

Closed-loop tension control, length control, and speed control systems aimed at ensuring the identical nature of constituent yarns from the innermost to the outermost layers of the warp beam, and minimizing the variation between beams characterizes new-generation warping machines. Direct beaming systems are preferably spindle-driven and equipped with thyristor controls for continuously varying spindle speed with a buildup of beam diameter as well as for bringing the system to an immediate halt in the event of a thread break. Minimization of a length of free warp sheet between the last guide roll and the winding-on point of the winding drum, automatic insertion of leases in individual sections, and leveling-off the variation in average tension between individual sections through feedback control systems are additional features of modern sectional warping systems.

10.8.3 Modernization in Sizing

The sizing process plays a central role in overall performance of a loom shed. Sizing is an energy-intensive process, and it is also a polluting one, especially so because of desizing. Innovations in this process have therefore been plenty, and many new methods of sizing are still being tried out. The transition from the double cylinder to the multicylinder and then to split drying systems has resulted in an improvement not only in productivity, but also in the quality of protection provided by size film to yarn. Similarly, a wetting of the warp sheet prior to its entry into the sow box, application of multiple sow boxes, and multiple immersions within the same sow box are practices aimed at regulating the wet pickup for various types of warp sheet without sacrificing the machine speed. The widespread acceptance of high-pressure squeezing in conjunction with modified sizing ingredients, which can be made into high-concentration but low-viscosity emulsions, is aimed at an optimum anchorage of size film in the body of yarn while lowering energy consumption considerably. Drying by hot air and infrared ray has been tried out to replace the steam-heated cylinder drying process, especially when contact drying is detrimental for the yarn. However, low efficiency at the desired machine speed and operational difficulties have limited their application. As the application medium of conventional sizing material is water, which is becoming scarce, and as the same has to be evaporated costing energy, which is expensive, conceptually different techniques, such as hot melt sizing, solvent sizing, or cold sizing, have been explored with limited success. The sizing ingredient for the hot melt process should flow very easily in the molten state and solidify very quickly after a kiss roll has applied a layer of the same to the yarn. It also should have reasonably good adhesion to the yarn as no additional anchorage in the yarn body can be generated. Solvent sizing has many practical limitations, most important being the recovery of the carcinogenic solvent itself. The cold sizing process bypasses the need of size cooking and the size application is also in a cold state, and hence some energy is saved but still requires that the aqueous medium be evaporated. In all these cases, the polluting desizing process poses a major challenge. Evidently, there is scope for innovation of a sizing process that would be low on demand for water and energy and pose minimal threat to the environment.

10.9 Noise Generation

A loom shed is a very noisy place, and hence efforts are made by loom designers to reduce noise to an acceptable level.

The picking and beating up processes of a shuttle loom are major sources of noise. The striking action of a picking cam against a plate or bowl followed by multiple impacts among various elements of the picking mechanism and the loud impact of the shuttle with elements of the shuttle box result in sounds of high intensity and frequency. Similarly, when a dense fabric is woven, the heavy impact of the reed with the fell of the fabric creates a sound of high intensity although of a much lower frequency. It may be noted that impact is the common feature in all the aforesaid sources of sound. Superimposed on these major periodic sounds would be those made by rotating meshed gears and reciprocating elements of the sley drive, take-up motion, dobby, Jacquard, etc.

The picking motion in shuttleless looms, such as air or water jet, do not exhibit any moving mechanical elements, and no impactful action is encountered in a rapier system. Gripper systems rely though on mechanical elements and impactful actions. However, its picking system is encased in an oil bath, thus damping out sound considerably, and checking of the small gripper, being much more controlled than shuttle checking, is not expected to create comparable sound. Sound created by the beating action in shuttleless looms would be, on the other hand, of much higher frequency, owing to higher loom rpm. Individual drives to most motions, such as shedding, sley drive, take up and let off as well as encasement of most moving elements should result in a damped sound of lower intensity.

Sound is transported through a medium by means of its rarefaction and compression. The energy transported in the process per unit time by unit area of the compressing and expanding medium provides a measure of the intensity of sound. As energy per unit time is equivalent to power in watts, the unit of sound intensity is expressed in watts/m^2. The sound power of an average whisper is 0.1 µW. The air pressure fluctuation created by sound is measured in Pascal. A normal human being can sense sound pressure varying in the range 20 µPa to 20 Pa. A logarithmic decibel scale (dB) is used to measure sound with reference to the hearing ability of human beings. A sound level of 0 dB is assigned to a sound intensity of 10^{-12} W/m^2, the threshold of hearing. A 10^1 times rise in this intensity would be recorded as 10 dB, and a 10^2 times rise would correspond to 20 dB and so on. Thus, a 100-dB sound would possess the intensity of $10^{10} \times 10^{-12}$ W/m^2. Expressing the sound level in dB scale is a convenient way of compressing the scale of numbers associated with variation in pressure (from 20 µPa to 20 Pa) or of power (from 10^{-12} to 1 watt) into a manageable range from 10 log (10^0) dB = 0 dB to 10 log (10^{12}) dB = 120 dB.

As a certain quantity of energy distributed over a larger area becomes less intense, the intensity of sound also weakens with distance traveled. In fact, there is an inverse square relationship between the intensity of sound and location of its origin. This aspect presents a practical problem in measuring noise created by a machine. Noise recorded at varying distances from a loom would be different. Indeed noise measured at various locations but at the same distance from the center of a loom may also be different as the different sources of noise are located at different points within a machine.

Similarly, noise in a loom shed of 50 looms would have a different level of intensity as compared to one from within a shed of 500 looms. This issue is further complicated by the effect of nature of the surrounding space in a loom shed.

Another aspect that has a bearing on noise measurement is the loudness of sound as perceived by human beings at varying frequencies. Sound in the frequency range of 1000 to 5000 Hz appears louder than that outside this range although the entire audible range spans over 20 to 20,000 Hz. Broadband sound-measuring instruments, employing condenser microphones, indicate weighted average (dBA) of all audible frequencies with the lowest audible sound that the human ear can detect as the reference point for determining the decibel level of noise so as to mimic the human ear to a degree. For suppression of noise through engineering control, octave band analyzers are used to identify critical frequencies at which the majority of noise is generated.

Approximate measurements suggest that shuttle looms create noise in the range of 100 to 110 dBA, and gripper and jet looms result in noise of 95 and 92 dBA, respectively. Incidentally, an eight-hour exposure to noise level of 90 dBA leads to hearing damage as well as other health repercussions, such as cardiovascular effects, etc. A stricter regimen, such as that of the U.S. Navy medical department, prescribes that a person, exposed for more than 2 days of time weighted average per month to a noise level of 84 dBA, would be considered to be at risk.

Evidently, a modern loom shed employing state-of-the-art single phase biaxial shuttleless weaving machines suffers from undesirable and unacceptable levels of noise pollution. Quite surprisingly, however, scientific studies aimed at exploring avenues to bring down such noise levels to within tolerable limits are hard to come by in accessible literature.

10.10 Techno-Economics

The techno-economics of a production system is aimed at objectively evaluating its technical feasibility and economic viability for a certain product type in a given market during a specific period of time. Even in today's globalized market, demand for a particular commodity varies widely from location to location primarily due to varying consumer habits and degree of affluence. Similarly, habits and demands of the same group of homogenous consumers may change considerably over a period of time owing to rapid changes in extraneous conditions. This is compounded by varying degrees of continuous technical development of all production systems. As a result, it is impossible to specifically rank different production systems in techno-economical terms in an absolute sense.

Technical superiority of shuttleless looms vis-à-vis shuttle looms have been covered in detail in Sections 10.1 through 10.7. Similarly, the comparative advantage of one type of shuttleless loom over another type has also been outlined. However, an entrepreneur is primarily concerned about the return that the investment in a particular production system generates. This factor, namely return on investment, depends on a number of other factors, such as capital cost of equipment, labor cost, energy cost, sales cost of manufactured products, bank rates, building cost, maintenance cost, raw material cost, fixed overhead cost, etc. Some of these aspects will be briefly discussed in the following. At this point, it is worth noting that, in spite of the technical superiority of shuttleless looms over shuttle looms, strong and firm selvedges can be produced only on the latter. Hence, for some products, one may be forced to opt for a technically and economically less effective system.

Capital cost, that is, cost of machinery and allied equipment is a major factor governing choice of type of loom. A plain shuttle loom is the cheapest, and projectile looms have been for quite some time the most expensive in India. Amongst shuttleless looms, the water jet system along with a water softening plant may turn out to be the cheapest, dictated partly by its limited market demand owing to the inability to weave wide fabrics and process hydrophilic materials.

In general, cost of a loom goes up with reed width. A simple linear relationship between cost (C) and reed width (R) can be stated as follows:

$$C = a + bR$$

The constants a and b depend on the type of loom.

The cost of a loom of a given type increases with weft insertion rate (W), and the latter increases with reed width approximately as given by the expression

$$W = kR^\alpha$$

Derivation of this expression is based on an empirical relationship between loom rpm (n) and reed width, which can be stated as

$$n = kR^{1-\alpha}$$

For the typical case of α being equal to 0.5, the relationship between W and R reduces to

$$W = kR^{0.5}$$

Behavior of the ratio (W/C) as function of reed width expresses the manner in which an incremental rise in reed width benefits investors purely from the cost point of view. As long as the value of this ratio keeps on growing with rising reed width, investors get a more than proportionate increase in production for the extra investment made in a broader machine. However, when this ratio stops increasing and, in fact, starts dropping, the extra investment for a broader machine becomes economically less viable. Hence, if this ratio is differentiated with respect to reed width and the first differential is equated to zero, then one can derive one condition for optimum reed width.

$$W/C = Z = kR^{0.5}/(a + bR) \tag{10.1}$$

$$dZ/dR = [(d/dR)kR^{0.5}] (a + bR)^{-1} + [kR^{0.5}] [(d/dR) (a + bR)^{-1}]$$
$$= [ka - kbR]/[2(a + bR)^2 R^{0.5}] \tag{10.2}$$

The RHS of the equation vanishes for $R = (a/b)$ yielding the value of the reed width at which an investor would have highest cost versus production benefit. Such an expression depends, however, entirely on the empirical equations governing the relationship between cost and reed width on one hand and that between loom rpm and reed width on the other. In situations in which $\alpha \neq 0.5$ or in which relationships between cost and width are nonlinear and/or discontinuous, the expression of optimum reed width would assume some other form.

Next to the capital cost, the cost of labor and cost of energy are of immense importance. Actual power consumed by looms of a certain type depends on its weft insertion rate. However, a more efficiently designed loom of similar type may not consume proportionately greater power for an incremental rise in weft insertion rate. For example, reduction of gripper mass by using carbon fiber–reinforced material as opposed to using steel would lead to lower power consumption for comparable weft insertion rates. Hence, the ratio (kWh/WIR) may be employed as an indicator of degree of sophistication of a loom within its own type. From the narrow point of view of economics of production, however, a better proposition would be ratio (kilowatt hour per kilogram of fabric produced). Table 10.1 lists typical values of the latter type. Purely from this narrow viewpoint, a water jet loom scores over all other types of loom, and its counterpart, the air jet loom, should be the least preferred one. The commercial popularity of the air jet and rapier looms, the

TABLE 10.1

Energy Consumption Pattern of Different Types of Loom

Type	kWh/kg
Shuttle loom	1.8
Gripper loom	1.3
Rapier loom	2.5
Air jet loom	3.2
Water jet loom	0.9

two highest energy-guzzling systems, reveal the complexity of economics of woven fabric production.

The cost of labor rises with economic well-being as the latter leads to greater material expectations of the working population and, as a consequence, also leads to an altered demographic profile. Shuttleless looms are less labor-intensive than shuttle looms, but operating a modern weaving shed still needs a sizeable number of manual operators. This fact alone has caused a global shift of this industry from developed economies to the developing ones. Such a shift will continue in the foreseeable future until a major technical innovation renders the cost of manual intervention in the manufacture of textile fabrics fairly insignificant.

Under the situation prevailing as of now, woven fabrics are manufactured in developing economies, and good-quality weaving machines are manufactured in developed economies. Hence, more often than not, shuttleless machines need to be imported by the woven fabric manufacturer. The cost of import in terms of various duties payable to the government adds to the overall cost of equipment. These machines need a steady power supply and good backup systems, such as very well prepared warp, highly skilled manpower, a well-constructed building with proper illumination and humidity control, etc. All of these are not easy to come by in developing economies and add to the cost of the product. The least demanding and most flexible system would, as a consequence, put the least economic burden on the investor and, therefore, would commercially be the most preferred one. Certain case studies report the lowest cost of production, highest cash inflow, highest return on investment, and steady long-term profitability of air jet looms functioning under Indian conditions in production of a range of commercially popular fabrics. Realization of these commercial aspects has prompted the discriminating weaving machine manufacturer to shift attention from raising the weft insertion ratio of the future generation of looms to making them more flexible and versatile.

11

Nonconventional Weaving Systems

Multiphase weaving machines and narrow fabric looms are being categorized here as nonconventional weaving machines, which, just like conventional weaving machines, also produce biaxial woven fabrics. These machines possess, however, some distinctive features that differentiate them from conventional flat biaxial weaving machines described in the foregoing chapters. Some of these machines, such as the circular looms and narrow fabric weaving looms, are encountered in commercial production systems while others are yet to be commercially successful.

11.1 Multiphase Weaving

The term "phase" is employed in conventional technical literature to indicate relative timing of events occurring within one operational cycle. The most commonly encountered terminology of three-phase electrical supply systems to AC motors refers to three conductors carrying voltage waveforms that are offset in time by $2\pi/3$ radians. As a result, current in each conductor reaches peak value sequentially. In a single phase system, on the other hand, there would be one single peak within one cycle. The specific event in these examples is voltage (or current), and a single phase is synonymous with one single peak within a cycle while three-phase exhibits three peaks. Extending this concept to the weaving process, a loom in which only one pick of weft is inserted in one revolution of the main shaft can be termed as a single phase loom whereas, in a multiphase loom, more than one pick is inserted in one loom cycle. However, insertion of one pick of weft, unlike the occurrence of peak voltage in one cycle, is not an instantaneous event. Its instantaneous nature can best be described by, for example, by the entry point of the weft carrier into the shed or exit point from the shed. Similarly, the instantaneous moment of the shedding process can be the crossing of healds and that of the beat up by transformation of the last pick of weft into the cloth fell. In a single-phase weaving machine, these defining moments of shedding, picking, and beating processes occur only once in each loom cycle and that too with considerable phase difference although, in multiphase machines, these moments occur a number of times within one loom cycle and more interestingly, in a way, simultaneously. Expressed more explicitly, a large number of weaving cycles—each comprising three primary motions—take place within one machine cycle of a multiphase machine in such a way that the instantaneous beating up of one weaving cycle may coincide with the instantaneous picking of the following weaving cycle and with instantaneous shedding of the previous weaving cycle.

According to Greenwood (1980), however, there is some ambiguity in definition of the term multiphase as many authors define phase number (n) as the number of shuttles inserting weft simultaneously. In this sense, if v is the shuttle velocity, then the WIR can be expressed as

$$W = nv$$

In the conventional single-phase loom, a shuttle remains in the warp shed for a fraction of the loom cycle (e.g., 140° and, hence, approximately 0.4th part of a loom cycle), which would mean that the phase number is less than one (in this case 0.4), and therefore, a single phase loom should be termed as a 0.4-phase loom!

Greenwood goes on to establish that, for all types of loom, the weft insertion rate increases with increasing reed width and approaches a limiting value as the reed width approaches infinity. If the scatter in velocity of the weft carrier is denoted by dv, then the expression of the limiting value of weft insertion rate is given by

$$W_L = v(v + dv)/dv \tag{11.1}$$

If the weft carrier is driven positively, then its velocity should be constant, and $dv = 0$. This would suggest that for looms with positively driven weft carriers, the weft insertion rate approaches infinity as reed width approaches infinity. In certain types of multiphase looms, as would be described in the following, the weft carrier is driven positively. Hence, for such looms, the weft insertion rate would increase proportionately with reed width. This is an important feature of a class of multiphase looms.

There are two classes of multiphase looms, namely circular and flatbed looms. The flatbed multiphase looms are again of two types, namely the wave

FIGURE 11.1
Shed formation on a circular loom.

(or ripple) shed and sequential (or multilinear) shed looms. On circular looms, a large number of shuttles travel along the periphery of a circle through an equal number of sheds. These sheds are formed by warp yarns arranged in a circular manner (Figure 11.1). On wave shed looms, a large number of shuttles travel along a liner path through an equal number of sheds, which develop along the weft direction of a flat loom, and on multilinear shed flat looms, the weft carriers, such as the stream of an air jet, propel a number of weft yarns simultaneously into separate sheds formed along the warp direction.

11.1.1 Circular Weaving

The characteristic features of circular looms are the curved shuttle housing a large, low wind–angled, cross-wound tubular package for side withdrawal (Figures 11.2 and 11.3), a circular reed ring guiding a multitude of curved

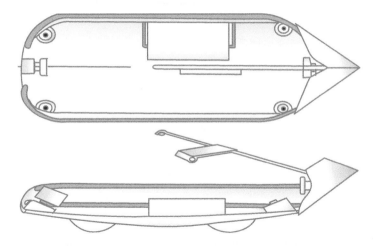

FIGURE 11.2
Schematic view of a shuttle for circular loom.

FIGURE 11.3
Tubular package within warp shed on a circular loom.

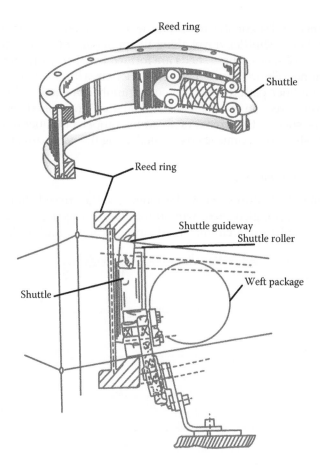

FIGURE 11.4
Shuttle in a circular reed ring. Lohia brochure.

shuttles running on rollers (Figure 11.4), cam-operated heald segments (Figure 11.5), a roller or magnet drive to shuttles (Figure 11.6), and a spur or needle-driven beat up of an inserted pick (Figure 11.7).

The plan and side views of the body of a shuttle are shown in Figure 11.2. From the side view depicted in the lower half of the diagram, it is observed that the shuttle has a curvature conforming to that of the reed ring (Figure 11.4), which, in turn, depends on loom radius. The shuttle base in the diagram is equipped with four antifriction rollers although, at times, only two very smooth bands along its length suffice for facing and sliding along the reed surface. Four more rollers, two near the upper wall and two near the lower wall of shuttle, ensure firm but rolling contact with the inclined upper and lower surfaces of the reed ring. These contacts come into play when a centripetal force acts on the shuttle during its rotation about the machine

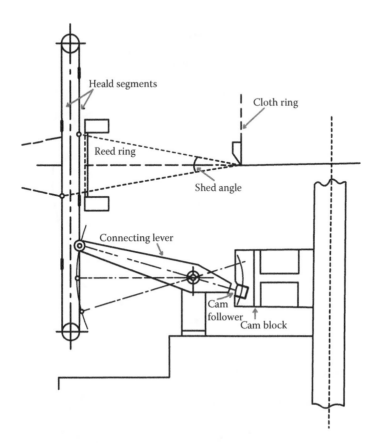

FIGURE 11.5
Drive to heald on a circular loom. Lohia brochure.

center. As is shown in Figure 11.3, the tubular weft package juts out of the shuttle body and occupies most of the shed space. A plan view of a circular machine exhibits, therefore, a multiple number of such weft tubes arranged along a circular ring juxtaposed between the reed ring and the periphery of the resultant tubular fabric. Shuttles remain firmly housed within the space between the reed ring and the sley race while their spurs stick out guiding the weft yarn close to the fabric fell. Repair of the weft break is carried out without removing either the shuttle or the package from within the shed while a replenishment of the package can be carried out either manually or automatically without disturbing the shuttle.

Warp yarns can be fed to the shedding zone either from a multiple number of beams arranged on suitable frames or from packages mounted on creels located around the weaving machine. These yarns are guided into heald segments that are arranged along a circular track around a reed ring (Figure 11.1). Such segmentation permits the warp sheet to be broken into narrow

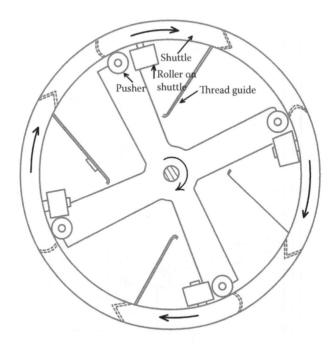

FIGURE 11.6
Mechanical drive to shuttle on a circular loom.

FIGURE 11.7
Beating up by a spur on a circular loom.

bands of width b. These heald segments are driven individually and at a phase difference with respect to neighboring heald segments (Figure 11.1) so that the narrow bands of warp they control can be formed into crisscrossing wave lines, which move in the weft direction at the same speed as that of the shuttles. Shuttles therefore perpetually enter new sheds, leaving behind old ones. An example of a cam drive to a pair of heald segments is depicted in Figure 11.5. The pair of heald segments is connected through a common heald-reversing motion so that when one moves up, the other automatically moves down. The rear heald segment is firmly linked to one end of a connecting lever, the other end of which carries a follower that runs in the track of a cam block. The cam block, in turn, is driven from the machine main shaft. Heald segments may also get individual drive from cam blocks. Such individual drives are required for weaves repeating on more than two ends, such as twill.

Propulsion of the shuttle can take place by mechanical or by electromagnetic means. The principle of the mechanical method has been depicted in Figure 11.6. In the side view of the shuttle shown in Figure 11.2, an additional roller at the rear end is seen as mounted on top of the roller that guides the shuttle along the sley race. This additional roller is engaged by a pusher, which is mounted on one of the radial arms extending from the central shaft shown in Figure 11.6. On the machine, there are as many arms with pushers as the number of shuttles. The thread guide shown sticking out of the shuttle is the spur that also pushes the inserted weft yarn into the cloth fell. The yarns of the bottom shed line are bound to come in between the pushers and rollers on the shuttle at an instant when the shed line changes from bottom to top. The manner of coupling of the three elements is such that the radial force that pushes the shuttle along its path also helps to push the yarns of the bottom shed line upward. In such an event, yarns of the lower shed line can, of course, suffer abrasive damage. Such a situation is avoided if it is ensured that instantaneous relative velocity between the pusher, the roller, and the yarns remains zero. A synchronization of shed movement, shuttle speed, and the roller size and speed as well as their locations are vital for minimizing abrasive wear of the yarns.

The electromagnetic method is superior to the mechanical method as it is contactless in its functioning, and hence, threads are not strained. The principle involves positioning a series of electromagnets under the shuttle path while shuttles move through the rakes, similar to guides used on gripper shuttle projectile systems, thus preventing any shearing action on the bottom shed line as well as accidental contact between the shuttles and magnets. Through a switching device, the electromagnets are energized in such a way that each shuttle experiences simultaneously a push and pull. Electromagnets may be replaced by permanent magnets to reduce investment cost in which case these magnets need to be mechanically moved along a circular track.

A reed on a circular loom does not take any part in the beating up process. Beating up is carried out in crossed shed either by an external element or by a spur on the tip of shuttle thread guide. In the former solution, an assembly of an inclined array of needles rotate about their axis in such a manner that a small part of each inclined assembly dips for a very brief period into gaps between a limited number of warp yarns pushing the interlacing segment of weft yarn to the fell of the cloth. Evidently, in addition to an axial rotation, such assemblies have to move with the shuttles along a circular path. In the other method, the tip of the shuttle thread guide is equipped with a smooth nose that pushes against the pick inserted by the preceding shuttle. In either event, beating up takes place in a crossed shed.

Circular looms are primarily employed for production of tubular sacks and hoses from PP/HDPE tapes. These machines come in a range of diameters to produce tubes varying between 16 and 54 cm on the lower range, and the larger diameter machines produce tubes varying in diameter between 108 and 175 cm. The bigger machines employ around 10 to 12 shuttles at a time, and the ones with smaller diameter come with four to six shuttles.

The actual diameter of a circular machine, measured at the reed ring, is much higher than that of the resultant fabric that forms around the cloth ring (Figures 11.1 and 11.3). The distance between the reed ring and the cloth ring is governed by a simple trigonometric relationship. Let

the height of shuttle base = h

the shed angle at the cloth ring = α radians

The radial distance between the reed ring and the cloth ring works out to be $[(h/2)/\tan(\alpha/2)]$. Thus, for a shuttle base height of 10 cm and a shed angle of 20°, the radial distance between the reed ring and the cloth ring would be 28 cm. Hence, if a cloth tube of diameter 150 cm is to be produced by said shuttle, then a machine diameter of nearly 2.1 m would come into question.

The tubular weft packages are fairly big with, for example, a 200-mm traverse and a 115- to 120-mm outer diameter. For such packages, the height of the shuttle base has to be around 130 to 140 mm, resulting in approximately a 1.8- to 2-m loss in space along the machine diameter. This loss can be reduced by opting for a higher shed angle, which, in turn, puts warp threads to a greater strain level.

The rpm of the main shaft of the machine multiplied by the number of shuttles yields the value of the weft insertion rates, which, for circular looms, can vary between 350 to 1100 m/min. For a multiphase loom, these values are very modest as high masses of shuttle (2.5 kg and above) and weft package (1.5 kg and above) coupled with the requirement of side end withdrawal of the weft yarn prove serious limitations. A large weft package size is dictated by the requirement of acceptable machine downtime governed by the frequency of package change, which, in turn, dictates the size of shuttle. The technical advantage of a circular loom over a flat shuttleless loom is hence restricted entirely to the form of the specialized product, namely a tube without a side seam.

The larger tubes produced by a circular loom are employed as flexible intermediate bulk containers (FIBC) as packaging for cattle feed and husk and even as tarpaulins after slitting and processing, and the narrower tubes can be used for packaging food grains, sugar, flour, cement, and similar granular and powdery materials. Some narrow tubes made of leno construction can be employed for packaging fruits and vegetables that need a high degree of ventilation. Leno constructions are produced on circular looms employing two healds, which are operated by an additional cam system that impart lateral motion to one set of the heald segment.

11.1.2 Ripple Shed Weaving

A ripple or wave shed or a rectilinear multiphase loom works on similar principles as a circular loom but, as opposed to circular looms, can operate at very high WIR with fairly moderate shuttle speed. Even then, it has had no commercial success. Let

$$\text{Width of a ripple shed loom} = R \text{ m}$$

$$\text{Length of a shuttle} = \ell \text{ cm}$$

$$\text{Width of a heald segment} = b \text{ cm}$$

$$\text{Semi-wave length of ripple shed} = a \text{ cm}$$

$$\text{Maximum shed depth} = h \text{ cm}$$

$$\text{Speed of shuttle} = v \text{ m/s}$$

Then

$$\text{Number of sheds across shed} = 100R/a$$

$$\text{Number of shuttles in shed at any instant} = n = 100R/a$$

Hollstein (1978) has shown that under the condition that a shed segment is fully open when the shuttle tip is about to enter it and this segment moves to a closed position when the tail of the shuttle just leaves it then time in milliseconds for shed change is given by $20(\ell + b)/v$. In this expression, no time has been allowed for dwell of the shed, which is, however, necessary for shed formation with heald segments. Hence, the time in milliseconds required by the shuttle to move through the distance $2a$ is equal to $40(\ell + b)/v$, which, in turn, is the time period for one shed wave. Assuming a sinusoidal waveform of the shed, the displacement function $s(t)$ of healds can be stated as follows:

$$s(t) = h\sin[(\pi v \times 10^3)/20(\ell + b)]t \tag{11.2}$$

The shuttle height affects amplitude h, and a smaller and faster shuttle and smaller width of heald segment reduces the period. A smaller shuttle would, in turn, reduce the wavelength $2a$, leading to an increase in value of n.

Hence, the key to increasing WIR in ripple shed looms lies in minimizing values of ℓ and b.

As $2(\ell + b)\ 10^{-2}$ m of weft is inserted by each shuttle before a shed line changes from one extreme position to another and as n such shuttles are involved in this process at any instant, then the length of weft inserted between a shed change is equal to $2n(\ell + b) \times 10^{-2}$ m. The number of such shed changes per second being equal to $[(v \times 10^{3})/20(\ell + b)]$ for a ripple shed loom

$$
\begin{aligned}
\text{WIR} &= [2n(\ell+b)\times 10^{-2}][(v\times 10^{3})/20(\ell+b)]\\
&= nv \text{ m/s}\\
&= 60\,nv \text{ m/min}\\
&= 60(100Rv/a) \text{ m/min}
\end{aligned}
\tag{11.3}
$$

The expression of WIR underlines the exciting possibility of achieving an extremely high value of WIR by increasing the reed width and reducing the semi-wavelength of the shed even while maintaining a very modest value of shuttle speed. For example, a 5-m-wide loom with a value of $a = 50$ cm and $v = 15$ m/s, equal to that of a conventional flying shuttle, yields a WIR value of 9000 m/min. Quite remarkably, the loom rpm does not play any role in the WIR of a ripple shed loom, theoretically permitting thereby an unfettered increase in the value of reed width. As opposed to a circular loom, the shuttle of a ripple shed loom needs to carry sufficient yarn for one pick only. Hence, its size can remain fairly small, thereby permitting very high values of n and v.

Principles of shedding, picking, and beating up are similar to those observed in circular looms although there are differences in detail.

Shedding of heald segments can be carried out by a mounting series of tappets on a tappet shaft with the required phase difference. Each tappet system operates on one narrow heald segment while neighboring heald segments can be similarly operated by other systems of tappets set up with the requisite phase difference.

Picking a shuttle may be carried out mechanically by ensuring that a roller mounted on the shuttle gets pushed by another roller, which is carried on a slider that slides on a reed ring. The slider is mounted on a conveyor belt, which carries a number of such sliders to impart the required push to the respective shuttles. The other method of picking is by magnetic principle wherein weft carriers are carried by a conveyor into the shed where they are acted on by respective electromagnets. These electromagnets are fitted to a moving tape.

A novel system of beating up by employing a rotary reed was demonstrated in the Czechoslovakian Kontis machine. In this solution, shuttles carrying weft yarns are pushed mechanically from right to left by the action of pairs of rollers. A reed is constructed not by a set of parallel wires but by a set of parallel thin circular lamella. Each lamella sports a number of curved radial slots at one end of each of which is located an upturned nose. Neighboring lamella are arranged in the cylindrical reed with such a phase difference that slots of the assembled lamella form helical grooves in the reed, the number of helices equaling the number of curved slots in each lamella. The lamella of the cylindrical reed rotate between adjacent warp yarns in such a way that the inserted picks are laid securely deep within its helical grooves. Evidently, the number of helical grooves in the rotary reed has to equal the number of shuttles inserting picks at any point of time. At the point of beat-up, relevant segments of each pick are guided out of respective grooves by the curved slots of rotating lamella and get pushed into the cloth fell by upturned noses, which stick each lamella at the respective locations.

The weft carrier on a rectilinear multiphase loom carries the weft for one pick only and hence has to be refilled after leaving the warp sheet and before being brought back into the picking mode. After having inserted the weft, the carriers are moved by a conveyer system back to the picking side via a filling station. A filling station may have a multitude of winding heads in different architecture, the description of which is not being gone into.

In spite of great expectations from these machines in the '70s and '80s, they did not prove to be commercially successful on account of problems associated with following issues.

1. Repair of a broken pick
2. Maintaining uniform and the same tension in all picks
3. Maintaining uniform beat-up force across the fabric width
4. Speed limitation imposed by inertia of weft carrier and of filling stations

11.1.3 Wave Shed Weaving

The limitations of a ripple shed loom were elegantly resolved in the wave shed loom M8300 developed in the 1990s by Sulzer Textil. Functional principles of this system are illustrated in the following:

A rotor constitutes the central element of this system shown in Figure 11.8. A warp sheet (5, 6) is fed onto the surface of this rotor via a set of warp-positioning elements (4) and is stretched through gaps between beat-up elements (2) along the rotor surface. A large number of equidistant rows of these beat-up elements are arranged on the rotor surface. The warp sheet is split into multiple numbers of sheds along its length by the shed forming elements (3). Picks are inserted by air jet streams through the channels provided by shed-forming elements, and after a nearly 180° wrap, the warp sheet lands as woven fabric (1) onto the fabric support bar (7).

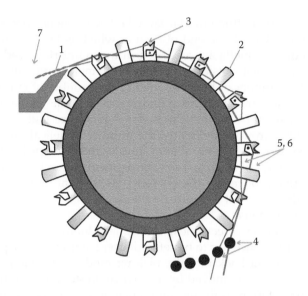

FIGURE 11.8
Side view of rotor shedding on a wave shed loom. (From Adanur, S., *Handbook of Weaving*, 2000, CRC Press.)

The warp-positioning elements can be moved laterally (Figure 11.9), shifting, in the process, the position of the warp yarns they control, thus moving them either away from or aligning them along the path of the shed-forming elements (3), which follow the band of the warp sheet landing onto the rotor. A warp yarn aligned along a corresponding shed-forming element would

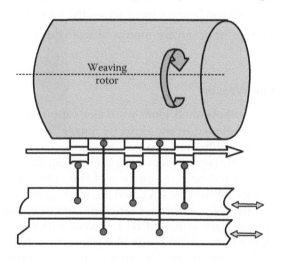

FIGURE 11.9
Mechanism for warp selection on a wave shed loom. (From Adanur, S., *Handbook of Weaving*, 2000, CRC Press.)

be pushed up to the top shed line, and a warp yarn, which is moved away would occupy the bottom line. A number of rows of shed-forming elements are arranged along the width of the rotor, resulting in a succession of sheds along its circumference. Every row of shed-forming elements is followed by a row of much taller beat-up elements. A closer view of the shed-forming and beat-up elements is provided in Figure 11.10. Picks inserted in channels of shed-forming elements are carried forward with the respective sheds until the warp sheet is eased off the rotor by the fabric support bar. As a result, the corresponding shed-forming elements move away from the warp sheet and continue their journey along a circular path, allowing the inserted picks to escape through the respective openings just ahead of the fabric support bar. Subsequently, picks are beaten up by the taller beat-up elements whose tips move along a higher arc, enabling them to reach out to the cloth fell.

A schematic view of the air jet picking system, equipped with relay jets mounted conveniently on the rotor, is depicted in Figure 11.11. Each pick of weft follows a tortuous path in space, necessitated by the continuous movement of the shed along an arc. Accordingly, the drum-shaped controller mounted by the side of the rotor plays the central role of keeping the shed-forming channels fed with pick of weft at the right speed and by the right amount.

Adanur (2000) reports that a typical M8300 machine can operate with three bar air pressure propelling picks at a speed of about 22 m/s across a 1.9-m-wide rotor inserting 2800 picks/min at a WIR of 5400 m/min. A warp density of up to 32 ends in one centimeter can be woven on such systems, which are less noisy and less energy-guzzling than comparable single-phase looms and occupy less space while yielding a much lower weaving cost.

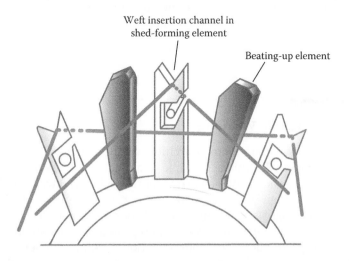

FIGURE 11.10

Close-up view of beating and shedding elements on a wave shed loom. (From Adanur, S., *Handbook of Weaving*, 2000, CRC Press.)

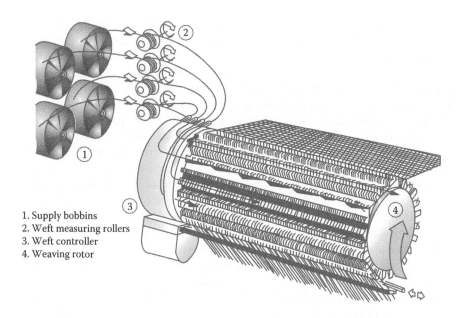

1. Supply bobbins
2. Weft measuring rollers
3. Weft controller
4. Weaving rotor

FIGURE 11.11
Air jet weft propulsion on a wave shed loom. (From Adanur, S., *Handbook of Weaving*, 2000, CRC Press.)

In spite of the many advantages of the M8300 wave shed loom over single-phase looms as well as over ripple-shed multiphase looms, commercial success has eluded this technology so much so that it is no longer being displayed in international exhibitions. One reason could be that its product spectrum is limited to a very narrow domain without being in any way unique to this system. Greater versatility is evidently the direction of its future development.

11.2 Narrow Fabric Weaving

A narrow fabric weaving loom is architecturally different from a conventional weaving loom. It exhibits multiple weaving heads organized along the loom width. Each head weaves a narrow fabric having a conventional woven selvedge on one edge and a knitted selvedge on the other. Heald shafts are made of very light and strong material such as carbon fiber–reinforced plastic in order that inertia and fatigue to repeated stresses are minimized, permitting a high frequency of loom operation. Picking is generally carried out by the tip of a swiveling needle head that operates like a Gabler rapier tracing out a wide arc in space to accommodate which the reed also sweeps along a wide arc without any perceptible dwell. Such picking and beating motions exert considerable dynamic strain on the weft and abrasive strain on warp yarns. The woven fabric passes through the gap between a pair of tape plates and is taken up by suitable take-up systems. The tape plates maintain a positive control on the entire width of the fabric fell.

A typical setup of multiple heads on a narrow fabric weaving loom is depicted in Figure 11.12. Each weaving head is complete with respect to its own shedding, picking, and beating-up systems while a common take-up and let-off system serves all the heads. This implies that it is not possible to weave different fabrics on different heads of the same loom. It also implies that interruption of the weaving process on any head brings the rest of the heads to a simultaneous halt.

The number of heads on a commercial unit may vary between two and 10, and the reed width of a head can vary between 27 and 210 mm. The 210-mm-wide configuration comes in a set of two heads, and the 27-mm-wide configuration may sport as many as 10 heads although a four-head, 27-mm configuration as shown in Figure 11.12 is quite common. A typical configuration for weaving seat belts can be a 46-mm head machine with six heads. The rpm of a loom operated under commercial conditions can be around 1000 although much higher values in the range of 2000–3000 can be quoted by machine manufacturers from the point of view of machine dynamics. The determining factor of machine speed is quality of warp and weft yarns, and complex weaves can slow down a machine.

Viewed as a system by itself, a narrow fabric weaving loom has to be classified as a single-phase system although it exhibits some attributes of a multiphase loom on account of its facility for inserting multiple pieces of weft simultaneously. Thus, on a 10-head system working at 1000 rpm and weaving tapes of 25 mm width on each head, a 250 m length of pick is inserted every minute by picking needles moving individually at an average speed of approximately 90 m/min. The low overall weft insertion rate (WIR) of such a loom is accounted for by the very small effective fabric width of only 0.25 m. However,

FIGURE 11.12
View of the weaving heads of a narrow fabric loom.

the ratio of overall WIR to an individual weft carrier velocity of such a loom works out to be around 2.7, which is as much as five times higher than even that of a modern single-phase air jet loom. A multiphase loom with a theoretical WIR of 6000 with 50 simultaneous weft insertions yields a much higher value of 50, demonstrating its clear superiority over all forms of single-phase looms.

A close-up view of a single head is shown in Figure 11.13. Weft drawn in through the eye of a curved needle that moves along an arc is inserted in the form of a loop into the shed. The needle tip carrying a weft loop near its extreme point of throw past the right-hand selvedge of fabric can be seen in Figure 11.14. From Figures 11.12, 11.13, and 11.14, it is observed that one end of the picking needle is mounted on a picking arm the other end of which is fastened to an upright rocking shaft. As a result of the rocking motion of the upright shaft, the L-shaped assembly of the picking arm and needle is rotated to and fro, which results in a large rapier like in-out sweep of the needle tip along an arc. An eye on the needle tip through which weft yarn is threaded pushes the weft bodily into the shed, forcing the yarn to move rapidly through the narrow opening of the eye at a speed almost twice that

FIGURE 11.13
Close-up view of beginning of weft insertion on a narrow fabric loom.

FIGURE 11.14
Close-up view of an inserted weft loop on a narrow fabric loom.

of the needle. This is bound to abrade the yarn surface and, depending on the nature of the yarn, may even wear out the needle eye. Hence, the surface of the needle eye needs to be very hard and very smooth, which can be generated such as through the plasma coating of the basic spring steel material.

With the needle eye positioned just outside the edge of the knitted selvedge (Figure 11.14), one of the two arms of weft loop occupies an inclined position inside the warp shed. One end of this arm is firmly anchored to the woven selvedge end, and its other end passes under the needle and rises up through the needle eye. The second and upper arm of the weft loop passes above and along the needle out of the shed opening and toward the weft supply zone. An additional warp yarn, termed a catch thread, is guided by a finger toward the knitted selvedge just as a latch needle moves forward through the gap between the picking needle and the arm of weft loop anchored to the woven selvedge (Figure 11.14). The latch needle pulls the catch-thread through this gap, securing, in the process, the tip of the weft loop to the selvedge. During its return journey, the needle eye slides across the upper arm of the weft loop and moves out of the warp shed and also away from the path of the closing-in reed, leaving behind a double pick in the shed.

The loop made from the catch thread is knitted into a pillar stitch, thus providing rigidity to the concerned selvedge. Occasionally, an additional lock thread, guided by a static thread guide, is also employed along with the catch thread for the purpose of improving the appearance and firmness of selvedge. The two types of threads are brought in through separate tensioning systems to the fabric edge in the form of an additional shed with the lock thread usually kept at a lower tension and occupying the bent configuration of an upper shed line while the catch thread remains straight along the lower shed line. The weft needle passes under the lock thread but above the catch thread, and the latch needle, while passing through the gap between the two arms of weft loop, passes above both lock and catch threads. The catch thread wraps around the lower arm of the weft loop, and the lock thread wraps around its upper arm during the withdrawal motion of the latch needle. This results in an equal degree of restraint to the two arms of the double pick and a very sharp selvedge line. Clearly, thread tensioning as well as the exact setting of the path of the two threads and the latch needle becomes quite critical, and therefore, working with one single catch thread is a much easier proposition.

Latch needles may be crank-driven or cam-driven. A cam-driven needle enjoys a more precise displacement profile. The latch needle pulls in the catch (and lock) thread through a loop from the previous cycle and casts off the old loop across a stripper plate, the side view of which is shown in Figure 11.15. It is an additional thin strip fastened to the side of the pair of tape plates, through which the woven fabric passes to the take-up zone. The side view of tape plates is indicated in the diagram by discontinuous lines. The stripper plates near the right selvedge of fabric are also clearly visible in Figures 11.13 and 11.14.

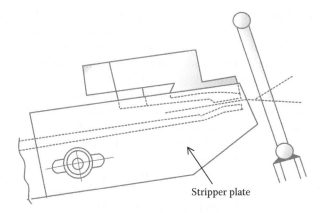

FIGURE 11.15
View of stripper plate for knitted selvedge on a narrow fabric loom.

Each heald is linked to a shedding lever (Figure 11.16), and shedding levers equaling the number of healds are mounted parallel to each other within a frame. Each shedding lever is fulcrummed at its rear end on the mounting block, which is fastened rigidly to the machine frame while the tip of each lever moves up and down in a slot provided in a guide bracket. An antifriction bowl is mounted on each shedding lever, which is acted on by a pattern chain. A pattern chain is constructed by joining links suitably together for providing the desired upward displacement profile to the antifriction bowl in keeping with the desired weave. The downward motion of the shedding lever is realized through extension springs. Pattern chains, one for each heald, are mounted on pattern drums, which are, in turn, mounted on a common shaft. Each drum serves one weaving head and can support a large number of pattern chains, equal to the maximum number of healds, which

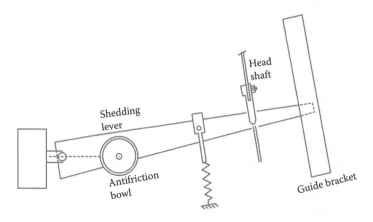

FIGURE 11.16
Elements of mechanical shedding motion on a narrow fabric loom.

can be operated on a given machine. For example, a machine with a 16-heald capacity would have pattern drums each of which can support 16 pattern chains. The chains are maintained under tension by means of suitable tensioning guides. Such cumbersome mechanical systems have been replaced in modern machines by electronic systems.

A tape plate secures the woven fabric just beyond the cloth fell for preventing any vertical movement. A modified form of tape plate lips provides secure gripping of the fabric fell for heavy fabrics, which are prone to bumping. Beyond the tape plate, a take-up system designed along similar lines as on conventional weaving machines pulls the woven fabric steadily away from the weaving zone and deposits the product into a can.

Three different types of take-up system are shown in Figure 11.17. Under the effect of spring tension, a nip roller presses against the take-up roller, which is driven positively with a controlled angular speed, resulting in a steady withdrawal of woven fabric from the weaving zone. Depending on the nature of the product, a varying extent of wrap between woven fabric and the rollers in the take-up system can be realized by choosing from one of the three configurations.

Warp can be prepared on narrow beams, but it is more convenient and economical to have the warp packages on a creel when a long run of the same product is planned. Warp packages for all weaving heads of a machine are mounted on the same creel, and segments of the sheet are drawn around rubber-covered brake drums so as to even out tension differences between individual warp yarns before being fed to the weaving zone. Brake drums are pulled around by tension in the warp sheet while overcoming frictional resistance generated by a suitably mounted dead weight system.

Narrow fabric weaving looms equipped with a needle weft insertion system result in fabrics with a woven selvedge along one of its edges and a knitted one on the other. The properties of a structurally and materially different knitted selvedge on one edge would be different from that of the woven edge, and this difference may cause problems in some technical applications in which emphasis is on homogeneity. Thus, for example, problems

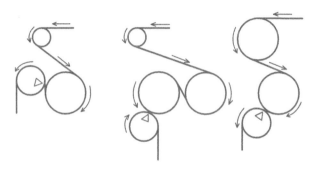

FIGURE 11.17
Principle of fabric take-up systems on a narrow fabric loom.

of unevenness owing to variation in thickness and tensile properties of the two edges would be encountered in applications requiring precise multiple wrapping of such a narrow fabric on an object. Besides the problem of dissimilar selvedges, narrow weaving looms equipped with a needle weft insertion system cannot be employed for generating seamless tubes, a form of product that has many technical end uses. For such and similar other specific technical applications, narrow fabric looms with a shuttle propulsion system are specially manufactured. Such looms may exhibit multiple layers of shuttle on two picking sides and also may be equipped with shedding systems that can split a warp line into sheds at different levels, permitting weaving separate fabric layers that get jointed together in a seamless manner through the weaving process itself. Width of individual layers as well as the extent and locations of their joints with neighboring layers can be controlled as per requirement.

References

Adanur S (2000). *Handbook of Weaving*, CRC Press Publication, Boca Raton, FL, USA.

Greenwood K (1980). The weft insertion rate of multi-section looms with flying shuttles, *Journal of Textile Institute*, 71, No. 3, 147–164.

Hollstein H (1978). *Fertigungstechnik Weberei*, Vol. 1, VEB Fachbuchverlag, Leipzig.

12

Formation of Weft-Knitted Fabrics

12.1 Introduction

Knitted fabrics are known to have existed even before 256 A.D., evidenced by samples found in Syria and Egypt. This art of producing fabrics was introduced to Europe by the Arabs.

The first knitting machine was invented in 1589—incidentally, 200 years prior to the French revolution—by Rev. William Lee of Nottingham. In hand knitting, yarn is looped around pin(s) in the manner shown in Figure 12.1. The resultant fabric, which is a matrix of rows and columns of loops, is formed by creating a single element in each complete cycle of operation. Hence, if a fabric needs to have 100 loops in each row, then 100 cycles of operation would be needed for one row. Subsequently, loops of the row just completed would be transferred one after the other to another pin, and in the process, new elements are generated for the next row. The machine invented by W. Lee could, however, generate one complete row in each cycle of operation. As shown in Figure 12.2, such a machine has needles, and each needle supports a column of the matrix, referred to in the foregoing. In each cycle of operation, all needles execute similar motion and produce a loop each. It is no wonder then that this method of production is immensely more productive than hand knitting. Such a qualitative change in method of production did upset the prevailing system considerably, and so the Rev. Lee was hounded out of his native land. It took nearly two more centuries before any further development in mechanization of the knitting process could take place. The real surge in development of this technology is closely linked with commercial availability of man-made filament yarns. Thus, the period subsequent to the second World War witnessed tremendous growth, so much so that, currently, international production of knitted fabrics matches the weaving method of fabric production.

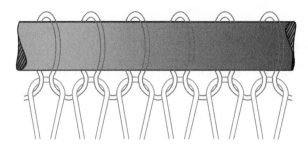

FIGURE 12.1
Knitting with pins.

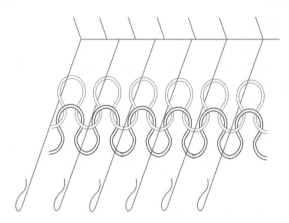

FIGURE 12.2
Principle of mechanized knitting.

12.2 Basic Concepts

The process of knitting, as practiced commercially, can be broadly classified into the following groups:

There are, in principle, two broad classes of knitting processes, namely *weft knitting* (Figure 12.3) and *warp knitting* (Figure 12.4). The overall direction of the yarn path in the weft-knitting process is along the fabric width. This is analogous to that of a pick of weft in the weaving process. In the warp-knitting process, the general direction of the yarn path is along the fabric length, similar to that of the warp in weaving. Moreover, weft knitting can be carried out from one single spool, analogous to a package of weft. For warp knitting, however, one needs a warp beam.

A *wale* is a column of loops intermeshed with each other along the length direction of a knitted fabric. In Figures 12.3 and 12.4, four wales or wale lines are shown. The distance between the centers of neighboring wale lines is known as *wale spacing* (*w*). The number of wales in the unit width of fabric is

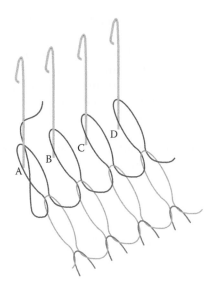

FIGURE 12.3
Principle of weft knitting.

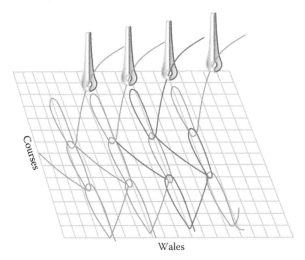

FIGURE 12.4
Principle of warp knitting.

designated as *W*. A *course* is a row of loops produced by adjacent needles during one knitting cycle. In Figures 12.3 and 12.4, three courses are shown. The distance between courses of two successive needle loops on the same wale line is known as *course spacing* (*c*). The number of courses in a unit length of fabric is designated as *C*. The product of the number of wales per unit width and courses per unit length of fabric equals the number of loops per unit area of the fabric. This is designated as *stitch density* (*S*). Stitch density happens to be a more accurate measurement than the linear measurement of the courses and wales of a knitted fabric because tension acting in one direction of a knitted fabric may, for example, produce low values in courses/ unit length and high values for wales/unit width. However, when multiplied together, this effect cancels out to a degree. The value of *S* is therefore more representative of the dimension of a knitted fabric as compared to its value of *C* or *W*.

The production process of knitted fabrics can be represented by the flow diagrams in Figures 12.5 and 12.6. Both warp- and weft-knitting processes are evidently much simpler than the weaving process.

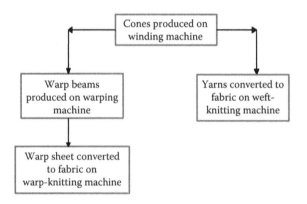

FIGURE 12.5
Sequence of steps in production of knitted fabrics.

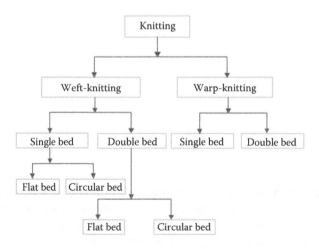

FIGURE 12.6
Types of basic knitting processes.

A knitted fabric can be defined as a two-dimensional continuum, produced by looping around (intralooping and interlooping) of individual yarns or sets of yarns. The building block of a knitted fabric is the *loop*. In a woven fabric, yarns are held in place owing to inter-yarn friction. In a knitted fabric, the looping of yarns results in a positive binding. Thus, if the coefficient of friction between yarns is hypothetically reduced to zero, then a woven fabric would disintegrate on its own but not a knitted fabric.

Loops are stitches hanging on to one another (Figure 12.7) and held in shape on the upper and lower interlacing zones by other loops. The last row of loops knitted in a fabric is only bound at its lower zones and can be referred to as a row of half loops.

Each loop consists of a crown (a), two arms (b), and two feet (c) (Figure 12.7). Two zones of interlacement around the base with a loop located below and another two zones of interlacement around the crown with a loop located above hold a loop in place. These zones of interlacements can be termed as binding points if contact zones are simplified to points. If the feet (c) of a loop lie above the binding points at the base, and the arms (b) lie below (Figure 12.8), then the loop is being viewed from the *technical back* side. If, on the other hand, the feet (c) are below the binding points and the legs (b) are above, then the *technical front* of the loop (Figure 12.9) is on view. Incidentally, the crown (a) is known as the needle loop. Moreover, the two elements "c," if joined together on the same side of the loop, would result in a mirror image of element (a). This so-joined element is also known as a sinker loop.

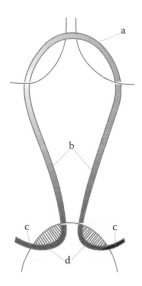

FIGURE 12.7
Front view of a weft-knitted loop.

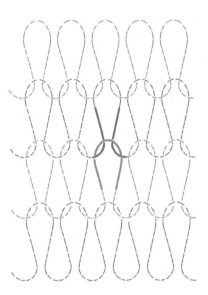

FIGURE 12.8
View of technical back side of weft-knitted loops.

FIGURE 12.9
View of technical front side of weft-knitted loops.

12.3 Process of Loop Formation

As shown in Figure 12.10a, a straight yarn segment (1) is brought close to a column of loops (2) supported by a needle (the black dot in the center of loop already formed). In the next phase (Figure 12.10b), the yarn segment (1) is bent into a shape resembling a loop. This new yarn and the column of (half) loops are then brought closer such that one would be on top of the other

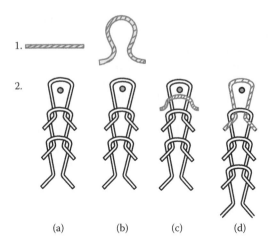

FIGURE 12.10
Process of loop formation. (a) Yarn feeding. (b) Shaping new yarn to loop form. (c) Interlooping.
(d) Casting off.

(Figure 12.10c). Subsequently, the needle pulls this new loop, crown first, through the half loop. At the same time, it releases/casts off/knocks over the half loop so that the same hangs with its crown supported around the base of the new half loop, whose crown is now held in the hook of the needle (Figure 12.10d). Simultaneously, the old half loop gets converted into a complete loop. On comparing with Figure 12.8, it is observed that the technical back side of loops is shown in Figure 12.10. In fact, from the arrangement of new yarn and half loops in Figure 12.10c, it is apparent that the needle must have penetrated the loop it has been supporting from the back of the plane of the loop to its front side, then caught the new yarn and drawn it to the back of the plane of the half loop. This process is also known as stitching to the back, and the resultant loop is termed as having been stitched to the back. Hence, a loop stitched to the back reveals its technical back side to the observer.

Similarly, if Figures 12.10c and 12.10d are turned around about their axis of symmetry by π radians, one would observe the needle penetrating the half loop from its front to the back side and pulling the yarn from the back to the front side. This is equivalent to stitching a loop to the front, and the resultant appearance of the loop would be similar to that shown in Figure 12.9. Thus, a loop stitched to the front reveals its technical front side to the observer.

Figures 12.8 and 12.9 represent the same construction, viewed from its two technically different sides. A fabric on a knitting machine would exhibit one of the two surfaces to the operator, and its nature would depend on the relative spatial disposition of the operator, the half loop, and the new yarn. If the half loop is between the operator and the new yarn, then the resultant fabric would exhibit its technical front side to the operator. If, on the other hand, the new yarn is between the half loop and the operator, then the fabric would exhibit its technical back side. However, if there is a combination of these two cases, that is, of the new yarn being in front of some of the half loops and being at the back of some other half loops, the resultant fabric would exhibit an admixture of technical front and technical back surfaces.

Until now, only weft-knitted loops have been considered. The warp-knitted loops, shown in Figure 12.11, display a different configuration although the basic concepts are common. Here, the sinker loops, instead of being symmetrically dispositioned about the loop axis, are situated on the same side of the loop. Moreover, these loops may (Figure 12.11a and 12.11b) or may not (Figure 12.11c) cross each other. The former is termed a *closed loop* and the latter an

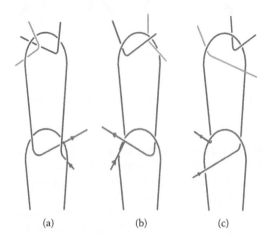

(a)   (b)   (c)

FIGURE 12.11
Views of warp-knitted loops. (a) Technical front side of a closed loop. (b) Technical back side of a closed loop. (c) Technical back side of an open loop.

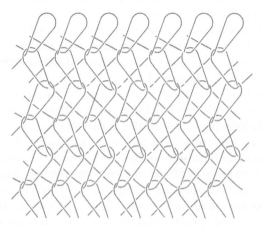

FIGURE 12.12
View of technical front side of warp-knitted loops.

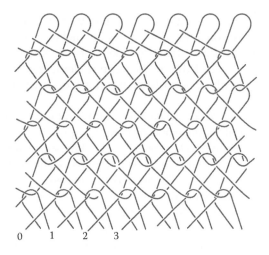

FIGURE 12.13
View of technical back side of warp-knitted loops.

open loop. One recognizes the technical back side (Figure 12.11b and 12.11c) and technical front side (Figure 12.11a) of those loops easily. The resultant fabrics from the loop element shown in Figure 12.11a is shown in Figure 12.12 whereas the one corresponding to Figure 12.11b is shown in Figure 12.13. In a warp-knitted loop, the element equivalent to the sinker loop is known as the *underlap,* and it appears always on the technical back side of the fabric.

12.4 Basic Weft Knits

12.4.1 Plain or Single Jersey

The constructions, illustrated in Figures 12.3, 12.8, and 12.9, are formed by a matrix of similar loops. Hence, on one face of the resultant fabric, all loops exhibit a technical back side while, on the other face, only technical front sides are on view. This is the simplest of all weft knits produced on machines employing only one set of needles. Hence, this knit is termed as plain or single jersey or plain knit.

12.4.1.1 Derivatives of Single Jersey

A loop is a stitch exhibiting four binding or interlacement zones: two around the needle loop and two around the base. If, however, the two zones of interlacement around its base are done away with, then a new structural element, namely the *tuck,* is formed. Furthermore, if all the four binding zones are removed, then, evidently, a straight segment of yarn, namely the *float*, would materialize. One may consider tuck and float stitches as derivatives of loop stitches. Thus, the three basic structural elements of a knitted fabric are loop, tuck, and float. It is noticed that no structural element can be formed with interlacement zones only around the base of a loop if the half loop is kept out of purview.

A great many derivatives of single jersey knit can be developed by combining the three structural elements judiciously. A hypothetical knit involving all the three elements is shown in Figure 12.14. This knit shows five wale lines and seven full course lines and, hence, a total of 35 loops (or stitches), all of which are showing the technical back side. The stitch on the extreme right-hand top corner is a loop; the ones in the center are a tuck and float, respectively, at the top and bottom, and the one at the left top corner is a half loop. The overall direction of the yarn path in the fabric is shown by the arrow whereby the arrowhead would lie, alternately, to the right and to the left of the line segment as one moves from one course to the next. One notices that the float stretches across three wale lines in the third course from the bottom and, hence, would be counted as three float stitches while the tuck is limited to one wale line in the fifth course. Thus, in the depicted construction, there are 31 loops, three floats, and one tuck, making up a total of 35 structural elements. It is noticed that the tuck and float stitches do not appear in isolation; indeed, they are always accompanied by the conventional loops, which differ in length and shape from the rest of the loops. A float in a woven fabric, however, is not accompanied by a stitch! In this sense, knitted constructions differ fundamentally from woven constructions.

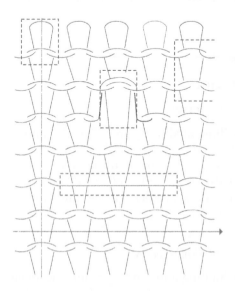

FIGURE 12.14
View of basic structural elements of knitted construction.

12.4.1.1.1 Characteristics of the Float Stitch

It is observed that a float stretches across the technical back side of a knitted fabric and, hence, would be completely hidden from the view from the front side. Thus, if one wants to deliberately hide yarns from the front side as, for example, while generating a design with yarns of many colors, one could take recourse to floating the undesirable colors at the desired locations.

The float, being a straight segment of yarn, imparts an additional rigidity to the fabric along the course direction. It would also tend to pull in adjacent wale lines close together, tending to make the fabric narrower. The float also tends to make the fabric appear thinner at the affected zones.

A float is, however, a loose yarn segment on the technical back side of a single jersey fabric and is therefore prone to snagging caused by any sharp object or abrasive surface. This would distort and damage the fabric. Hence, the actual length of a float is always restricted to very short stretches. If, however, design considerations demand a long float, then it should be tucked into the structure at suitable intervals, limiting the snagging potential to an acceptable level while, at the same time, preventing the yarn from being accessible to view on the technical front side as the tuck stitch also is visible primarily from the technical back side only.

12.4.1.1.2 Characteristics of the Tuck Stitch

A tuck stitch exhibits two free arms that spread away from the concerned wale line. The neighboring wale lines therefore would be pushed away from their normal location in the resultant fabric. This results in localized openings on two sides of a tuck stitch and in a reduction in the number of wales per unit length. Effectively, the resultant fabric becomes more porous and wider.

The needle loop of a tuck stitch is always accompanied by the needle loop of a complete loop. Indeed, if a tuck stitch is extended over n courses, then there would be $n + 1$ needle loops bunched up together. This collection of yarns at one location makes the fabric thicker.

12.4.1.1.3 Effect of Float and Tuck Stitches on Fabric Properties

Turning attention to the four extended loops accompanying the three floats and a tuck stitch in Figure 12.14, one may make a quick estimate of their length difference vis-à-vis the other loops in the construction and hazard a fair guess about the resultant difference in areal density (mass/unit area) of the fabric. Let us allot one unit to the needle loop, one unit each to the two arms, and one unit to the sinker loop, making up a total of four length units for a normal loop. Thus, a fabric made of 35 loops would have a total of 4×35, that is, 140 length units of yarn for a construction consisting only of loops. If, however, three float stitches replace three normal loops and if every float stitch is also assigned one unit, and the extended arms of the distorted loops are assigned two units each, then the total units of yarn length work out to be $4 \times 29 + 3 \times 6 + 3 \times 1 = 137$ units. Recalling that a float stitch makes the fabric narrower, which would hence occupy less area, one may infer that introduction of float stitches may not change the areal density of the fabric significantly. Let the tuck stitch be considered now, and let a total of three length units be assigned to the tuck stitch. Introduction of every tuck stitch would therefore increase the total yarn length by one unit. However, the tuck stitch makes the fabric wider, and therefore, this increase in length unit may as well not make a significant difference to the areal density.

One can summarize the effect of introduction of derivative stitches to plain jersey knit by the following statements:

- A tuck stitch makes the fabric wider, thicker, and more porous.
- A float stitch makes the fabric narrower, thinner, and more rigid in the course direction.
- The effect of tuck and float stitches on the areal density of fabric would depend critically on the actual change in fabric width although, presumably, the effect may be marginal.

12.4.1.1.4 *Popular Single Jersey Derivatives*

Combining loop and tuck stitches, one can develop knits, such as cross tuck (both single and double), Lacoste, and crepe, and cross miss, bird's eye, mock rib, and twill can be developed by combining loop and float stitches. Some twill knits can also be generated by combining all the three types of stitches (see Figures 12A.1 and 12A.3b in the Appendix).

12.4.2 Instability and Asymmetry of Plain Loop

12.4.2.1 *Curling of Fabric Edges*

Figure 12.15 shows a model of the technical front side of a single jersey loop, made by using flexible cables pinned onto a flat board (Doyle, 1952). The adjoining straight lines show the X, Y, and Z axes. Owing to the binding zones and also due to the elasticity of the material, the loop exhibits a pronounced three-dimensional configuration. Hence, the axis of yarn constituting the loop can be projected on the XY, YZ, and XZ planes. By dropping normal line segments from the yarn axis on the three planes, the corresponding contours on the respective planes have been generated. Due to its elastic properties, a looped yarn would try to regain the undeformed shape, that is, its original straight form. This causes generation of reaction couples, which tend to straighten up the bent form of projected contours. These couples have been indicated on the XZ and YZ planes. The couple on the XZ plane would tend to undo the concave curvature of crowns of adjacent loops along a course line, resulting in curling of the fabric across the wale lines (along courses) in the manner shown in Figure 12.16, amounting to a rolling up of the fabric into a tubular

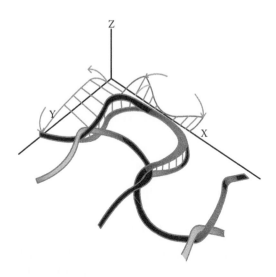

FIGURE 12.15
Model of single-jersey loop. (From Doyle, P. J., *Journal of Textile Institute*, 43, 19–35, 1952, CRC Press.)

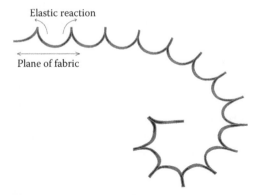

FIGURE 12.16
Curling along course line. (From Doyle, P. J., *Journal of Textile Institute*, 43, 19–35, 1952, CRC Press.)

form whereby the direction of curling is from the technical front side to the technical back side along the course lines. Similarly, the couple on the YZ plane would result in curling of the fabric along the wale lines (across courses) from the technical back to the technical front in the manner shown in Figure 12.17. Thus, a fabric sample composed of very few wale lines and a relatively large number of course lines (say six wale lines and 20 course lines) would roll up into a tube exhibiting the technical front side to the viewer. Conversely, a strip made up of a large number of wale lines and relatively few course lines would roll up into a tube exhibiting the technical back side to the viewer. However,

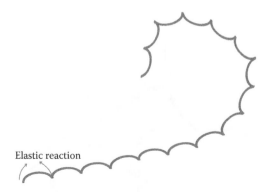

FIGURE 12.17
Curling along wale line. (From Doyle, P. J., *Journal of Textile Institute*, 43, 19–35, 1952, CRC Press.)

if a large rectangular piece of plain single jersey fabric is kept on a table free of any constraint with the technical front side exposed to the viewer, then the side edges parallel to the wale lines would tend to curl inward, and the top and bottom edges would curl outward. Obviously, this opposite tendency of neighboring edges would not permit a rectangular fabric to roll up into a ball. But this lively behavior of fabric edges would be a hindrance to further downstream processing (for example, in sewing the cut edges).

12.4.2.2 Spirality of Wale Line

The view of an ideally symmetric loop is depicted in Figure 12.7. The two halves of the loop about its axis of symmetry along the wale line are mirror images of each other. Such a situation may occur in practice when monofilaments or flat multifilaments are converted into loops. However, a loop made of spun yarn invariably becomes asymmetric as shown by the image in Figure 12.18. The

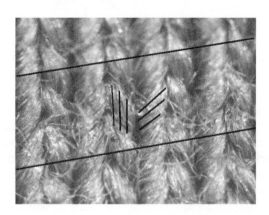

FIGURE 12.18
View of an asymmetric loop.

technical front side of four wale lines and five course lines of a plain knit are shown in this figure. The binding points around the crown of each loop are parallel to the course lines, indicated by two thick lines drawn across the figure. Apparently, the course line is not normal to the wale lines, and hence, the line joining the binding points also is not normal to the wale line. If then the course line is held parallel to the base line of the frame, the wale lines would appear inclined to the right within the frame. If the fabric is produced in a tubular form, such wale lines would appear spiraling around the surface of the tube. This phenomenon, termed the "spirality" of a wale line, is a typical problem of plain knits made of spun yarns, the reason for which is explained in Figure 12.19.

The nomenclature related to the direction of twist in a spun yarn is illustrated in Figure 12.19a. The circle with a dot in the center refers to a force normal to the plane of the paper or screen and directed toward the viewer, and the circle with a cross in the center refers to a force normal to the plane of the paper or screen but directed away from the viewer. The two opposite forces when applied at the two end points of a straight line constitute a couple. The systems of couples for generating a Z-twist and a S-twist in spun yarns are shown in Figure 12.19a.

The conversion of a Z-twisted yarn into a symmetrical half loop is shown in Figure 12.19b. Two systems of couples act on the two interlacement zones at the base of the half loop. The first system is represented by the pair of opposite forces acting on the two end points of the line segments AB and CD, respectively, accounting for the 3-D shape of loops, which finally leads to curling of the fabric edges. The other system of couples acts at the points A, B, C, and D in such a way that the line AB itself is subjected to a S-twisting couple and the line segment CD is subjected to a Z-twisting one. A blow

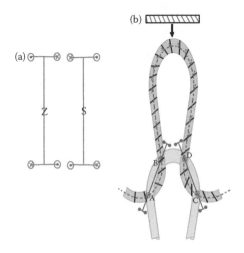

FIGURE 12.19
(a) Formation of S- and Z-twists. (b) Twisting and detwisting of loop arms.

up of this system of couples on the interlacement zones is illustrated in Figure 12.20. Evidently, these couples are responsible for effectively ply-twisting the interlacing yarn segments. As a result, the left arm of the half loop is subjected to an S-twist, and the right arm receives a Z-twist. If the yarn itself is Z-twisted, as is the case in Figure 12.19b, the left arm of the half loop would be untwisted to a certain extent and the right arm would be over-twisted. Hence, the rigidity of the left arm would be reduced, and that of the right arm increased. The first system of couples acting on the interlacement zones would, accordingly, be able to bend the two arms out of the fabric plane (along the YZ plane) to a varying extent, the more rigid one bending less and the more pliable one bending more. Similarly, the in-plane bending (along the XZ plane) of the two arms would also be different, the more rigid one bending less and the more pliable one bending more. Viewed along the XZ plane only then, the loop would exhibit a tilt to the right to accommodate the more rigid right arm. Viewed along the YZ plane, two distinctly separate convex curves would account for the projection of the two arms, and along the XZ plane, the concave projection should show a clockwise angular displacement with respect to the plane of symmetry. The loop evidently becomes asymmetric. Indeed, the 3-D configuration of the yarn of the new loop becomes quite convoluted with the plane of the loop exhibiting a rotation about the wale line while the wale line itself assumes a spiral path.

As a result of the second system of couples, then the left arm of a loop made of Z-twisted yarn would show less twist than the right arm, and the entire wale line would bend to the right of the line of symmetry. These observations are confirmed by Figure 12.18. If the wale line is held upright in such a situation, the course line would exhibit a positive slope as depicted in Figure 12.21. The corresponding angle of inclination is termed the "spirality"

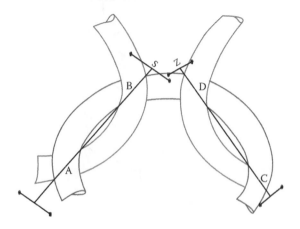

FIGURE 12.20
Couples acting in loop-interlacement zones.

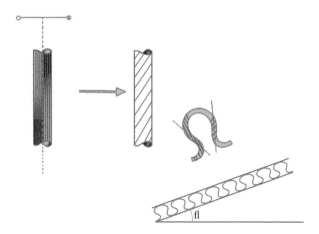

FIGURE 12.21
Spirality due to Z-twisted yarn.

angle. If, on the other hand, a plain knitted fabric is made of S-twisted yarn, a negative spirality as per Figure 12.22 would result.

The spirality of the wale line can be suppressed by a number of measures, the most elegant of which involves choosing the direction of rotation of a circular knitting machine in keeping with the direction of yarn twist. This is explained in Figures 12.23 and 12.24.

The needle bed of a circular knitting machine (Section 12.6.1) rotates either in the clockwise or in the counterclockwise direction. As a result, the tubular fabric produced by the machine also rotates in the same direction as the machine. During rotation of the machine, loops are formed sequentially on adjacent needles, and the resultant tubular fabric is pulled downward by

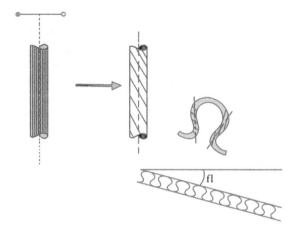

FIGURE 12.22
Spirality due to S-twisted yarn.

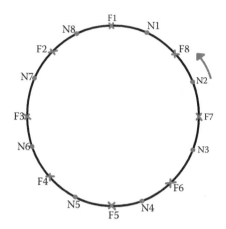

After one M/C revolution

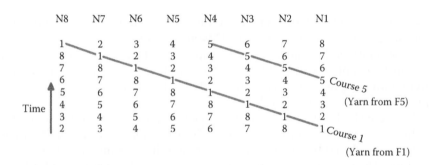

FIGURE 12.23
Mechanics of course line skewness.

the take-up system. Hence, the course line follows a helical path around the fabric tube. This is explained in some detail in the following.

The circle in Figure 12.23 represents the periphery of the cylinder of a circular knitting machine around which eight feeders—F1 to F8—and eight needles—N1 to N8—are arranged. The cylinder rotates in the counterclockwise direction, carrying the needles sequentially past the eight feeders, each of which knits loops in its wale line in the process. Hence, in one rotation of the cylinder, eight loops would be added to each of the eight wale lines held by the eight needles. The matrix under the circle outlines the sequence in which yarns from the different feeders, represented simply by the corresponding feeder number, are linked together in the individual wale lines.

As the cylinder rotates, the needle N1 approaches the feeder F1 while N2 approaches F8, N3 approaches F7, and so on. Thus, after a lapse of the first unit time, the eight needles N1 to N8 have knitted loops from feeders 1, 8, 7, ... and 2 as shown in the bottom row of the matrix. These loops

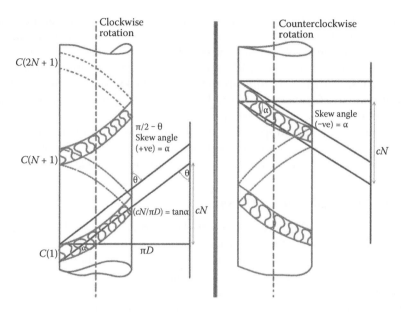

FIGURE 12.24
Direction of machine rotation–dependent course skewness.

are, however, not linked to each other as the corresponding yarns are from eight different sources. In the second unit time, the needle N1 approaches the feeder F2 while N2 approaches F1, N3 approaches F8, and so on. The resulting sequence of loops on each of the wale lines is shown in the second row from the bottom of the matrix. In this manner, the entire 8 × 8 matrix can be constructed. The path of a particular course resulting out of yarn from a particular feeder is found out by joining up the corresponding number in each of the wale lines. This has been illustrated in the matrix by joining the numbers 5 and 1. Clearly, a negative slope is formed by the course lines.

The helix angle of the course line or simply its skewness gets accentuated by the presence of a multitude of feeders on a circular machine. The larger the number of feeders, the greater would be the deviation of a course line from its ideal path, normal to the wale line. Referring to Figure 12.24, one can derive the following relationship:

$$\text{Angle of skew of course line} = \alpha = \tan^{-1}[cN/\pi D] \qquad (12.1)$$

where
D = diameter of fabric tube
c = course spacing
N = number of feeders
θ = helix angle of course line

For a clockwise rotating machine, the skew angle is positive, and a counterclockwise rotating machine generates a negative angle of skew. It follows, therefore, that the positive angle of spirality of the wale line resulting out of a Z-twisted yarn (Figure 12.18) can be suppressed by knitting such yarns on a circular machine rotating in the counterclockwise direction, and the spirality due to a S-twisted yarn can be effectively suppressed on a machine rotating in the clockwise direction.

The other options for suppressing spirality involve choosing balanced plied yarns or choosing yarn without any twist. However, such options restrict the freedom of choice of yarn and should be considered as the last resort.

12.4.2.3 Stabilization of Knitted Structure

The instability of a plain loop, manifested by curling, is caused by the elastic energy of recovery in the constituent yarns. An obvious solution is stress decay, achievable either by (1) full relaxation of natural fibers or by (2) heat setting of thermoplastic materials. Such solutions are however not applicable to materials like PAN. The other solution is to neutralize the reaction couples in the YZ and XZ planes by structural modification. Thus, if adjacent to the clockwise moment generated in a loop another loop with a counterclockwise moment can be knitted, then the two opposing moments would nullify each other's destabilizing effect on the resultant fabric. Rib and purl structures are constructed incorporating such an approach.

12.4.3 Rib Knit

The simplest rib knit is shown in Figure 12.25. It has alternate wales stitched to the front and is known as a 1 × 1 rib. The projection of the technical front side

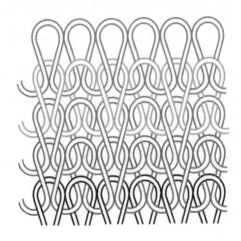

FIGURE 12.25
Loop interlacement in 1 × 1 rib.

of a course of loops on the XZ plane, in a manner similar to that of Figure 12.15, would exhibit curved lines alternately concave and convex, thus neutralizing the reaction couples of each. This would overcome the curling tendency along the course lines. Considering the curling tendency along the wale lines, it is observed that if the projection of the first wale line on the YZ plane is composed solely of convex lines then the next wale line would display only concave ones. Thus, the tendency of the first wale line to curl inward is compensated by the opposite tendency of the adjoining wale line. Hence, in this structural solution, the neutralizing of reaction couples along a course line is achieved internally whereas the wale lines neutralize the curling tendency of each other.

A noticeable difference between a plain knit and 1×1 rib knits lies in the spatial configuration of sinker loops. The sinker loop connecting a plain wale (stitched to the front) with a rib wale (stitched to the back) would have a sigmoidal configuration instead of a semicircle as is the case with plain knit when the projection on the XZ plane is taken into account. Figure 12.26 illustrates this point stage wise. Figure 12.26a depicts the projected view of a plain loop on the XZ plane. Two neighboring needle loops are connected by an exactly similar sinker loop via similar but much smaller semicircles turned upside down. Each small upturned semicircle is made up of the projection of one arm of the corresponding loop. Comparing Figure 12.26 with Figure 12.15, the reaction forces acting at the binding points can easily be recognized. The corresponding configuration of a rib sinker loop and the

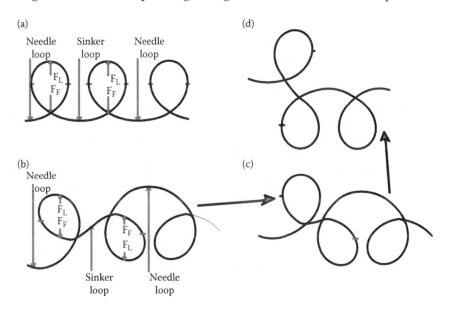

FIGURE 12.26
(a) Projection of plain loop on XZ plane. (b) Projection of machine state 1×1 rib loop on XZ plane. (c) Onset of shrinkage of relaxed 1×1 rib loop. (d) Final relaxed configuration of 1×1 rib loop.

associated reaction forces are depicted in Figure 12.26b. The couple made up of a pair of reaction forces on the base of the loops of adjoining wale lines would cause the rib and the plain wale lines to rotate around the Y-axis (Figure 12.26c) accompanied by a lateral shift toward each other necessitated by the constant length of the sinker loop. The final appearance of the adjoining plain and rib wale lines would be similar to that of Figure 12.26d. Because of this shift of wale lines along and across the plane of symmetry, a rib acquires a vertical cord-like appearance, displaying in the relaxed state only the technical front side on both faces of fabric.

A close look at the sinker loops of Figures 12.8 and 12.26 also reveals the cause behind the difference in unraveling behavior of plain and rib knit fabrics. The plain knit fabric can be unraveled from two edges, namely the courses knitted first and knitted last. The corresponding yarns sticking out of the two edges can be pulled out as the loops of each wale line would simply rotate bodily about the respective course lines without any resistance. However, in the case of the rib knit, such a process is possible only from the course knitted last. Any attempt to pull out yarn from the course knitted first would be thwarted by the interlocking sinker loops, which cannot simply rotate bodily around the course lines. The starting course of a rib knit thus provides a stable end edge for parts of garments.

The simplest rib fabric is made of 1 × 1 rib knit. Relaxed 1 × 1 rib is theoretically twice as thick and half the width of an equivalent plain fabric, but it has twice as much width-wise recoverable stretch. In practice, a 1 × 1 rib normally relaxes by approximately 30% compared to its knitting width. This characteristic, together with its stable starting course makes the rib particularly suitable for segments of garments, such as necklines of pullovers, cuffs of sleeves, rib borders for waistlines, etc. These elements need to grow considerably to a varying extent in the width direction when the garment is pulled on the body and, when in place, need to grip the body part securely.

There are a range of ribs apart from the 1 × 1 rib, the first figure in the designation indicating the number of adjacent plain wales and the second figure the number of adjacent rib wales. Single or the simple ribs, such as 2/1, 3/1, etc., have more than one adjacent plain wales but only one rib wale in between. Broad ribs have a number of adjacent ribs as well as plain wales, for example, a 6/3 derby rib. It may be mentioned here that when a larger number of similar wale lines are knitted adjacent to each other, in either plain or rib wale lines, the resultant fabric becomes thicker and broader. This results out of the curling tendency along the course line of adjacent similar wale lines, accounting for larger thickness whereby the lesser number of rib sinker loops make the fabric shrink less in the course direction. A thicker fabric would also be warmer as it would trap more air.

12.4.4 Purl Knit

Just as a rib construction involves neutralizing reaction couples along a course line, a purl construction neutralizes reaction couples along a wale line. To that end, it is necessary to have loops stitched both to the front as well as to the back in each wale line. Hence, some course lines would be knitted to the front and some to the back.

On projecting this structure onto the YZ-plane, one would encounter a similar phenomenon as when a rib construction is projected onto the XZ-plane. Hence, the fabric would shrink lengthwise due to the courses, stitched to the front and back, moving away from each other across the plane of symmetry and, at the same time, closer to each other along the fabric plane.

Considering Figure 12.27, a 1 × 1 purl construction, it is noticed that the successive pair of binding zones across needle loops along the same wale line is subjected to reaction forces of opposing sense. Hence, successive course lines would tend to rotate in opposite directions about their axis of symmetry, in the process drawing close to each other, making the resultant fabric twice as thick as plain knit in the relaxed state. This would also cause the needle and sinker loops to literally jut out of the fabric plane and be more visible on both faces than the rest of the body of the loop. One would then have literally technical back sides on both faces of the fabric. The lateral stretch of 1 × 1 purl is same as that of plain knit, but its lengthwise elasticity is almost double. It can be unroved from both ends because all the knitted courses are individually quite the same as plain knits.

Derivatives of purl knits, such as moss stitch and basket stitch, can be generated by skillfully combining, in each course, stitches to the front and to the back (see figures in the Appendix).

FIGURE 12.27
Loop interlacement in 1 × 1 purl.

12.4.5 Interlock Knit

This knit is generated by interlocking or intermeshing two 1 × 1 rib knits. In Figure 12.28, a course of 1 × 1 rib in white has been intermeshed with a similar course in black. Similarly, the 1 × 1 rib course using speckled yarn has been interlocked with a corresponding course generated from white yarn. It is observed from this figure that

- Each wale line is actually composed of two columns of loops whereby the top column is knitted to the front and the lower column is knitted to the back. Hence, the fabric would, similar to a rib fabric, exhibit only front-stitched loops on both faces of fabric. However, unlike rib fabrics, the surface won't exhibit cords and won't show the back side of the loops even when stretched laterally.
- The intermeshed rib courses cross over across corresponding sinker loops in the manner illustrated in Figure 12.29.
- Because the interlocking rib courses are produced one after the other, there is a vertical shift between the two columns of loop of the same wale line.
- Because of the interlocking, the adjacent plain and rib wale lines cannot collapse as much as in the case of 1 × 1 ribs because the corresponding space is occupied by loops of the crossing course. Hence, this fabric would not shrink as much as the parent 1 × 1 rib.

Eight lock is a 2 × 2 version of interlock. Such a knit involves the interlocking of two 2 × 2 ribs. Similarly, 3 × 3 and 4 × 4 interlocked structures can also be produced.

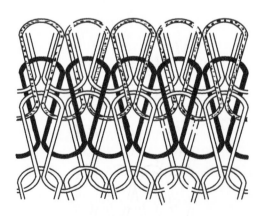

FIGURE 12.28
Loop interlacement in interlock.

FIGURE 12.29
Crossing of sinker loops in interlock.

12.5 Conventions for Representation of Weft-Knitted Stitches

It is already apparent that pictorial diagrams of looping of yarn are both difficult to develop and difficult to understand when a certain degree of complexity is incorporated in the knit. Hence, instead of actually drawing the stitches physically, conventions are made use of for indicating the kind of knit.

Both point and square papers may be employed for this purpose as illustrated in Figure 12.30. In the case of a point paper, a point represents a needle. The lines, both curved and straight ones, drawn around or adjacent to a needle, represent the yarn path and indicate the type of stitch. A square block, on the other hand, needs to be filled up with various symbols for representing the different types of stitches.

Examples of some stitches based on the four basic weft knits, namely plain, rib, purl, and interlock, employing three structural elements, namely loop, tuck, and float, are illustrated in the appendix using both conventions, namely the point paper and the square paper.

The matrix of points on a point paper represents a matrix of needles. A row of needles when supplied suitably with yarns results in a knitted course. Hence, a row of points on a point paper can, in effect, be equated to one course, and a column of points can be deemed to represent a wale line. Accordingly, the matrix of points is, in effect, a matrix of stitches. In a plain knit and its derivatives, all stitches are knitted either to the front or to the back. Hence, all such stitches are, in a way, similar in a course. In the other three variants, however, there is usually an admixture of stitches knitted to the back and to the front in each course. Hence, one needs to provide two rows of points for representing each course of such constructions, one row for representing stitches knitted to the front, and the other row represent stitches knitted to the back. Therefore, depending on the kind of construction, namely single jersey or double jersey,

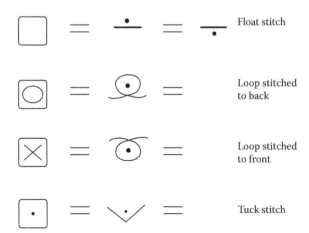

FIGURE 12.30
Convention of representation of stitches.

one chooses either one row of points or two rows of points for representation of stitches pertaining to one course of the related knit.

The arrangement of needles on the needle bed(s) (Section 12.6.1) needs to be reflected in the selection of points of each course. This refers to the set out and the gating of needles. If all the grooves on a needle bed are equipped with needles, then all the points in a row need to be selected. If, however, some needles are removed, then the corresponding points on the point paper also need to be ignored. Similarly, there are differences in the mutual arrangement of needle beds for rib, interlock, and purl knitting; for rib knitting, the two beds are so arranged that their needles are mutually offset, and those for interlock knitting are arranged face to face. In purl knitting, the same needle can occupy either of two grooves located face to face in the two beds. Hence, the grooves would have to be face to face. Evidently, understanding these and related aspects demand a grasp on the interaction between needles, their beds, and the resultant fabric constructions. This issue is discussed in the following.

12.6 Systems of the Basic Weft-Knitting Machines

12.6.1 Needles and Beds

The bearded needle was the first to be produced around 1689. It is the simplest in structure, being made from a single piece of wire, at times drawn to a thickness as low as 0.1 to 0.2 mm. The bearded needle (Figure 12.31) is made up of five parts, namely (1) shank, (2) stem, (3) hook, (4) beard, and (5) eye.

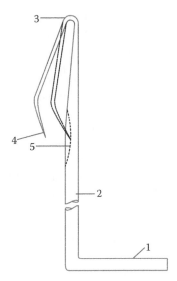

FIGURE 12.31
View of a bearded needle.

The shank is meant for mounting the needle on a rigid body. Any movement of the rigid body would therefore be exactly followed by the needle. Usually the rigid body is a bar, which is provided with matching holes on one of its faces in which shanks of bearded needles are pushed. A group of such needles mounted on a bar is finally clamped on the bar by suitable fasteners. Alternately, a group of such needles may be cast in a small block. Such blocks are, in turn, screwed on to the moving bars. Bearded needles do not thus move independently of each other. All needles mounted on a bar move together in a group.

The stem is the thickest part of the needle and supports a wale line. A cross-section of the stem is designed to counter the bending couple exerted by tension in the wale line. The yarn constituting a wale line slides up and down along the stem, which therefore needs to have the required smoothness as well as hardness to resist wear and tear.

The hook of a bearded needle encloses a finite space in which yarn of the new loop is housed. Evidently, there has to be a correspondence between the space available within a hook and the thickness of yarn that a hook can convert into a loop. During the formation of tuck stitches, more than one yarn would be within the hook. Hence, capacity of the hook can become a constraint to the number of tucks that can be formed over consecutive courses.

The beard is an extension of the hook and is the thinnest portion of the entire needle. The kink in the front of a hook is the starting point of the beard, which terminates in a sharp point at its free end. Geometry of the bearded part may be approximated to that of a cone. If a light pressure normal to the stem is applied on the kinked portion of the hook, then the tip of the beard would move in and touch the body of the stem at a specially designed location, called the eye of the stem.

The eye of the stem is a depression that permits the beard tip to sink into body of the stem, creating, in the process, a very smooth and continuous outer surface joining the hook and the stem. With tip of the beard sunk into the eye, a loop under tension can slide easily along the stem and onto the outer wall of the hook. If the tip of the beard is away from the eye, then the sliding loop, instead of riding along the outer wall of the hook, would land up within the space enclosed by the hook.

A latch needle (Figure 12.32a, b, and c) is made up of (1) the hook, (2) the latch, (3) the stem, and (4) the butt. The hook is designed to catch feed yarn from a supply package and pull it close to the wale line supported by the needle and subsequently pull it through, either from the back or from the front of the wale line. The latch is a freely fulcrummed body, which can swing up and down about a rivet mounted suitably on the needle. The rivet passes through the base of the latch, which is slotted in a groove on the needle body. The free end of a latch ends in a spoon-like object that is designed to securely enclose the outer surface of the tip of the needle hook when the latch is located at its uppermost position. Hence, a latch needle is relatively massive compared to a bearded needle. When the latch swings to its uppermost position, the space under the needle becomes closed, and when it swings to its lowermost position, the space under the needle becomes open. The

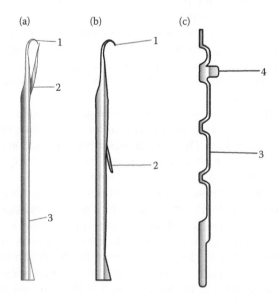

FIGURE 12.32
Views of a latch needle. (a) Latch in closed position. (b) Latch in open position. (c) A meandered stem.

closed space situation is required for the old loop to slide along the stem onto the outer wall of the needle latch and the hook for getting cast off, and the open space situation is necessary for enabling the loop to slide away from within the hook down onto the stem and beyond the grip of the latch, a process termed as clearing. It can be observed from Figure 12.32a and 12.32b that, for both the cast-off and clearing processes, the circumference of the loop has to be larger than when the loop is simply sliding along the working part of the stem. The design of the latch as well as its fulcrum, along with the design of the needle hook, are therefore crucial for enabling a knitting machine employing a latch needle to knit small loops without damaging or breaking the yarn.

The butt of a latch needle is a projection on the needle stem, which is meant to be lodged securely in the groove of a suitable cam track. The butt faces the same direction as the needle hook. The contour of a suitable cam track is followed by the butt during the operation of a knitting machine, thus imparting the corresponding motion to the needle. During continuous operation of a knitting machine, the needle butt is subjected to large impactful forces in addition to considerable frictional and thermal strains as it is forced to follow a path dictated by the profiles of the two opposite walls of cam track. The design of the cam track as well as its suitable and continuous lubrication is very important not only for quality knitting, but also for a long life of the very expensive latch needles.

The latch needle appears very robust owing to its massive nature. However, it is prone to damage of the latch and breakage of the hook and butt. For example, if, for some reason, the latch gets slightly bent, then its spoon won't sit properly around the tip of the hook and thereby would strain the loop being cast off. The hook and the butt often break due to impulsive shocks associated with the motion of a needle as well as the formation of a loop. A solution to such a problem is attempted through the meandered shape of a needle stem (Figure 12.32c). The body of such a needle has some ability to absorb shockwaves that propagate along the needle axis.

The advantage of a latch needle over a bearded needle is primarily in its self-acting nature. The operation of a latch is carried out by the sliding of a loop along the length of the needle body. The loop flips the latch closed during the cast-off motion and flips it open during the clearing motion. Thus, for creating the open and closed hook space, no additional element is required by a latch needle. On the other hand, a bearded needle needs a presser bar to press the tip of the beard into the needle eye for creating a closed hook space. This action of the presser bar also slows down the needle and therefore the machine considerably, accounting for a lower productivity of bearded needle systems. On the other hand, the throw of a latch needle, that is, its amplitude of movement in a cycle, is nearly double the length of the latch. This movement is required to ensure that a loop held within the hook can securely clear the latch and land on the stem. In a bearded needle, on the other hand, this amplitude is just slightly larger than the distance between the crown of the hook and the tip of the beard. Lower amplitude can result in a higher number of cycles and therefore higher machine productivity.

The compound needle (Figure 12.33a) does away with the problems associated with a latch. Such a needle is made up of two parts, namely the main grooved body (1) and the jack (2). Both components 1 and 2 are equipped with butts (4) and (5). Hence, the two components are driven independently by suitable grooved cams. The needle has a hook (3) and exhibits a grooved channel along its length. The jack is a thin and stiff wire, which can fit into and slide along the channel easily. The two components essentially move in opposite directions. At one extreme stage, the jack reaches its lowest position when the needle is at its highest. This results in an open space under the hook as shown in Figure 12.33a. The opposite occurs when, as shown in Figure 12.33b, a closed space under the hook is required. The total required displacement can be split equally between the hooked body and the jack. The total displacement itself being fairly low, the amplitude of the hooked body is effectively much lower than that of either the bearded or the latch needle. This system can therefore be operated at a much higher frequency, resulting in the highest productivity of the resultant knitting machine. In the absence of a latch, this needle is also less prone to damage than the latch needle while being as robust. The absence of a latch also ensures that the casting off and the clearing loops are not strained, permitting knitting of smaller loops and less elastic yarns.

Double-ended latch needles (Figure 12.34a) are, in essence, two latch needles joined end to end without their respective butts. Such a needle, shown

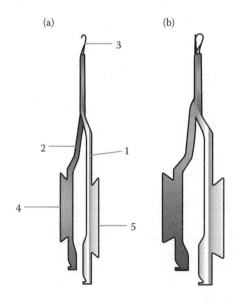

FIGURE 12.33
Views of a compound needle. (a) Throat space closed. (b) Throat space open.

(a)   (b)

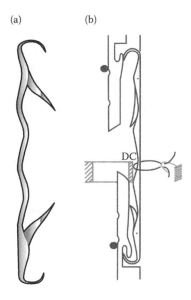

FIGURE 12.34
Views of a double-ended latch needle. (a) Needle in Free State. (b) Needle engaged with sliders on bed.

along with two sliders in Figure 12.34b, would hence be devoid of any butt but would have loop-forming elements on both ends. The sliders are equipped with butts, which follow the contour of respective groove cams. At any time, only one of the two sliders would engage with one of the two hooks of needle and impart it a to-and-fro motion while the hook at the other end of the needle remains free to catch yarn and form loops. In Figure 12.34b, the lower slider is engaged with the corresponding needle hook while the upper ones are free. A double-ended latch needle can therefore form loops at either of its two ends. The resultant wale line remains supported at the center of the kinked portion of the hook. These needles are used only on machines equipped to knit purl stitches.

Needles, whatever their type may be, need to be held in place in order that they may move along a predetermined path and maintain a fixed distance from each other. This is accomplished by either housing them in a bed or mounting them on bars.

When grooves are milled on a flat metallic plate or on the outer surface of a hollow right circular metallic cylinder at predetermined spacing corresponding to the thickness of the needle to be housed, then a bed is formed. A bed may be circular or flat. A circular bed may be a right circular hollow cylinder or a disk. In the former case, grooves are milled along the outer wall of the cylinder. In the case of a disk—commonly referred to as a dial—grooves are milled radially on the working surface. Obviously, the thickness of the wall separating two neighboring grooves would keep on diminishing as one approaches the center of the dial. Dials are hence milled along an outer annular ring only, leaving rest of the surface of a dial flat. On the other hand, if a hollow cylinder is cut along a generator and, so to say, opened out and pressed flat on a 2-D plane, then a flat bed is created. Knitting machines may thus be circular or flat, having single or double beds. Latch and compound needles can be housed in beds.

In a bar, on the other hand, no groove needs to be milled. The bearded needle is mounted on a bar by inserting its shank into a suitable hole, drilled on the surface of the bar. There are as many holes on the bar as the number of needles to be mounted on it. After mounting the required number of needles on the bar, small caps are screwed on side by side, thus rigidly clamping the needles on the bar. In the event of needle damage, the respective cap can be unscrewed, and the damaged needle can be taken out to be replaced by a new needle. Alternately, the needles can be cast in a block of lead in advance and the block then fastened on to the bar.

12.6.2 Sinkers

Sinkers are thin metallic strips made of pressed steel sheets in various shapes and sizes. A typical sinker employed on a circular weft-knitting machine is shown in Figure 12.35. Such a sinker is made up of a butt (1), nib (2), throat (3), and belly (4). Sinkers on weft-knitting machines are held upright in an operational state by being made to slide along grooves milled on the surface of an annular ring while an extension of this very ring acts as a wedge that fits the lower cavity of sinkers, preventing their sideways movement. Such a ring is mounted on top of a cylinder. Thus, the outer surface of a knitting cylinder houses knitting needles, the hooks and butts of which radiate outward while the top of the cylinder carries a ring supporting upright strips of sinkers, which operate through gaps between needles. The butt of a sinker is lodged in the groove of a cam, and a relative motion between the two results in a to-and-fro motion of the sinker. Thus, each needle is neighbored by a sinker, and they move at right angles to each other.

The nib, the throat, and the belly of a sinker exercise specific control on the sinker loop during the formation of a new loop as well as during the withdrawal of a knitted fabric from the zone of fabric formation.

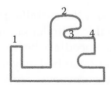

FIGURE 12.35
View of a sinker.

12.6.3 Knitting Cams

Knitting needles need to be moved in order that they may knit. This movement can be achieved through a group drive or through an individual drive. If needles are mounted on a bar and the bar is moved, then all needles move together. This is the principle of a group drive. On the other hand, if the needle butt is made to follow the contour of a cam, then a needle would also move independent of its neighbors in a manner dependent upon the profile of the cam and inertial forces. Warp knitting is invariably carried out through a group drive. On the other hand, an individual drive is extensively employed in weft knitting.

A typical cam, meant for producing interlock construction, is depicted in Figure 12.36a and 12.36b. Cam tracks have been traditionally generated by fixing small elements on the inside wall of a jacket in such a manner that along the length of the jacket a groove is created. These elements yield linear cam tracks, which cause very high shocks to needle butts during reversal of motion. Nonlinear cam tracks, which are manufactured as complete blocks on NC-machines, are employed in modern knitting machines. However, for explaining the basic concept of a knitting cam, a linear cam system has been chosen here.

The jackets shown in Figure 12.36 are made to bear upon needles whose butts are secured inside their grooves. The projected view of such jackets may be rectangular in nature or may appear as a segment of ring. Considering the rectangular view in Figure 12.36a, one can make out short vertical line segments representing butts of needles within two such grooves. This jacket has two cam tracks. Moreover, these two tracks, similar in all respects, exhibit a lateral shift with respect to each other. Each track is made up of an upthrow cam (12, 13), a swing cam (7, 8), a stitch cam (1, 2), and guard cams (14, 15, 16, 17, 18, 19, 20). By means of screws 3 and 4, the stitch cams may be moved up or down. The swing cams may also be made either to lie flush along the surface of the upthrow cams (12, 13) or remain in their raised position as shown in the figure. The corresponding needles, which can ply in these two tracks are shown on the right-hand side of this diagram. The distance between the hook and butt of one needle is shorter than that of the other. The former is known as a short (butt) needle and the latter as a long (butt) needle. The short needles engage in the cam track closer to the upper edge of the jacket, and the long needles engage with the farthest track. It is evident that the short needles and the long needles receive to-and-fro motion from their respective cam tracks and would knit at different time instants.

The other cam track shown in Figure 12.36b on a jacket of the shape of the ring has upthrow cams (1, 2), swing cams (3, 4), stitch cams (5, 6), and guard cams (7–14). Short needles ply in the lower track while the long needles are guided by the upper track. The needles, in addition to being of different lengths, may exhibit a difference in butt heights as well. Such

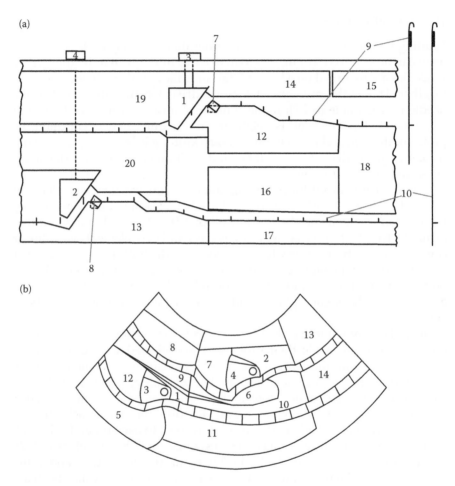

FIGURE 12.36
Views of (a) cylinder and (b) dial cams for interlock knit.

needles are termed as low butt needles or high butt needles as the case may be. A high butt needle would be more fully inside the cam track groove as compared to a low butt needle. Hence, if some cam elements (e.g., the swing cam) are thinner than other elements, then the depth of the track at the corresponding location may not be sufficient to guide the low butt needles whereas the high butt ones would be acted on by the cam track. This results in selection of certain needles to knit differently from other needles for the purpose of producing knits that are derivatives of the basic ones.

12.7 Sequence of Loop Formation

12.7.1 Single Bed

12.7.1.1 Loop Formation on Single Flat Bed

The sequence is depicted in Figure 12.37. At the outset (Figure 12.37a), the latch needle (1), housed in a bed with verge (3) is supporting a wale line (2). The motion given to its butt makes it move from left to right. In the process, the loop resting within the needle throat and prevented from moving with the needle due to the take-down force acting downward (shown by the arrow), slides along the inner wall of the needle and flips the latch open (Figure 12.37b). The loop rides along the inner wall of the latch and slips off onto the needle stem (Figure 12.37c). This process is also termed as clearing. During its further outward movement, the needle comes across the feeder. Feeders on weft-knitting machines guide the yarn along course lines and hence must move from needle to needle; alternately, the needles must move past a feeder one by one. The yarn fed to the needle is gripped as the needle starts its return journey (Figure 12.37d). The loop slides below the outer wall of the latch. During the further return journey of the needle, the new yarn is pulled closer to the loop, which now slides along the outer wall of the latch after having flipped it closed (Figure 12.37e). The needle in Figure 12.37f has reached

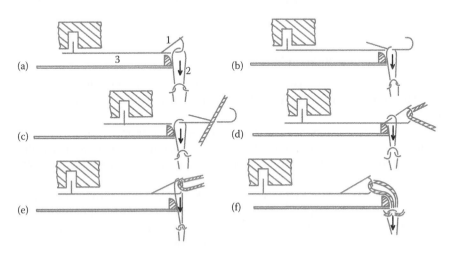

FIGURE 12.37
Sequence of loop formation on single flat-bed knitting machine. (a) Needle in starting position. (b) Needle in forward motion. (c) Needle in clearing position. (d) Feed yarn caught by needle. (e) Needle in casting off position. (f) Needle with a new loop.

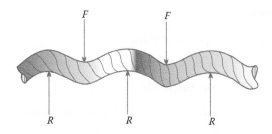

FIGURE 12.38
Loop-making forces on a yarn segment.

its rearmost position, in the process pulling the new yarn through the old loops and casting off the latter. As the needle starts afresh its forward journey, the take-down mechanism assists in rotating the new loop by $\pi/2$ radians, adding in the process a length equal to course spacing to the fabric.

Figure 12.37e and f shows that the new yarn has been bent into the loop shape. However, for bending a yarn, it is necessary to provide two supports or restrictions while applying the bending force (F) along a line intermediate between these restrictions (Figure 12.38). In the present case, the verges (3) of the needle bed provide restrictions to the yarn, in the process supplying the force R, and a needle thus pulls the yarn within the gap between neighboring verges by supplying force F for producing a loop.

12.7.1.2 Loop Formation on Single Circular Bed

Such machines may be classified into two kinds, namely verge top and sinker top. The sequence of loop formation on the sinker top machine has been illustrated in Figure 12.39.

The needle (1), which is housed in tricks of cylinder, moves in the vertical plane. Every needle is neighbored by a sinker (2), which moves along the horizontal plane. The sinker loop is gripped by the sinker throat (Figure 12.39a). Hence, when the needle moves up, the tendency of the fabric to follow the needle is countered by the grip on the sinker loop by the sinker throat. As opposed to a flat-bed machine, only a fraction of the fabric take-down force acts on the loop held by the needle. Moreover, the speed of the needle in such machines is generally much higher than that of flat-bed machines. Hence, it becomes necessary to grip the edge of the fabric securely so as to prevent it from following the needle.

The needle in Figure 12.39b has reached its uppermost point and encountered the feeder. The sinker is observed to be retracting while the needle moves down. This backward movement of the sinker continues, and by the

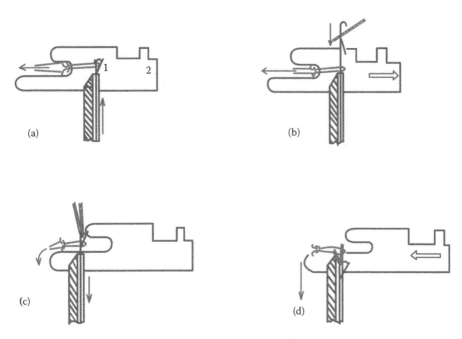

FIGURE 12.39
Sequence of loop formation on sinker top circular knitting machine. (a) Needle and sinker in
starting position. (b) Needle about to catch feed yarn. (c) Needle nearing casting off position.
(d) Needle with a new loop.

time the needle hook is almost at the level of the sinkers, the sinker belly
is exposed to receive the yarn being pulled in (Figure 12.39c). Hence, the
restriction to further downward movement of yarn is provided by the sinker
belly. If the sinker would not have moved back, then the restriction would
have been provided by the nib or shoulder of the sinker. In such an event,
the sinker loop would not have come under the grip of the sinker throat in
the succeeding cycle. In Figure 12.39d, the needle has reached its lowermost
position below the sinker line, and the sinker is in the process of moving to
the left so as to grip the sinker loop. The fabric take-down force helps in pulling the course knitted downward.

The interface between the fabric and the needles on single bed machines
is illustrated in Figure 12.40. If the fabric comes out in tubular form, then the
technical front side is presented to the viewer. On the other hand, the technical back side is offered to the observer when knitting takes place on a single
flat bed. Hence, from the viewpoint of the observer, a single cylinder knits
loops stitched to the front, and a single flat bed knits loops stitched to the back.

The action of the stitch cam on the butts of needles in pulling new yarn
below the verge line in a phased manner is outlined in Figure 12.41. It is
noted that, at any instant, a finitely small number of needles pulls the feed

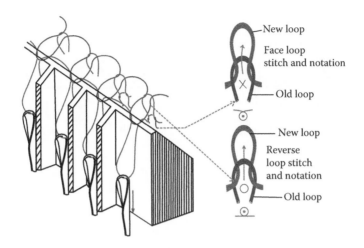

FIGURE 12.40
Interface between fabric and needles on single-bed knitting machine.

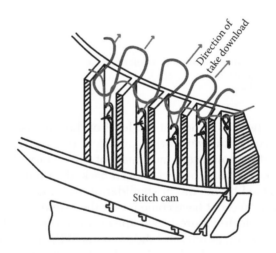

FIGURE 12.41
Action of stitch cam on needle butts.

yarn below the verge line and that too in a phased manner. In the figure, only two needles on the right-hand side are pulling the yarn down whereas the others are rising up. Hence, applying Amonton's law and knowing the yarn input tension as well as the wrap of the feed yarns around the needles, verges, and cast-off loops, it is possible to calculate the maximum tension in the yarn while being formed into a loop and ensure that the same remains within acceptable limits.

12.7.2 Double Bed

12.7.2.1 Rib Gating

A double-bed machine, illustrated in Figure 12.42, may be flat or circular. If flat, the beds are arranged at 45° to the horizon such that the vertical sectional view of the pair of beds shows an inverted V. Such beds are termed V-beds. A dial and cylinder machine or a cylinder–cylinder machine constitutes the circular double-bed machines on which rib and other double-bed gating are possible. In flat as well as in dial and cylinder machines, the two beds are arranged normal to each other so that when the corresponding needles of the two beds move out, they cut across a common line of displacement and create a suitable lap in which the new yarn to be knitted is laid by the feeder. Great care needs to be taken while feeding a new yarn to needles on a double-cylinder machine.

As explained in Section 12.4.2 (Figure 12.25), rib fabrics are made up of wale lines, some of which must be knitted on a bed in which needles stitch loops to the front (cylinders or front bed) and others on another bed (dial or back bed or another inverted cylinder). The neutral plane across which needles of the two beds move (either from the front to the back of the plane or from the back to the front of the plane) is formed by the tensioned edge of the fabric between the two beds.

The term gating refers to the arrangement of needles on the beds. Figure 12.42 shows a 1 × 1 rib gating in which each cylinder needle is neighbored by a dial needle, and they are offset with respect to each other in such

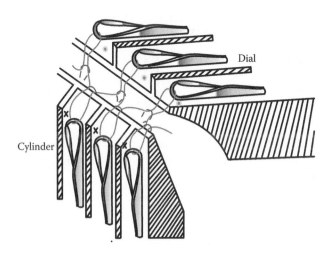

FIGURE 12.42
Double-bed rib-gated knitting machine.

a way that when both sets move forward to catch yarn, each cylinder needle passes through the gap between two neighboring dial needles.

The corresponding sequence of loop formation is illustrated in Figure 12.43. It is noticed that although the dial needle moves out first (Figure 12.43b), the cylinder needle generates a loop earlier (Figure 12.43f). It is, of course, possible for both dial and cylinder needles to form loops simultaneously. This feature of simultaneous or phased loop formation is governed by timing. In general, the dial forms a loop later than the cylinder (delayed timing). Synchronized timing makes the system highly sensitive to fluctuations in input parameters, and the yarn is subjected to very large strain.

It is observed that there is no sinker in the loop-formation system, but each bed displays prominently its respective verges. These verges are used for casting off the old loop. The tendency of individual loops to move up with their needles gets neutralized by their opposing senses. Thus, for example, when the cylinder loop tends to go up in the vertical direction, the dial loops would tend to move forward horizontally. In the process, the fabric as a whole would try to move along the resultant direction, which is at an angle to each bed. This direction would be matching that of the fabric edge that happens to be under take-down tension. Hence, the fabric won't move when the needles move out to their yarn-catching positions.

The yarn is fed above the latch and within the hook of each needle. Adjustment of the feeder with respect to the needle is critical. If the dial needle were to catch the yarn earlier than the cylinder needle, then the feeding plate has to bring the yarn onto the dial hook while ensuring that the

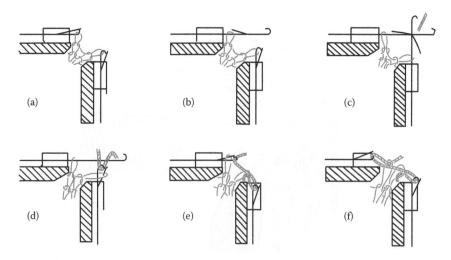

FIGURE 12.43
Sequence of rib-loop formation. (a) Needles at starting position. (b) Dial needle clears. (c) Cylinder needle clears. (d) Feed yarn caught by cylinder needle. (e) Cylider needle casts off and dial needle starts to form loop. (f) Dial needle casts off.

yarn does not slip below some of the following needles. This is obviously a more difficult proposition.

12.7.2.2 Interlock Gating

Referring to Figure 12.28 as well as Section 12.7.2.1, it is obvious that in order to have two columns of loop in the same wale line, whereby one column is knitted to the front and the other is knitted to the back, it is necessary to have dial and cylinder needles along the same displacement line. However, in order that they do not collide, one uses long and short (butt) needles (Figure 12.44) in conjunction with a cam setup outlined in Figure 12.36. However, Figures 12.36 and 12.44 are not compatible because the cam system of Figure 12.36 would make long needles of one bed and short needles of the other work together whereas the individual 1 × 1 ribs shown in Figure 12.44 are products of the simultaneous working together of short needles of both dial and needle followed by the working together of long needles of both beds. Thus, either the needle arrangement of Figure 12.44 needs to be changed or the cam system modified. The arrangement of needles in Figure 12.44 is known as interlock gating characterized by needles of two beds occupying the same displacement line. With an interlock gating, two interlocking ribs are knitted with a phase difference by two neighboring feeders. By making alternate feeders idle, it is possible to knit a 1 × 1 rib on an interlock machine. Similarly an eight lock gating made up of pairs of long and short needles in each bed can be made to knit a 2 × 2 rib.

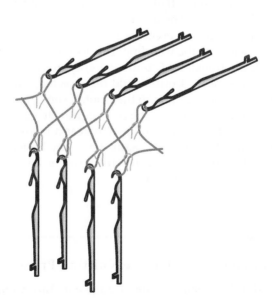

FIGURE 12.44
Interlock gating.

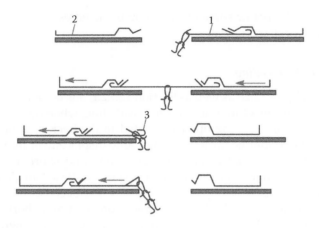

FIGURE 12.45
Loop formation with purl gating.

12.7.2.3 Purl Gating

Double-ended latch needles are employed in purl gating, and tricks of two beds are aligned in such a manner that a needle can easily be transferred from one bed to the other. In Figure 12.45, the movement of the needle (1) is controlled by the sliders (2). If the needle keeps on knitting on the bed on the right-hand side only, then loops would be stitched to the front, a direction opposite to when the needle knits loops stitched to the back on the bed on the left-hand side. As shown in the figure, the double-headed latch needle is first transferred to the left-hand bed by the corresponding slider. Subsequently, the needle remains on this bed and executes a complete loop-forming cycle, adding, in the process, a loop to the wale line. This loop is stitched to the back.

If all the needles are transferred simultaneously from one bed to the other after every course, then a 1 × 1 purl would result as loops would be stitched in alternate courses to the back and to the front. If some needles are allowed to remain in one bed and the rest of the needles transferred to the other bed, then a rib (or its derivative) would follow whereas a plain knit is generated if all needles are made to knit on the same bed. Any hosiery article having a combination of rib, plain, and purl structures can be knitted effectively on double-bed machines with purl gating.

12.8 Guidelines on Control of the Knitting Process

12.8.1 Relationship between Machine Gauge and Yarn Count

Each knitting machine is characterized by its "gauge," which represents the number of needles in the unit length of the machine bed. Usually the unit of length is 1 inch. However, for straight bar weft-knitting machines, the unit of length is 1.5 inches. Similarly, for Raschel machines, the unit of length is 2 inches. The term "cut" is also used at times to indicate the number of needles in a circumferential length of cylinder. Empirical formulae, a typical set of which is listed below, can be used as a guideline for choosing a yarn count suitable for a weft-knitting machine of a given type and gauge. Evidently, depending on the conditions at hand, one can choose a yarn count greater or smaller than that given by these formulae. If yarn count is expressed in English cotton count N_e, then the following relationships between count and gauge would hold.

Circular bed: For single jersey, $N_e = (\text{Gauge}^2)/20$

For rib, $N_e = (\text{Gauge}^2)/6$

For interlock, $N_e = (\text{Gauge}^2)/9.6$

Flat bed: For single bed, $N_e = (\text{Gauge}^2)/15$

For rib double bed, $N_e = (\text{Gauge}^2)/12.5$

It would appear from the above formulae that flat-bed machines need finer yarns compared to circular-bed machines. However, the difference is really due to the different nature of yarns employed. Flat-bed machines are normally produced in coarse gauges to convert coarse and bulky yarns into products, such as a pullover. Such yarns of the same English count would show higher thickness values. It is therefore more advantageous to express the machine gauge–yarn count relationship as a function of the raw material in question.

Restricting attention to circular knitting machines, it is observed that, for the same machine gauge, the finest yarn would be required to knit a rib, and the coarsest yarn is employed for the single jersey. This is due to the difference in availability of space between neighboring needles through which the knitting yarn must be manipulated into corresponding loops. A single-jersey machine has only one single row of needles, and a double-jersey machine has twice that number. Hence, a much thinner yarn and a smaller needle can effectively operate in the reduced space of a double-jersey machine of the same gauge as that of a single-jersey machine. Moreover, recalling the interlock gating, one observes that only alternate needles operate from any one bed at a time; hence, the effective available space in an interlock machine of the same gauge is higher than that of a rib machine, permitting knitting thicker yarns on the interlock machine as compared to a rib machine of the same gauge.

12.8.2 Control of Loop Length

Although a loop stitch is not the only structural element of a knitted fabric, control of its length is vital for controlling fabric quality. The loop length is decided once for all on the machine through the interaction of factors such as yarn input tension, stitch cam setting, machine gauge, and coefficient of friction of yarn with yarn and with metal as well as cam contour and take-down tension on the fabric. However, ignoring the other factors and restricting attention to the most basic ones, namely, machine gauge and stitch cam setting, it is possible to make a theoretical estimation of loop length with the help of Figure 12.46. In this figure, the two hatched rectangles represent a section of two neighboring sinkers. The needle hook is situated at a depth of h below the line formed by joining the top of all sinkers (sinker line) when the corresponding needle butt is at the lowermost position in the cam track formed by the tip of the stitch cam. This position of needle hook is called the knitting point. At the knitting point, the needle hook holds the longest length of yarn below the sinker line. Considering the geometry, it is observed that an equilateral triangle with a base nearly equal to the pitch of the machine (a = inverse of gauge) and sides each equal to half of the theoretical loop length (ℓ_t) is held at knitting point.

Thus,

$$(\ell_t/2)^2 = (a/2)^2 + h^2$$

or

$$\ell_t = (a^2 + 4h^2)^{0.5} \tag{12.2}$$

The control of loop length on a given machine thus primarily boils down to control of the value of cam settings. On a multi-feeder machine, it is important to have the same cam setting on all feeders. Similarly, reproducibility

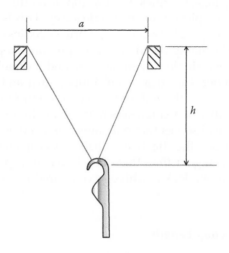

FIGURE 12.46
Theoretical estimate of loop length.

of a fabric depends primarily on the ability to reproduce a cam setting. It must however be mentioned that the actual loop length in a knitted fabric is usually lower than the theoretical loop length owing to a complex system of forward and backward flow of yarn within the knitting zone.

Next to the cam setting, which can be termed as a machine variable, tension in yarn flowing into the knitting zone plays a decisive role in deciding the loop length. The magnitude of yarn tension is, in turn, affected by the settings of the yarn tensioner, the path that the yarn has to take from the supply package to the knitting zone and the coefficient of friction of yarn with elements in its path. On a multi-feeder machine, it is absolutely essential to be able to duplicate all these variables without any variation on all the knitting systems, a task that is almost impossible. Hence, there is a very high probability of a certain variation in loop length at different locations on a weft-knitted fabric. This variation would be lower in fabrics produced on modern machines employing good-quality yarn. On most modern machines, the critical settings can be monitored and set more precisely, and good-quality yarns guarantee greater uniformity of properties.

12.8.3 Productivity of Knitting Machines

Productivity of circular weft-knitting machines can be increased by increasing the number of feeders per inch diameter.

If

$$n = \text{no. of feeders on a machine}$$

$$C = \text{courses/inch of fabric being produced}$$

$$r = \text{machine rpm}$$

and

$$E = \text{percentage machine efficiency}$$

Then the rate of production in m/h can be expressed as

$$Q = \frac{1.524Ern}{100C} \tag{12.3}$$

Moreover, if

$$d = \text{the machine diameter in inches}$$

$$g = \text{the machine gauge}$$

$$t = \text{yarn count in Tex system}$$

and

$$\ell = \text{loop length in millimeters,}$$

Then the rate of production in kg/h can be expressed as

$$P = 0.6\pi E dg n r \ell t \times 10^{-9} \tag{12.4}$$

Evidently, an increase in machine rpm and the number of feeders can lead to a higher rate of production provided the efficiency remains same. But a higher number of feeders would mean handling a higher number of yarns, leading to a higher breakage rate. This leads to a drop in efficiency. Similarly, higher r would mean greater yarn tension (variations) and higher wear and tear on the machine, leading to a possible lowering of E. A typical example may be cited for underlining the critical interaction between n, r, E, and Q in which Q for the combination $n = 48$, $r = 36$, and $E = 50$ is higher than Q for $n = 48$, $r = 18$, and $E = 76$, which, in turn, is again higher than $n = 96$, $r = 18$, and $E = 47$. It is thus necessary to find out the optimum combination of n, r, and E for each quality before a long production run is undertaken. A larger number of feeders also gives rise to what is known as drop or skewness. This refers to the deviation of courses from the horizontal line if the wales are held vertical. This inclination measured in radians would be given by

$$\alpha = \tan^{-1}(nc/\pi d) \tag{12.5}$$

where
  d = machine diameter
  n = number of feeders and
  c = course spacing

Hence, for coarse fabrics knitted from thicker yarns with large course spacing, skewness is more marked. The effect of skewness is noticeable when yarns of more than one color are employed. By reducing the number of feeders or/and increasing the course density, skewness can be reduced.

For increasing the number of feeders while keeping the machine diameter constant, one has to redesign the cam track such that every knitting cycle takes less circumferential space of the cylinder or dial. This can be done if either the cam angles are made higher or the throw of the cam is shortened. The former gives rise to large needle–cam forces and jerks. The latter is achieved by, for example, employing a compound needle. It is also possible to increase the number of feeders by reducing the amount of idle run of needles between two successive feeding stations.

Another method of increasing production on a machine of the same gauge and type is to employ the coarsest possible yarn. A coarser yarn is generally knitted to a larger loop length, causing thereby a drop in the number of courses per inch. An increase in yarn Tex and loop length leads to a higher value of P, and a drop in course density raises the value of Q. Hence, employing a coarser yarn leads to a rise in both volume and mass of production per unit time. On the other hand, machines of finer gauge invariably lead to a drop in production as the count of yarn in a Tex system has to be lower for such machines. Machines of finer gauge are invariably more expensive and employ a larger number of knitting elements, leading to higher running costs. It is thus inferred that finer-gauge machines, which are more expensive than equivalent coarser gauge ones and which need finer and therefore more expensive yarns, should be used for producing fabrics of superior quality that would fetch higher returns.

The aforementioned technical aspects have a bearing on the economics of knitting, partly explaining the reasons behind a preponderance of coarse-gauge and single-jersey machines in the industry.

As against circular weft-knitting machines, the flat-bed weft-knitting machines, generally employing to-and fro action of feeders, cannot be run at very high speeds. Various kinds of patterning possibilities is the main strong point of flat-bed knitting (such as racking of beds, unrestricted positioning of yarn carriers, possibility of programmed variation of loop length even within the same course, multiple number of beds, etc.).

12.9 Relationship between Geometry and Properties of a Loop

12.9.1 Importance of Loop Length and Loop Shape

The property of a fabric is a function of the properties of the raw material employed and of the structure used in making the fabric. The property of a structure can be related to the geometry of spatial configuration of its elements.

Geometry is defined as properties and relationships of magnitudes (as lines, surfaces, solids) in space. Deriving relationships between critical geometrical parameters of a structure helps in characterizing geometry. Owing to the complexity of the knitted structures, only the simplest ones have been subjected to similar treatment.

Loop length and shape of loop are two important geometrical parameters. A loop may be thought of as a diamond having sides of length unity (say, 1 inch). Thus, it has a total length of four units and contributes 1.41 units each to the length and width of fabric. If the length of loop is increased to eight units (i.e., doubled), then it would contribute to two units each to the length and width of the fabric. In the first case, it would occupy one square unit of space in the fabric, and in the second instance, it would occupy four times as much. Hence, the loop length is directly related to length, width, and area of the resultant fabric.

In place of the diamond-shaped loop, one may construct hinged lattices and mount a loop on such a lattice. The lattice may be easily deformed both lengthwise and widthwise, implying that even while loop length remains unchanged, dimensions of the resultant fabric can be different. Hence, the shape of the loop along with its length determines the fabric dimensions. Loop length also influences the mechanical property of fabric. On application of the same load to loops of the same length but made of yarns of different thicknesses, one observes that the loop made of the thinnest yarns deforms to the largest extent, and that made of the thickest yarn deforms the least. Consequently, if the loop length is reduced proportionate to yarn diameter, all loops would deform by the same extent. One can then infer the following:

a. If two fabrics are knitted to the same loop length from different yarn counts, that containing the finer count can be more easily distorted.

b. For the same count of yarn, fabric made up of a larger loop length will be more susceptible to distortion than one containing a shorter loop length.

12.9.2 Geometry of Weft-Knitted Loop

J. Chamberlain (1949) attempted to relate the length of a plain loop (ℓ) to its dimension. He assumed that a loop would be made up of circular arcs and straight lines. Moreover, the loops of adjacent wales and courses would be in contact whereby the position of maximum width of one loop would coincide with the position of minimum width of the interlacing loop. This would, in effect, indicate a jammed structure. Accordingly, if d is the yarn diameter, then one can derive the following geometrical relations from Figure 12.47.

$$\text{Wale spacing } w = 4d$$
$$\text{Course spacing } c = BD = (ED^2 - EB^2)^{0.5}$$
$$= 3.4643\,d.$$

This leads to the expression of shape factor and of the loop length

$$\text{Shape factor} = w/c = 1.17$$
$$\text{Loop length} = \ell = 2PQ + 2PS \qquad (12.6)$$
$$= 16.6d$$

Thus, for a geometrically jammed loop, its length is uniquely decided by the diameter of yarn employed. It also has a unique shape. As the loop

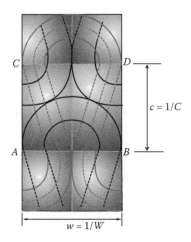

FIGURE 12.47
Geometrical model of a jammed loop.

length and loop shape are the determining parameters of a knitted fabric, Chamberlain's model appears to suggest that the properties of a knitted fabric are uniquely determined by the diameter of yarn employed to knit the fabric. This is, of course, very far from reality.

Peirce (1947) extended this model by introducing curvature of the loop across the plane of fabric and removing the condition of jamming. His equation reads

$$\text{loop length } \ell = w + 2c + 5.94d \tag{12.7}$$

These pioneering efforts of both Peirce and Chamberlain were based on an assumed configuration of the loop, underlining the postulate that the role of yarn in governing fabric dimension is determined solely by its dimensional properties. Subsequent work by various research workers has proved conclusively that yarn elastic properties play a significant role in determining the relationship between the geometric parameters of loop and its dimension.

A contribution of great significance was made by Munden (1959), who observed that a fabric in a zero strain state has loops of unique configuration such that constants of proportionality relating dimensions with loop length at a minimum energy level are also unique. These constants are given as K_C, K_W, and K_S whereby

$$\text{course spacing } c = \ell/K_C \tag{12.8}$$

$$\text{wale spacing } w = \ell/K_W \tag{12.9}$$

$$\text{area occupied by one loop} = cw = \ell^2/K_C K_W = \ell^2/K_S$$

$$\text{stitch density} = K_S/\ell^2 \tag{12.10}$$

$$\text{shape factor} = w/c = K_C/K_W \tag{12.11}$$

Extensive work has been carried out by various workers to establish the validity of these expressions as well as magnitudes of these K values, and their results suggest that they vary in magnitude considerably.

Notwithstanding the inherent simplicity of the set of formulae, proposed by Munden (1959), practical data, in a broad sense, exhibit a degree of agreement. Thus, for example, a nearly hyperbolic relationship is obtained between stitch density and loop length over a very wide range of commercial fabrics (Figure 12.48). Hence, these simple formulae may be conveniently put to use for deriving the relationship between other important variables of practical significance.

12.9.3 Some Useful Expressions

12.9.3.1 Fabric Areal Density

Where

g = weight in gram of 1 meter of yarn

S = stitch density (number of loops per square inch)

ℓ = loop length in inches

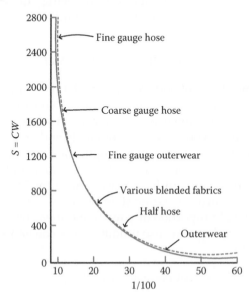

FIGURE 12.48

Functional relationship between stitch density and loop length. (From Doyle, P. J., *Journal of Textile Institute*, 44, 561–578, 1953, CRC Press.)

The fabric areal density in terms of grams per square meter (GSM) can be expressed as

$$GSM = [\ell/(39.37)][S(39.37)^2]g$$
$$= S\ell g(39.37)$$

Substituting the expression of stitch density S in terms of the relevant dimensional constant K_S, the expression reduces to

$$GSM = K_S \ell g(39.37)/\ell^2 = (39.37)K_S g/\ell \qquad (12.12)$$

Hence, the smaller the loop length, the higher would be the fabric areal density.

12.9.3.2 Fabric Width

Fabric width can be expressed as the product of the number of wale lines in a fabric and wale spacing. If n_W is the number of wale lines, then

$$\text{fabric width} = n_W w \text{ inches}$$
$$= n_W \ell/K_W \text{ inches} \qquad (12.13)$$

As both K_W and n_W can be treated as constants, it follows that larger loop length yields wider fabric.

12.9.3.3 Fabric Length

Length of a knitted fabric is the product of the number of courses in the fabric (n_C) and the course spacing (c). Hence,

$$\text{fabric length in inches} = n_C c \qquad (12.14)$$
$$= n_C \ell/K_C$$

As both K_C and n_C can be treated as constant, it follows that larger loop length yields longer fabric.

12.9.3.4 Tightness Factor of Fabric (TF)

The fractional area covered by a knitted loop can be expressed as the ratio of area covered by yarn constituting the loop and the total area actually occupied by the loop. Thus,

$$\text{fractional area covered by a loop} = [\ell d]/(\ell^2/K_S)$$
$$= dks/\ell$$

Now, switching over to the SI system and expressing the units of length in cm, one can state that the yarn diameter d in cm is equal to

$$d = \text{constant}(\text{tex}^{0.5})$$

Hence, the fractional area covered by a knitted loop is equal to

$$[\text{Constant } K_s]\text{tex}^{0.5}/\ell.$$

The value of the fractional area is less than unity. To obtain an equivalent value larger than unity, the expression of the fractional area is divided by the quantity within the third bracket, yielding a modified variable termed as the tightness factor.

$$\text{TF} = \text{tex}^{0.5}/\ell \tag{12.15}$$

In this expression, the unit of the loop length ℓ is in cm.

References

Chamberlain J (1949). The geometry of a knitted fabric in relaxed condition, *Hosiery Yarns and Fabrics*, 2, 106–108.

Doyle P J (1952). Some fundamental properties of hosiery yarns and their relation to the mechanical characteristics of knitted fabrics, *Journal of Textile Institute*, 43, 19–35.

Doyle P J (1953). Fundamental aspects of the design of knitted fabrics, *Journal of Textile Institute*, 44, 561–578.

Munden D L (1959). The geometry and dimensional properties of plain knitted fabrics, *Journal of Textile Institute*, 50, T448–T471.

Peirce F T (1947). Geometrical principles applicable to the design of functional fabrics, *Textile Research Journal*, 17, 123–147.

Further Readings

Iyer C, Mammal B, and Schach W (1995). *Circular Knitting*, Meisenbach GmbH, Bamberg.

Raz S (1993). *Flat Knitting Technology*, Universal Maschinenfabrik, Westhausen, Germany.

Spencer D (2005). *Knitting Technology*, Woodhead Publishing Limited, Cambridge, UK.

Appendix

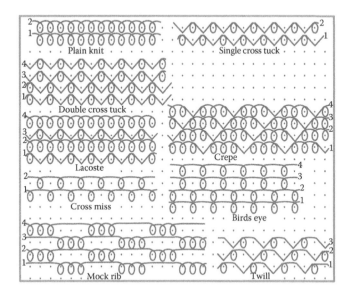

FIGURE 12A.1
Dot matrix representation of some single-jersey derivatives.

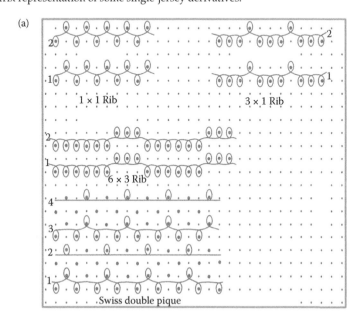

FIGURE 12A.2
(a) Dot matrix representation of some rib derivatives.

(b)

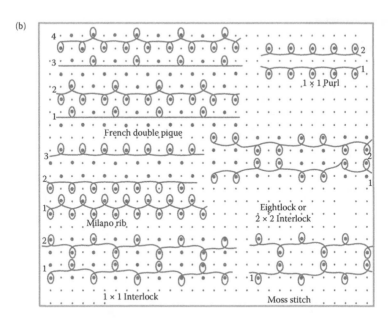

(c)

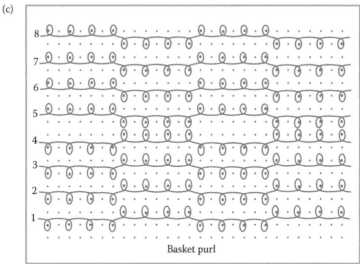

FIGURE 12A.2 (Continued)
(b) Dot matrix representation of some interlock derivatives. (c) Dot matrix representation of a purl derivative.

(a)

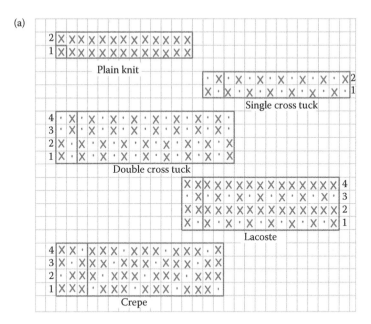

(b)

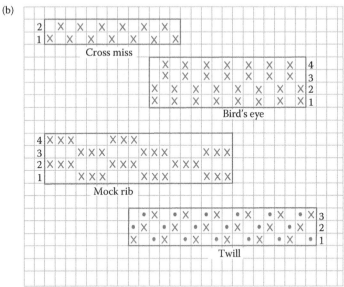

FIGURE 12A.3
(a) Square paper representation of some single-jersey derivatives. (b) Square paper representation of some single-jersey derivatives.

(a)

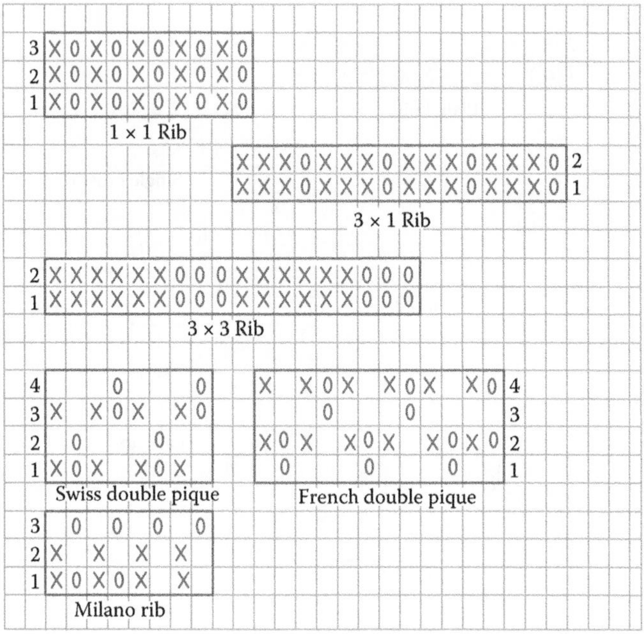

1 × 1 Rib

3 × 1 Rib

3 × 3 Rib

Swiss double pique

French double pique

Milano rib

(b)

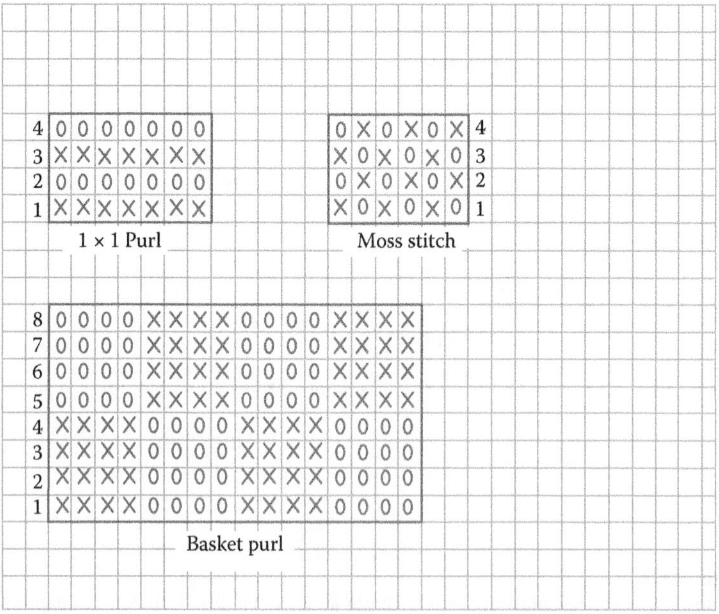

1 × 1 Purl

Moss stitch

Basket purl

FIGURE 12A.4

(a) Square paper representation of some rib derivatives. (b) Square paper representation of some purl derivatives.

13

Formation of Warp-Knitted Fabrics

13.1 Warp-Knitting Machines

13.1.1 Broad Classification

All commercial warp-knitting machines are of the flat-bed type, and the needles are group-driven. Warp-knitting machines with circular beds do exist but are rarely encountered in practice. Indeed, the very advantage of a warp-knitting system over other systems of fabric production is, to a large extent, obviated in circular warp-knitting machines. Hence, this system is not discussed here.

Warp-knitting systems are broadly classified into tricot (pronounced as tree-coe) and Raschel. The word "tricot" is derived from the French word "tricoter," which literally means "to knit," and the word "Raschel" stands for the name of a person to whom the inventor dedicated his invention. There are some fundamental differences between the two types as result of which Raschel machines can be employed to produce a much wider range of household and technical textiles. Tricot machines, on the other hand, have a narrower palette of products to offer but can be employed for mass production of a basic range of goods. Essentially, the Raschel technique permits the combination of a much wider range of raw materials and constructions than tricot, and it is the route to production of three-dimensional and multi-axial warp knits, altogether a new class of textile fabrics.

13.1.2 Tricot Machine

The sectional view of a typical tricot machine is shown in Figure 13.1. The machine can be broadly divided into four sections. Section I contains the central drive to knitting elements, section II represents the knitting zone in which all knitting elements converge and interact for converting yarns into knitted fabric, section III forms the yarn feeding zone, and section IV is the cloth take-up zone.

The driving mechanism is usually housed in an enclosure filled with machine oil. A central crankshaft, driven by the main motor is linked to

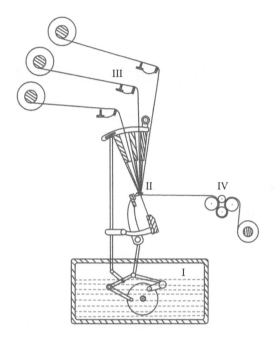

FIGURE 13.1
Sectional view of a tricot machine.

reciprocating rods, which jut out through the top plate of the enclosure. One set of rods is linked to the needle bar, one set to the sinker bar, and another set is linked to the guide bars. The needle bar carries needles and performs the function of a needle bed. The sinker and guide bars carry sinkers and guides, shown in Figures 13.2 through 13.5.

The sinker of a tricot machine (Figure 13.2) has a slightly different contour compared to that of a weft-knitting machine (Figure 12.35, Chapter 12). The function of the throat, belly, and nib are exactly the same as described in Section 12.6.2. The butt of Figure 12.35 is replaced by a foot in Figure 13.2,

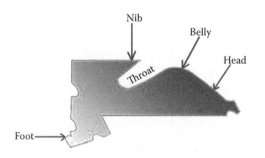

FIGURE 13.2
Sinker of a tricot machine.

FIGURE 13.3
Sinkers cast into a block.

and a prominent protrusion below the belly, termed the head, is an additional feature in the latter.

The sinkers in a tricot machine, just as the needles, are mounted on a bar and therefore group-driven. In order to facilitate the process of secure and exact mounting, a certain number of sinkers that would occupy 1 inch of space on the machine, that is, equal to the gauge of the machine, are arranged in a mold and cast in lead. Such a cast is shown on Figure 13.3. The throat, belly, and nib of each sinker are clearly visible, and the channel formed by neighboring sinkers stands out clearly in the diagram. The head of each sinker is encased in a well-rounded front, and their feet are gripped by the cast securely. Each block of a cast exhibits a hole through which a suitable bolt can be inserted for being fastened to the sinker bar. Thus, an 84-inch-wide bar can hold 84 such blocks side by side. Such a method of mounting ensures that all sinkers are equidistant, and in the event of necessity of replacement of a damaged sinker, the corresponding block can be taken out quickly and replaced by a fresh block.

A guide and the block in which it has been cast in a group, similar to the sinkers, are shown in Figures 13.4 and 13.5. Each guide, made of a pressed steel sheet, has a circular hole punched at its tip as shown in Figure 13.4. Guides are cast in alloys to a 1-inch width (Figure 13.5), and such blocks are mounted on bars, which are termed guide bars. On a tricot machine, there would be one needle bar and one sinker bar but usually more than one guide bars.

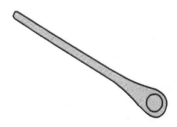

FIGURE 13.4
View of a guide.

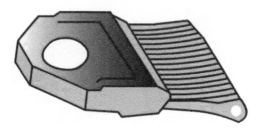

FIGURE 13.5
Guides cast into a block.

An analogy can be drawn between a guide bar of a warp-knitting machine and the heald of a weaving machine. Warp yarns on a weaving machine are drawn through heald eyes following a particular drafting plan, and the healds are moved along predetermined paths according to a lifting plan. Similarly, warp yarns on a warp-knitting machine are drawn through guides of different guide bars according to a predetermined plan, and guides are moved in a specific manner according to a lapping plan. Just as one warp yarn on a loom is drawn through one heald eye only, so is one warp yarn on a warp-knitting machine drawn through one guide only. Moreover, just as a loom exhibits more than one heald, so does a warp-knitting machine operate with more than one guide bar. Just as healds are numbered in a specific order, so also are the guide bars. Essentially then, the movement of warp yarns on a loom is controlled by healds, and on a warp-knitting machine, the same is controlled by guide bars. Moreover, just as a loom with a larger number of healds can be used to produce a wider palette of weaves, so can a warp-knitting machine with a larger number of guide bars be employed to generate a greater variety of knits.

Just like the sinkers and the guides, needles on a tricot machine are also mounted on a bar, termed as a needle bar. Such needles do not need butts for following the groove of a cam track. Needles may be either cast in lead blocks, which, in turn, are fastened to the needle bar, or they can also be directly and individually mounted in suitable grooves in the needle bar. Compound needles are commercially in vogue for modern tricot machines. Compound needles have two elements, namely the needle and the jack. Hence, such needle systems need two bars. The needles are directly mounted onto tricks cut in the needle bar, and jacks are cast in lead blocks of suitable width and then fastened to its respective bar. Bearded needles, which used to be very common on tricot machines, have been gradually phased out. These needles need to be repeatedly slowed down and speeded up within one cycle of operation, resulting in reduced productivity of corresponding machines.

Needle, guide, and sinker bars are supported by links and levers, which are finally connected to the three sets of rods sticking out of the oil-filled enclosure (section I of Figure 13.1). The rods reciprocate with the same frequency but different amplitudes. These motions are translated suitably by the links and levers so that the tip of the guide bars move along an arc; needles move up and down, and sinkers move to and fro.

Viewing Figure 13.1, one can locate three guide bars converging from the top to the knitting zone (section II) at which a needle from the needle bar also converges from the right and from below the fabric plane. The other element in this zone is the nearly horizontal sinker bar from which a sinker protrudes into the knitting zone. Such a machine would be termed a three-bar tricot, the term three-bar referring to the number of guide bars with which the machine is equipped. The three guide bars are supplied with three sets of warp sheet (section III), which come from three sets of warp beams. Sections of these two beams are visible on the top left-hand corner of the figure. The guide bars, as opposed to the healds of a loom, consume the warp sheet, usually at varying rates. Hence, although one may feed a large number of healds from the same warp beam on a loom, one would, in general, need an independent source of warp supply for different guide bars. This feature points to a greater degree of freedom and complexity of warp knit constructions vis-à-vis woven constructions.

The resultant knitted fabric is pulled away from the knitting zone along a horizontal plane by the take-up rollers (section IV). One may take note of the angle between the needles and the fabric; it is nearly 60° in the diagram. The fabric being produced from the warp sheet through the interaction of guides, needles, and sinkers is initially supported by the serrated discontinuous surface provided by the bellies of neighboring sinkers (Figure 13.3). This is very similar to the support that the sinkers on a circular knitting machine provide to a single-jersey weft-knitted fabric. The smooth and rounded front of the sinker blocks of a tricot machine provide an extension to this fabric support along which the knitted fabric is pulled away by the fabric take-up system. The angle between the plane of the knitted fabric and the needles is therefore dictated by the angle that the serrated plane provided by sinker bellies subtend at the needles. It may be recalled that the throats of neighboring sinkers play a crucial role in ensuring that the knitted fabric is kept in its position while the needles reciprocate. Such a role is possible only if the sinker loops connect the neighboring wale lines in a course. Clearly, in the absence of such loops, even for over a single course, the sinkers lose their effect, and fabric production becomes difficult. This forms an essential bottleneck in the tricot system. The Raschel system differs from tricot in this very particular aspect; absence of sinker loops linking adjoining wale lines is not an issue in Raschel fabric production.

13.1.3 Raschel Machine

The side view of a typical Raschel machine is depicted in Figure 13.6. In this figure, (1) is the needle, (2) the sinker, (3) the set of guide bars, (4) the trick plate, and (5) the knitted fabric. Raschel machines employ latch needles (and also compound needles), which are mounted on a bar, and there may

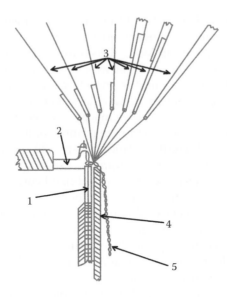

FIGURE 13.6
Side view of a Raschel machine.

be one needle bar (single-bed Raschel) or even two needle bars (double-bed Raschel) in a machine. The sinker performs the function of holding down the fabric during the upward journey of the needles but play no role in the casting-off process. For assisting in casting off, the trick plate is employed.

As can be observed from Figure 13.6, the fabric, coming out of the knitting zone, is taken down over the trick plate at a very acute angle such that the fabric and needle axes are nearly parallel to each other. During the downward motion of the needle, the new yarn is bent across the arms of the cast-off loop because the loop just cast off is bodily supported by the trick plate (Figures 13.7 and 13.8). Each newly cast-off loop in the body of the fabric is supported by the top edge of the trick plate and thus resists, as a group, any movement of the fabric with that of the needle moving downward. Thus the supports or restrictions necessary for bending a yarn (see Section 12.7.1.1 and Figure 12.38), which is provided by sinkers to sinker loops on a tricot machine, is provided in a Raschel machine by the edge of a trick plate, which supports both arms of loops just cast off. Thus, the role of sinker loops in the loop-formation process on a Raschel is obviated, permitting production of wale lines devoid of sinker loops. Many networks showing large openings can therefore be effectively produced on a Raschel. Needless to mention, the manner of casting off in tricot would not permit production of structures in which neighboring wale lines are not connected by sinker loops. However, net-like structures can also be produced on tricot.

Because the yarn sheet fed to the needles of a Raschel and the resultant knitted fabric are nearly on parallel planes, a very high take-down tension

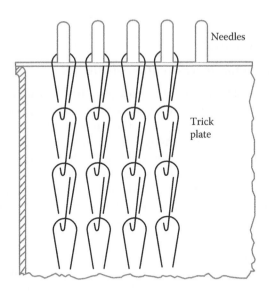

FIGURE 13.7
Trick plate supporting cast of loops.

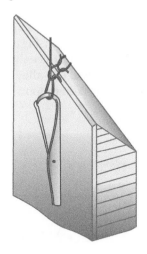

FIGURE 13.8
Side view of a trick plate supporting loops.

can be applied on fabric without loading the needles excessively. This feature also permits maintaining very straight and rigid wale lines on the machine, a necessary condition for producing nets with large openings.

As the fabric is pulled in the downward direction by the take-up motion, it is conceivable that two sets of needle bed and trick plate can be easily arranged back to back for the production of a double-layered knitted fabric.

13.2 Basic Warp Knits

13.2.1 Convention for Representation of Stitches

A point paper is used to represent warp knits whereby the columns of points represent wale lines and rows represent courses. Hence, points in the same column are a time-displaced view of the same needle. Gaps between columns, which can be termed as alleys, are numbered in the manner shown in Figure 13.9. For simple knits, one starts with the number zero and moves from left to right for onward numbering of the alleys.

A warp knit is described by curved lines drawn around points, which represent needles being lapped around by yarns. Each warp yarn is controlled by a guide, and each guide moves around a specified needle in such a manner that the corresponding yarn is wrapped (or lapped) around the needle axis, enabling the needle to catch the new yarn in its hook. The lapping motion in front of the needle, that is, on the side on which its hook is located, is known as *overlapping,* and the corresponding motion at the back of the needle is known as *underlapping.* The overlapping motion ensures that a needle catches the yarn, and the underlapping motion gives rise to underlaps in warp-knitted structures (see Figure 13.10b). Such a diagram showing the sequence of overlapping and underlapping is known as a lapping diagram.

A typical lapping diagram is shown in Figure 13.9. It consists of four needles/wale lines and three courses. The trace shown in the bold line is duplicated by those shown in dashed lines. The alleys, that is, the gaps between neighboring needles are numbered zero to four. Restricting attention to the

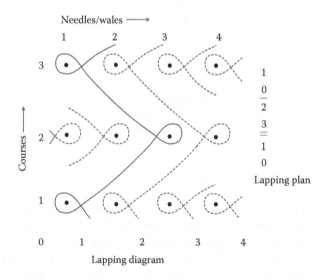

FIGURE 13.9
Lapping diagram and lapping plan.

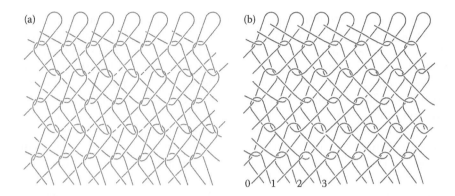

FIGURE 13.10
(a) Technical front and (b) technical back of warp-knit fabric.

trace made by the bold line, one observes that the point of initiation of the course 1 is in alley 1. This means that at the starting point, the guide or yarn under consideration is in alley number 1. The first effective motion of the guide or yarn in this course is from alley 1 to alley 0. This motion can be broken up into two straight segments as shown in Figure 13.11. The first segment S_I represents the motion of the guide from the back side of the needle to its front side, and the second segment represents the overlapping motion OL. Reverting back to Figure 13.9, it is observed that the guide or yarn comes back from the front of the needle 1 to its backside and continues its journey in the subsequent course across the backside of needles 1 and 2 and reaches the alley number 2. This is represented by the straight segments S_O and UL in Figure 13.11. In the second course, the guide/yarn moves from alley 2 to the front side of needle 3, again a motion made up of two segments S_I and OL. It is noticed that this OL has the same magnitude but an opposite direction compared to that of course 1. Subsequently, the guide or yarn moves to the back side of needle 3 and repeats its journey along the back side of needles to be back in alley 1. This motion can be split up into S_O and UL. This UL is also equal in magnitude but opposite in direction to the previous UL. The lapping motion of the third course is a repeat of that of the first course.

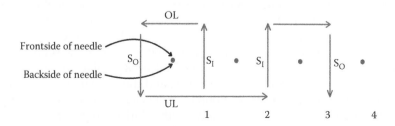

FIGURE 13.11
Break up of lapping movement into elements.

Now it is evident that the lapping motion of one guide does not make a fabric. However, when the lapping motion of an entire guide bar is considered, then the resultant fabric comes into effect. Figure 13.10 represents the resultant inter-looped construction arising out of the lapping plan of Figure 13.9. A close comparison of these two figures reveal the manner in which neighboring wale lines get interconnected by the sinker loops and loops in a wale line inter-loop for formation of the fabric. The inclination of courses alternately to the right and left clearly makes the resultant fabric highly unstable. Such constructions can only be employed to produce very cheap backing fabrics. In this sense, warp knitting is one step ahead of weaving insofar as the ability to generate simple constructions is concerned. With one heald, it is impossible to produce a woven fabric, and with one guide bar, a series of warp knits in which only the length of underlap UL differs can be produced.

From Figure 13.11, the following can be inferred:

- The motion S_I is always of the same magnitude and direction.
- The motion S_O is always of the same magnitude and direction.
- Motions S_I and S_O are always in opposite directions being however of the same magnitude.
- The motions S_I and S_O taken together can be equated to a swinging motion of guide and hence of the corresponding guide bar. Such a motion has indeed been described in Section 13.1.2 and is generated by the central drive given to the guide bars.
- The motions OL and UL are lateral motions given to the guide bars and determine the extent and direction of underlap as well as the needle on which the overlap takes place at any instant. Such motions are termed as *shogging*.
- The complete lapping diagram of a guide bar is thus a pictorial representation of its shogging and swinging motions in a chronological sequence.

Instead of a lapping diagram, one may also employ the lapping plan (or chain plan), which is a digitized form of the former. A lapping plan corresponding to the lapping diagram has been listed on the right-hand side of Figure 13.9. A pair of numbers, representing the overlapping motion, is separated by a straight line segment and is followed by another pair of numbers representing the next overlap. The numbers across a line segment logically represent the underlapping motion. In this particular construction, the underlap is across the back of two needles, and the overlap is always across one needle. Hence, such a construction is termed a 2 × 1 tricot. By varying the length of underlap, it is possible to create other knits. On paper, one can create knits with very long underlaps although practical limitations are encountered on the machine and in the actual fabric-formation process.

13.2.2 Single Bar Knits

13.2.2.1 Pillar Stitch

A pillar stitch (or chain stitch) is a stitch construction in which both overlaps and underlaps are always carried out across the same needle. As there are no lateral connections with the neighboring wales, only a link-up in the direction of the wales exists. Chains of disconnected wale lines result out of such construction, and a fabric therefore is not created (Figure 13.12). Illustrations of open and closed pillar stitches are illustrated in Figures 13.13 and 13.14.

As one can see from the chain, the smallest repeating unit of the open pillar stitch is extended over two courses, and that of the closed pillar stitch over one course. The smallest repeating unit is termed the repeat. There is,

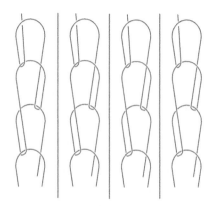

FIGURE 13.12
Pillar stitch.

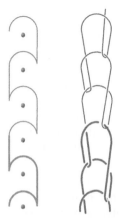

FIGURE 13.13
Open pillar stitch.

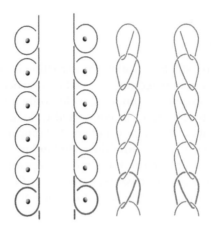

FIGURE 13.14
Closed pillar stitch.

however, a difference between a height repeat and a width repeat. The closed pillar stitch has a height and width repeat of one stitch each way. The width repeat of the open pillar stitch is one stitch, and the height repeat is two stitches.

In conjunction with other stitches, the pillar stitch can be employed for production of outerwear and for ribbed velour fabrics (corduroy). Even in these fabrics, the open pillar stitch is more popular as it provides the necessary longitudinal stability and runs freely. The chain movement locks down the long underlaps of other guide bars, giving the characteristic ribbed effect.

13.2.2.2 1 and 1 or Tricot Lap

If the laps are carried out in alternate overlap and underlap motions on two adjacent needles, then a tricot knit is produced that can consist of closed or open stitches (Figures 13.15 through 13.17). As opposed to the pillar stitch, the 1 and 1 (tricot) stitch results in a textile fabric because the underlaps not only connect courses, but also connect the wales. As a result of the inclined underlap, the stitch heads of each successive row of courses are pulled in opposite directions (Figure 13.16). The needle loops of tricot knits exhibit therefore alternate rows leaning to the left and to the right.

The closed tricot knitted fabric is very light; it can be stretched in all directions and is only used for fabrics that do not require a high degree of stability (e.g., fabrics for lamination, etc.). There are a considerable number of uses for the stitch, however, when it is employed in conjunction with other stitches or binding elements (locknit, velvet, etc.).

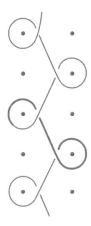

FIGURE 13.15
Lapping diagram of closed 1 × 1 tricot.

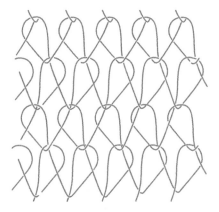

FIGURE 13.16
Technical back side of closed 1 × 1 tricot.

The open lap 1 × 1 stitch is shown in Figure 13.17. The fabric formed in this manner is very light, inconsistent in shape, and easily stretched and distorted.

13.2.2.3 2 × 1 Stitch

The 2 and 1 (or 2 × 1) stitch differs technically from tricot by the underlap, which is increased by one needle space so that each underlap in the fabric moves across bodily over one wale line. The longer underlap also alters the properties of the fabric. The resultant fabric offers more resistance to stretching and is denser than 1 and 1 tricot. As a stitch together with binding

FIGURE 13.17
Lapping diagram of open 1 × 1 tricot.

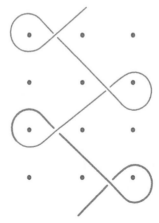

FIGURE 13.18
Lapping diagram of closed 2 × 1 construction.

elements and other stitch formations, the 2 and 1 construction can be widely used. The 2 and 1 stitch can also be generated with closed (Figure 13.18) and open (Figure 13.19) stitches.

13.2.2.4 3 × 1 and 4 × 1 Stitches

If the underlap is increased by one more needle space, the 2 and 1 stitch becomes 3 and 1, which, in conjunction with other stitches, such as pillar and tricot, will produce velvet or fabrics with less longitudinal and lateral stretch. The longer underlaps cross two wales on the back of the fabric, and, after a raising and shearing treatment, this fabric takes on a velvet appearance.

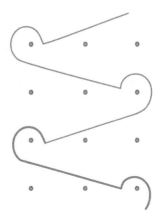

FIGURE 13.19
Lapping diagram of open 2 × 1 construction.

The 3 and 1 stitch, which can be generated with closed and open loops, has no importance as a single-bar product.

The 4 and 1 stitch has the same basic structure as the 3 and 1 except that the underlap is extended over one more needle space. The 4 and 1 stitch, therefore, has the largest underlap of all the common regular stitch constructions. The underlap, however, can be extended by a further one or two needles for certain pattern effects.

The 4 and 1 stitch has an unusually high longitudinal and lateral stability owing to the large underlaps when used in conjunction with the pillar stitch. In combination with the tricot stitch, underlaps of 4 and 1 remain loose (i.e., not bound in) on the reverse of the fabric and after raising a velvet-like finish is obtained.

13.2.2.5 Atlas Stitch

The Atlas stitch differs from those stitches dealt with up to now in that the laps are continued over two or more courses in one direction and then return in the opposite direction to the point from which they started. The position of the overlaps and underlaps are changed when the lapping movement changes so that diagonal stripes appear and cause differential light reflections. These stripes are a prominent feature of Atlas stitches, which can be subdivided into the following:

Closed Atlas (cross Atlas)
Open Atlas (common Atlas)
Closed, back-lapped Atlas
Open, back-lapped Atlas

The closed Atlas (Figures 13.20 and 13.21) is built up of underlaps and overlaps, which run in opposite directions and result in closed stitches. The turning course of the closed Atlas stitch is usually open. However, there are no objections to a closed Atlas with a closed return course as far as working and binding are concerned. The number of courses an Atlas moves before returning and the number of return courses is characteristic of the stitch and therefore must be indicated in the classification of stitch formation. The repeat of the four-course Atlas contains eight courses whereby the stitches of four rows at a time come alternately to the left and to the right. Half of the repeat of the Atlas stitch is called the reflector or mirror. The open Atlas (Figures 13.22 and 13.23) results in the lightest warp-knitted fabric

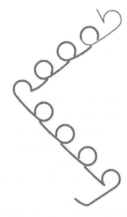

FIGURE 13.20
Lapping diagram of a four-course closed Atlas.

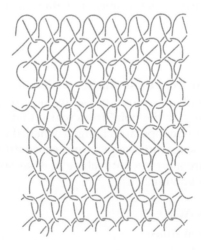

FIGURE 13.21
Technical back side of a four-course closed Atlas.

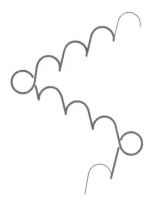

FIGURE 13.22
Lapping diagram of a four-course open Atlas.

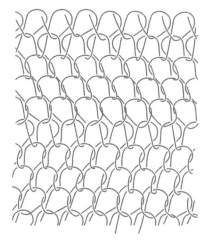

FIGURE 13.23
Technical back side of a four-course open Atlas.

and is more often in use than the closed Atlas. Apart from the closed return courses (which can also be open), this Atlas is built up solely from overlaps, which continue over several courses and then return. The stitch structure of this Atlas can be compared with a weft-knitted fabric distorted in its longitudinal direction.

If the layout of the closed Atlas is altered to lap alternate needles, the closed back-lapped Atlas is produced. Owing to the elongated underlap, the closed back-lapped Atlas is heavier than the closed Atlas. The back-lapped Atlas is recognized by the underlap, which stretches across one stitch on the technical back of the fabric in just the same way as the 2 and 1 stitch. The open Atlas can also be extended by one needle space by elongating the underlap. The resultant stitch is called the open back-lapped Atlas (Figure 13.24). The

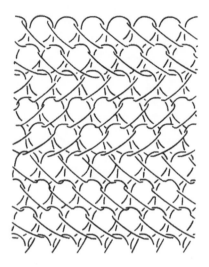

FIGURE 13.24
Technical back side of a three-course open back-lapped Atlas.

repeat of the three-course back-lapped Atlas contains six courses. All types of Atlas can be combined with one another or with all other types of stitches. However, the pillar stitch can only be worked successfully with the back-lapped types of Atlas.

13.2.3 Multi-Bar Warp Knit Constructions

As the simple stitches generated with one guide bar can only be used for certain technical purposes or for purposes without claim to high quality, several types of stitches are combined with each other into a range of knits for designing products with desirable properties. This is accomplished by employing more than one guide bar in a machine whereby each bar laps differently. The resultant products are superior in terms of stability, firmness, stretch, elasticity, pore size distribution, etc. Such fabrics become commercially useful and open up new vistas for application of textiles in elegantly solving many technical problems.

Working with several guide bars requires a distinct numbering system to prevent confusion. The numbering of the guide bars on tricot machines is done from the warp beams toward the operator, that is, the guide bar nearest to the warp beam is numbered bar I or the back guide bar, and consequently, the other guide bars set in front of this bar are numbered bar II, bar III, etc. (Figure 13.25). The bar closest to the operator is also termed as the front guide bar. The guide bars on the Raschel machine are numbered the other way around (Figure 13.26) because the pattern guide bars are usually fitted in from the front to the rear of the machine. However, the location of the front guide bar in both systems is closest to the operator, and yarns from

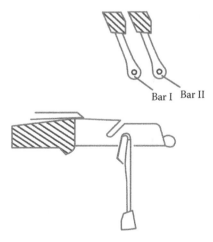

FIGURE 13.25
Numbering of guide bars on tricot machine.

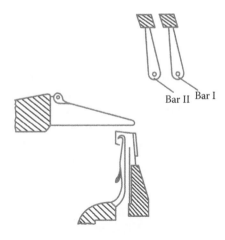

FIGURE 13.26
Numbering of guide bars on Raschel machine.

this bar occupy the outermost position on both surfaces of a warp-knitted fabric. This has been illustrated in Figure 13.27.

Figure 13.27a and b depict two extreme positions of a pair of guide bars in the course of their swinging motion. In Figure 13.27a, the guide bars are in the process of executing their overlapping motion, and in Figure 13.27b, they are executing the underlap motion. Two sets of yarns pass through two respective guide eyes and are linked to the knitted wale line held by the needle as shown in Figure 13.27b.

During the overlapping phase, the back guide bar (bar I) is farthest from the needle and, therefore, is located at a higher level in the swing of the arc. The corresponding yarn, therefore, lands at a higher level on the needle, and

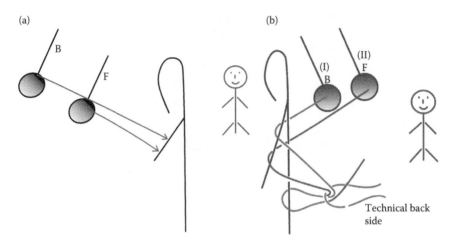

FIGURE 13.27
Positioning of front guide bar yarn on fabric surfaces. (a) Guides in overlapping mode.
(b) Guides in underlapping mode.

subsequently, during the casting-off process occupies a position closer to the
needle hook than the yarn from the front bar (bar II). Such a situation gives
rise to the effect of "plating" illustrated in Figure 13.28. In effect, a plated
loop, made up of two or more yarns occupying clearly differentiated positions in a needle hook, is a multilayered entity, showing up on the fabric's
technical face the yarn farthest from the needle hook. A knitted fabric made
of plated loops would hence also be multilayered. A warp-knitted fabric
would be accordingly always multilayered as it is made up of plated loops.
Furthermore, the technical front of a warp-knitted fabric would always show
on its technical face the yarn fed to the front bar.

FIGURE 13.28
Principle of plating.

During the underlapping phase depicted in Figure 13.27b, the front bar occupies a position farthest from the needle, and hence, the length of yarn dragged by the same across the back of one or more needles would lie on top of the underlaps resulting from other guide bars. As a result, the technical back side of a warp-knitted fabric would reveal the underlap of the front bar on top. The underlaps of the rest of the bars would lie below that of the front bar but would also be arranged in a layered manner, the last of the layer being made up of the yarn from the back guide bar.

One can create, therefore, a mental picture of the section of a warp-knitted fabric as being made up of an orderly layer of loops followed by an orderly layer of underlaps. The layers of loop start on the technical front side with the ones made from yarns of the front bar and ending with the ones from the back guide bar, and the layers of underlap start on the technical back side with the ones made from the front bar and ending up with those from the back guide bar. Thus, both the technical front and technical back side of a warp-knitted fabric would exhibit yarns of the front guide bar, and the yarns of the back guide bar would be at the core of the multilayered structure. This specific feature has a great relevance to engineering warp-knitted fabrics. The outer surfaces of a fabric are directly exposed to elements of nature and other corroding and abrading strains. Hence, the most suitable yarn to withstand such strains can be located on the two surfaces and therefore threaded through the front bar. The rest of the yarns from other guide bars would be progressively less exposed, and the ones at the core would be fully protected. Hence, the yarns at the core and thereabouts could be the ones with a relatively more vulnerable surface but possessing bulk properties that would impart desirable attributes to the product.

Central to the model of a warp-knitted fabric described in the foregoing is the phenomenon of the plating process. It may, however, be mentioned that such an eventuality may be deviated from if the relative location of the guides, the extent and direction of overlaps and underlaps of different bars, combination of open and closed loops, and yarn tensions or properties are suitably manipulated.

While indicating the position of the guide bars when applying various multi-bar stitch constructions, a further point must be mentioned concerning the type of lapping. When stipulating a particular knit produced from two or more guide bars, it is necessary to state which lapping movement is placed on which bar and also to state whether the bars move in unison or in opposition, that is, whether the underlapping and overlapping motions of the different bars are in the same direction or not.

Two lapping movements, such as 1 and 1 tricot and four-course Atlas are combined (Figure 13.29), cannot be considered as a whole but must be compared course for course. When the bars move together, the term "lapping in unison" is sometimes used together with "counter-lapping" to indicate movement in opposite directions.

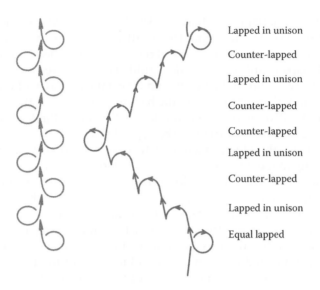

Lapped in unison

Counter-lapped

Lapped in unison

Counter-lapped

Counter-lapped

Lapped in unison

Counter-lapped

Lapped in unison

Equal lapped

FIGURE 13.29
Characterization of direction of lapping.

In order to avoid misunderstanding, all details necessary for the production of a construction should be indicated. Such details are

- Open or closed stitches
- Type of stitch and guide bar used
- Unison or counter-lapping motions

The stitch of the front bar is termed as cover, and hence, when classifying the constructions of the combined stitch formations, the cover is mentioned first followed by that of the others. Furthermore, when working with two or more bars, the thread run-in ratio of the various warp beams must also be indicated. The run-in ratio provides information on the relative consumption of threads by different guide bars. This thread run-in ratio is often not known and must therefore be estimated. By way of illustration, all overlaps can be assigned a value of two, underlap connections between two neighboring stitches in two consecutive courses the value one, and underlap connections between two neighboring stitches in the same course (two-needle overlap) 0.5. The thread proportions calculated for the overlaps (needle loops with two arms) and underlaps are entered in the layout or in the notation draft. The theoretical estimation of the run-in ratio serves as a guide for the setting of the beam drive. Slight alterations, depending on the fabric to be produced, can be carried out subsequently.

The effect of threading and shogging of a guide bar on fabric properties is governed by the following rules:

- The front bar affects primarily the widthwise behavior of warp-knitted fabric whereas the back bar affects the lengthwise behavior.
- A larger underlap of the front bar makes fabric more extensible than when the same underlap is assigned to the back bar.
- The greater the difference in the length of underlap of the two bars, the more extensible is the fabric if the difference is in favor of the front bar.
- Underlaps in the same direction tend to make the fabric shrink more.

Some simple two-bar knits are illustrated in Figures 13.30 and 13.31.

The two-bar tricot, depicted in Figure 13.30 has a run-in ratio of one and has poor cover. Longer underlap in the front bar is used to make locknit (Figure 13.31b). The run-in ratio is 3:4. The resultant fabric is soft, extensible, and exhibit tendency of curling at edges. The reverse locknit (Figure 13.31a), which is commercially produced in very large quantity by exchanging lapping plans of the two bars, exhibits a considerably reduced extensibility as well as greater stability as compared to locknit. The run-in ratio is 4:3.

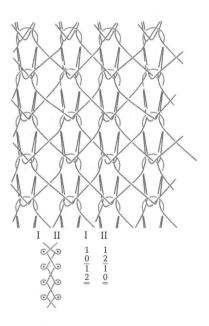

FIGURE 13.30
Two-bar tricot.

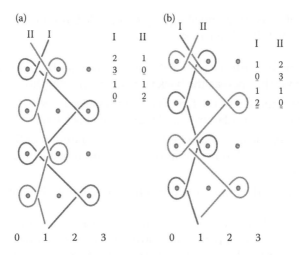

FIGURE 13.31
Lapping diagrams of (a) locknit and (b) reverse locknit.

13.3 Sequence of Loop Formation

13.3.1 Compound Needle on Tricot Machine

The sequence is shown in Figure 13.32 in eight stages. At the outset, the needle (1) is at its lowermost position, and the sinker (3) makes a forward motion to grip the sinker loops (underlaps) extending from the cast off loop (Figure 13.32a). The jack or tongue (2) is at its uppermost position, securely blocking the opening of the needle hook. The pair of guide bars (4) (front bar is II, and back bar is I) execute the underlap followed by a swinging-in motion (Figure 13.32b and c) as the needle moves up to a higher level and the jack moves down (Figure 13.32d). The sinker (Figure 13.32b and c) moves back a little to release tension in the fabric–yarn continuum and subsequently moves forward again (Figure 13.32d and e) to maintain control on the fabric edge and keep it in the desired position. The guide bars complete the overlap and swinging-out motion, thus effectively lapping the yarns around the stem of the needle hook. Subsequently, the jack rises and the needle descends (Figure 13.32e and f) while the sinker moves back (Figure 13.32f), gliding the fabric edge up onto its belly and away from its throat. This movement of the sinker permits the newly drawn-in loop to be pulled down across the sinker belly. As the needle and jack descend (Figure 13.32g) further toward the sinker line, the sinker keeps on moving back until the old loop is cast off. The cast-off loop passes into the fabric due to the action of the take-up mechanism at nearly a right angle to the needle (Figure 13.32h). The jack starts retracing its path away from the needle hook.

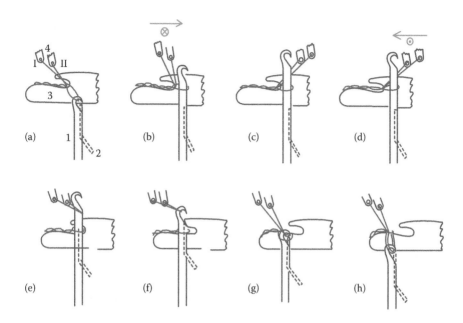

FIGURE 13.32
Loop formation sequence on a tricot machine with a compound needle. (a) Starting position. (b) Guides underlapping and swinging-in with needles moving up. (c) Guides in front of needle hook. (d) Guides overlapping and swinging out. (e) Feed yarns wrapped around needles. (f) Sinker moving back and needle moving to cast-off position. (g) Sinker at furthest position and needle casting-off. (h) Needle at knitting point with a new loop.

The needle movement is smooth and short. The movement of the jack or tongue supplements the motion of the needle such that the throat space gets closed during casting off (Figure 13.32e and f) but is open during the upward motion of the needle (Figure 13.32b, c, and d).

13.3.2 Latch Needle on Single-Bed Raschel

The sequence is shown in Figure 13.33 in six stages. Needles are mounted in slots on the needle bed, and the needle bed itself is located flush against the back side of the trick plate. The trick plate remains stationary, but the needle bed is moved up and down. The sinkers do not exhibit any throat as the tendency of the fabric to rise with the needles is nullified to a great extent by the alignment of and tension in the fabric under the take-down process. Nonetheless, the lower edges of the sinkers provide a degree of additional assistance to suppressing the fabric edge during the upward journey of the needles. Casting off is also carried out by the assistance of the trick plate. Hence, the sinker belly becomes redundant. Thus, the fabric edge is now free of any external constraint, and a continuum between the yarn sheet and the fabric is established more directly. The simplified sinker can indeed be further modified to carry

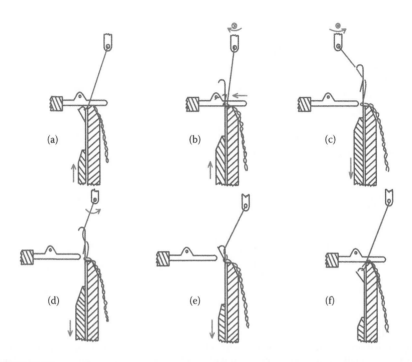

FIGURE 13.33

Sequence of loop formation on a Raschel. (a) Starting position. (b) Needle moving up while guide executes underlap and swinging-in motion. (c) Needle at top position while guide executes overlap and swinging-out motion. (d) Needle descending with feed yarn wrapped around. (e) Needle nearing cast-off position while sinker moves in. (f) Needle at knitting point with a new loop.

out some additional task as is exemplified in Figure 13.43b for insertion of weft across the fabric width, a powerful feature of the Raschel technology.

13.3.3 Latch Needle on Double-Bed Raschel

13.3.3.1 Special Features of Machine and Product

The arrangement of knitting elements on a double-bed Raschel machine is depicted in Figure 13.34. Two sets of trick plates (6) and needle beds (7) are arranged back to back, and the resultant fabric is pulled down in the gap between the two sets. In this figure, five guide bars are arranged for feeding yarns to the needles. However, the guide bars 1 and 2 feed one set of needles that may be termed the front bed, and the guides 4 and 5 feed the other set of needles in the back bed. Hence, two fabrics are formed separately on these two beds. The central guide bar (3) feeds yarn on both beds and links up the two fabric layers. Evidently, distance between the two sets of needle bed and trick plate affect the length of yarn segments connecting the two layers of fabric.

If these yarn segments are made of stiff yarns, then the resultant fabric would have a pronounced third dimension with a high compressional

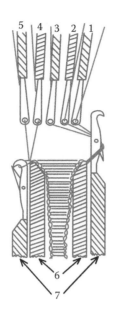

FIGURE 13.34
Arrangements of knitting elements on a double-bed Raschel.

resiliency. Such fabrics, known as spacer fabrics, are finding widespread use in diverse applications, such as compression bandages, body armor, boot soles, padding material in seats, geotextiles, etc. If, on the other hand, these connecting yarn segments are sliced at the center and the two fabrics separated downstream, then two cut pile fabrics would result.

The setup shown in Figure 13.34 exhibits an absence of sinker bar. The argument for the absence of sinkers is the same as that observed for double-jersey weft-knitting machines. But this absence of sinkers also means that such machines cannot be employed for production of fabric on only one of the two beds at a time. The guide at the center has to always connect the two fabric layers so that the tendency of one fabric layer to move up with the rising needles of that bed is nullified by the inability of the other fabric layer to move up due to the downward movement of the corresponding needles.

A related issue that affects the productivity of such machines results out of an unproductive phase in part of the lapping motion in each cycle. This is explained in Figure 13.35.

The diagram in Figure 13.35a shows only the guide bar that feeds the yarn on both beds in the course of its swinging-in and swinging-out motion with respect to the front needle bed that has cleared the old loop and is moving up. After the overlapping and swinging-out motion of the guide bar, as shown in Figure 13.35b, its location is above the needles of the back bed around which it has to lap the yarn next. In order to carry out this later exercise, this guide bar has now to swing back away from the top of the needles

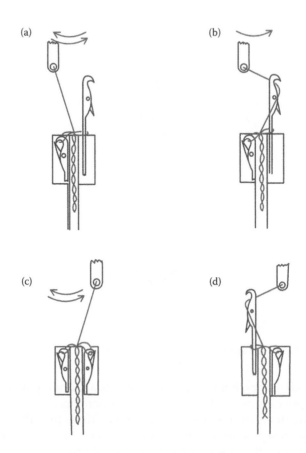

FIGURE 13.35
Unproductive lapping movement on a double-bed Raschel. (a) Lapping and shogging motion of guide. (b) Loop formation by front bed needle followed by idle swinging of guide. (c) Lapping and shogging motion of guide. (d) Loop formation by back bed needle.

of the back bed to those of the front bed, allowing the needles of the back bed to rise up, after which it can swing out again to the front of the needles of the back bed for overlapping. Hence, the needles of the front bed have to first move down and cast off, after which the guide bar can swing back and then swing out again. Thus, there arises a phase shown in Figure 13.35c in which the needles of both beds are occupying a position nearly at the same level as the trick plate and one swinging motion of the guide bar is not accompanied with either an underlap or an overlap. This is an idle and unproductive phase. Subsequently, this guide bar would execute one complete lapping motion around the needles of the back bed. Hence, one complete cycle of motion of this guide bar comprises one lapping motion around the needles of the front bed followed by an idle swinging motion, which is then followed by one complete lapping motion around the needles of the back bed. As the swinging motion of all the guide bars is derived from a central source, the entire bunch of guide bars execute this idle phase, causing an overall loss in production. Moreover, the guide bars meant to lap only on the back needle beds also take part in this extravagant swinging motion as do the guide bars meant to feed the needles of the front needle bed only. Evidently, a substantial amount of unproductive energy is also consumed simultaneously.

13.3.3.2 *Special Features of Lapping Diagrams and Lapping Plans*

The point paper representation of the lapping diagrams and lapping plans on a double-bed Raschel has to be interpreted differently compared to that of a tricot. To start with, just like the the gating of needles in a weft-knitted interlock knit, two beds with needles arranged face to face are to be provided for. Hence, two rows of points, one for the front bed (F) and the other for the back bed (B), are marked out as indicated in Figure 13.36. Here a pair of F and B rows effectively constitutes one complete course, although two complete lapping motions of each guide bar are consumed in generating the same. It may be recalled that in the representation of interlock knits (Figure 12A.2b) two separate pairs of rows of points, representing two courses constituting one effective interlock course, are used up. For double-bed Raschel constructions, however, only one pair of rows of points represent one effective course. Such a representation is evidently more compact, and therefore, a mental decoding is called for.

The representation of the lapping plan of a guide bar in Figure 13.36a is for four courses whereby the knit actually repeats on two courses only. The lapping plan corresponding to the diagram on the extreme left-hand side is written down in a manner slightly different from that in Figure 13.9 of a 2 × 1 tricot. First, the numbering of alleys is in an opposite sense. In fact, according to one convention, the alleys are numbered with even digits only, such as 2, 4, 6, and so on. Second, a pair of numbers is indicated against every row showing the overlap corresponding to that row. Hence, for the eight rows of lapping diagram, eight pairs of numbers are written down adjacent to the corresponding rows. This diagram is actually made up of two separate diagrams indicated on the right-hand side of the figure. Clearly, two separate fabrics are being formed on each of the two beds by this particular bar, which may be considered as the connecting guide bar (3) of Figure 13.34. Both these fabrics are of open loop 1 × 1 tricot stitch. There is, however, one important difference from the tricot fabrics formed on single-bed machines. The underlaps that directly connect adjacent wale lines in a single-bed tricot are, in this case, connecting the two fabrics formed on the two beds of a double-bed machine. They are oriented in the third dimension vis-à-vis the fabric plane.

Moving over to Figure 13.36b, one comes across a lapping plan that generates a 2 × 1 tricot on the back bed and a string of pillar stitches on the front bed. Unless there is some other knitting system that links these strings together at regular intervals, no fabric would be formed on the front bed. Hence, in effect, no fabric can be formed on a double-bed Raschel by the lapping plan described

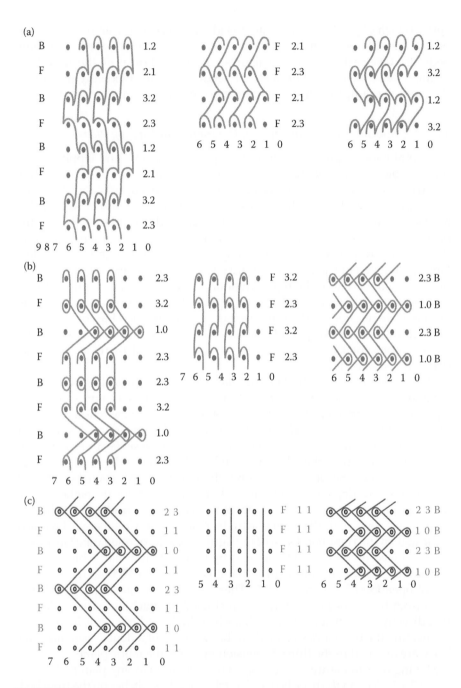

FIGURE 13.36
Principle of lapping diagram and lapping plan for a double-bed Raschel. (a) Open loop 1 × 1 tricot lap on both needle beds. (b) Open loop pillar on front bed and 2 × 1 closed tricot lap on back bed. (c) 2 × 1 closed tricot lap only on back bed.

in this figure unless it is accompanied by a suitable lapping by another guide bar. The lapping plan of Figure 13.36c, however, is of a guide bar that wraps around needles only of the back bed. The front bed is given a complete go-by. The lapping plans of Figure 13.36a and b belong to that of a connecting bar, and that of the Figure 13.36c belongs to one of the extreme bars.

13.4 Shogging Motion of Guide Bars

From the description so far of warp-knitting systems, it is evident that the range of warp-knit constructions on a machine is governed by (1) number of guide bars and (2) the possible combinations of shogging motions of the different guide bars.

Shogging motion is comprised of overlap and underlap motions. By and large, the extent of overlap motion is restricted to one needle although overlap across two needles is practically possible. Thus, for all practical purposes, the combination of underlapping motions (direction and magnitude) is central to generating different constructions.

In machines of the recent past, the shogging motion used to be carried out by mechanical systems. The two systems, namely the pattern wheel and the pattern chains, are briefly outlined in the following. In modern machines, however, electronic shogging motions are practiced, which is also briefly touched upon here.

13.4.1 Pattern Disk

A pattern disk has inclines of different slopes on its periphery. These inclines press against an antifriction shogging roller as the pattern disk revolves, and the resultant displacement of the antifriction roller is transmitted to the corresponding guide bar via a push rod. In a machine, there would be as many disks as the number of guide bars. All these disks are fastened together and work as a group. As a disk can push the guide bar only in one direction, the return motion has to be generated through the action of springs.

A pattern disk is actually a cam with inclines milled on its circumference according to the required lateral displacement profile of the guide bar. These inclines (Figure 13.37), which are required for overlapping and underlapping motions of the guide bar, are smoothly shaped and have a well-formed transition to and from each other. This ensures quiet and smooth running. For this reason, pattern disks are used on high-speed tricot machines. The stretch between two neighboring inclines is a period of dwell during which the guide bars execute a swinging motion through the alleys between needles.

The circumference of a pattern disk on tricot machines is usually divided into 48 equal parts, which, in turn, is divisible by the length of the pattern

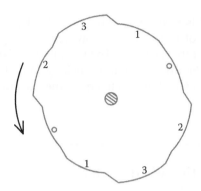

FIGURE 13.37
Pattern disk.

repeat. For execution of one cycle of lapping motion, usually two but, in some cases where the underlap stretches across a large number of needles, even three inclines may be required. If three inclines are needed for one course (three-phase), for example, first incline for overlapping, second incline for the first part of the underlap, and third incline for the second part of the underlap, then 16 courses are produced in one revolution of the pattern disk. If two inclines per course are employed (two-phase), then 24 courses can be accommodated around the circumference of the disk. Therefore, if the number of courses on the pattern disk circumference (16 or 24) is divisible by the courses in the pattern repeat, then it can be produced with a pattern disk.

The pattern disk (wheel) illustrated in Figure 13.37 would produce two courses of 2 × 1 tricot. The numbers 0, 1, 2, and 3 refer to those of the alleys between needles on a lapping plan. Thus, the height difference between each numbered concentric circle of Figure 13.37 must equal the distance between adjacent needles of the machine on which such a wheel can be employed.

A pattern wheel has therefore restricted utility because (1) the pattern cannot be changed to produce some other knits and (2) it is not interchangeable between machines of different kinds.

13.4.2 Pattern Chain

If the differences in height on a pattern wheel are created by a chain constructed of individual pattern links that are placed around a pattern drum and the links are connected by pins, then the resulting pattern range can be very large. A number of such chains, linked together laterally by pins and equaling the number of guide bars, is mounted on a pattern drum so that, when the drum rotates, the set of chains also rotates. However, the movement imparted in this manner to antifriction rollers bearing against the pattern chains is somewhat jerky as the links have to be ground at points where they connect. This grinding is generally straight although special sine-curve shaped links

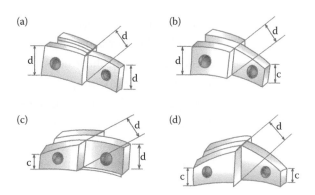

FIGURE 13.38
Links of pattern chain. (a) The A-link. (b) The B-link. (c) The C-link. (d) The D-link.

(Figure 13.38) can be used in certain instances. As a result, machines driven by pattern chains move more slowly than the ones driven by the pattern wheel. Just like the pattern wheel, the pattern chain is also negative in character.

The links of a pattern chain are made from hardened steel and have various shapes depending on the type of lap to be carried out. These are marked by the letters a, b, c, and d. The A-link (Figure 13.38) is without a slope, a B-link is ground to a slope on the end that first runs under the shogging roller. A C-link is ground on the opposite end, and a D-link is ground around on both ends. The links move fork end first. The links have different heights representing needle spaces, and on warp-knitting machines they run one needle space at a time being numbered 0, 1, 2, 3, 4, and upward.

Figure 13.39 represents a pattern chain for a three-phase displacement. The steps with the transition points for the underlaps and overlaps are marked from a to m. The run of the chain can be described as follows:

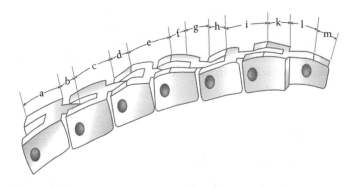

FIGURE 13.39
A three-phase pattern chain.

	a Swinging in of the guides
	b Guides shog for overlap
	c Swinging out of the guides
	d First shog for underlap
1st Course	e Stopping position of the guides
	f Second shog for underlap
	g Swinging in of the guides
	h Guides shog for overlap
	i Swinging out of the guides
2nd Course	k First shog for underlap
	l Stopping position of the guides
	m Second shog for underlap

Pattern chains enable a larger palette of patterns to be produced. However, the inventory of pattern chain links can also be appreciable (different heights and forms), and the precision of grinding would dictate the speed at which a machine can run.

13.4.3 Electronic Shogging

With the advent of electronics and the widespread use of stepper motors, the generation of shogging motion became liberated from an extremely cumbersome process of preparing sets of chains, driving a very heavy set of chains, and maintaining a large inventory of links as well as dependence on the skills of the chain maker. In an electronic system, the push rod linked to a guide bar receives motion directly from a stepper motor. Hence, this motion is positive in nature.

The stepper motor receives a command from a program stored in a microprocessor. In the microprocessor, a memory card is loaded, containing details of the pattern repeat of all guide bars. In addition, the card is also fed the yarn feed rate and the fabric take-up rate. These data can be stored in a PC and therefore can be retrieved when necessary. Reproduction of a sample is therefore very easy with such systems.

13.5 Some Important Warp Knits

The versatility of the principle of warp knitting is expressed by the ease with which fabrics of a wide range of properties can be generated very elegantly by some simple manipulations. By adjusting the extent and direction of underlap of different guide bars, the resultant fabric can be made highly stretchable or highly rigid. A rigidity in the machine and cross directions even higher than that of woven fabrics can be achieved by inserting straight warp and weft yarns and trapping them within the space between the underlaps and loops of the knitted structure. Indeed, the multiaxial warp-knitted fabrics possess extremely high rigidity even in bias directions. Such fabrics display planar isotropicity. Yarns that are not knittable but otherwise possess properties that are crucial for certain functions can be easily laid in without any damage either to the knitting elements or to the material itself. Similarly yarns with surface properties incompatible with the surrounding conditions can be hidden in the core layer of a multilayered warp-knit fabric. A range of pore sizes of given shape and dimension can be developed in the fabric by part-set threading of guide bars and intelligent selection of lapping plans. These and some other special features of warp-knitted constructions are briefly outlined in the following.

13.5.1 Single-Bed Knits

13.5.1.1 Nets

13.5.1.1.1 Nets on Tricot

It is possible to generate many net-like open work structures on tricot machines by the technique of part-set threading. In principle, it is necessary that, in each course, every needle receives at least one yarn so that knitting continues uninterrupted. When more than one guide bar is employed, it is possible to skip certain guides in each guide bar while drawing threads through them, a process termed as part-set threading. By choosing the lapping plan judiciously in conjunction with part-set threading, one can generate elegant net-like structures. The basic idea derives from the specific features of warp-knitted fabric formation, embodied in Figures 13.10 and 13.12. It is observed from Figure 13.10 that a loop may be made to pull away from the wale line if the relevant underlap is permitted a free play. Similarly, it is observed from Figure 13.12 that a gap between neighboring wale lines can be created by skipping underlaps. The corresponding length of opening can be controlled by varying the number of courses over which the underlap is skipped. An example is shown in Figure 13.40.

In the alleys 0, 1, 2, 3, etc., of the needle bar, the guides of the two guide bars I and II exhibit part-set threading indicated either by a + sign (yarn present) or a vertical line segment (no yarn). It is observed that guides of both bars carry yarns only in the odd-numbered alleys, and they are made to lap always in the "counter-lapping" mode. By choosing the length of underlaps properly, it is ensured that the yarns from guides of the two guide bars do not lap simultaneously on any needle. Hence, each needle gets only one yarn in every course, creating, in the process, loops that are inclined away from each other. As a result, the even-numbered alleys exhibit a crack during the first three courses,

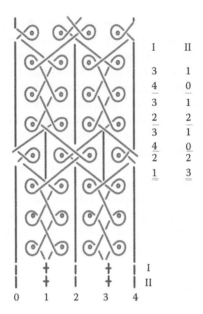

FIGURE 13.40
Lapping diagram and lapping plan of a net-like structure on a tricot machine.

indicated by the long line segments. These cracks are bridged in the fourth course by increasing the length of underlap of the two guide bars but resumed from the fifth course onward. Very short pinhole openings develop in the fourth course along the odd-numbered alleys as indicated by the short line segments. The resultant fabric would thus exhibit openings of two distinctly different shapes and dimensions distributed in a geometric pattern all over its surface.

13.5.1.1.2 Nets on Raschel

This is being covered in Section 13.5.1.3 under the broad category of weft-inserted structures.

13.5.1.2 Inlaid Structures

On warp-knitting machines, it is possible to produce laid-in (or inlaid) structures by making a guide bar execute (a) inlay lap, that is, underlap but no overlap and/or (b) mis-lap.

An inlay lap is one that passes backward and forward across the fabric, the corresponding yarn getting embedded in the structure by virtue of the fact that it is trapped between the face loop and the underlap of every stitch it crosses. Such a situation is illustrated in Figure 13.41 in which the back side of a tricot fabric, produced evidently by the front bar, traps the underlaps (0 0/3 3//) of the back bar. The inlay yarn may, by nature, be very difficult to knit, but its

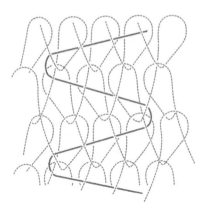

FIGURE 13.41
Lapping diagram of an inlay lap.

presence may be crucial for the functionality of the fabric. Such a yarn may, for example, be a stiff conducting element or even a highly stretchable material.

It is also possible to inlay a yarn along the warp direction (Figure 13.42). The filler or mis-lap is a line of thread that is laid in the alley (lengthwise) between two neighboring wales and trapped in place in a manner similar to the inlay lap. A three-bar laid-in structure is depicted in Figure 13.42, in which the front bar (bar III) knits a tricot, the second bar executes a mis-lap (0 0//), and the back bar (bar I) executes the inlay lap (0 0/4 4//). The mis-lapped yarn remains absolutely straight along the fabric length, and its tensile property can be fully exploited in the resultant fabric.

The structure shown in Figure 13.42 is suggestive of an extremely interesting possibility of construction of electronic circuits embedded in a

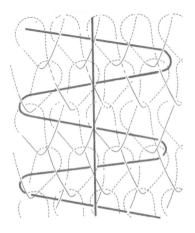

FIGURE 13.42
Lapping diagram of a mis-lap.

warp-knitted fabric if in conjunction with suitable conducting yarn material and part-set threading of guide bars, mis-laps and inlays are appropriately developed according to desired circuit diagrams.

13.5.1.3 Weft-Inserted Structures

A thread can be inlaid over a limited distance in a course due to limitations in the extent of shogging possible in one cycle of operation. However, just as a mis-lap can impart certain additional functionality to the product, a straight piece of yarn embedded in the fabric along its entire width imparts to the product an additional dimension. Such products fall under the category of weft-inserted structures.

The weft-insertion method employed on Raschel machines with the help of the magazine weft-insertion technique enables production of weft-inserted structures that have many technical and household end uses. The magazine, outlined in Figure 13.43a, consists of a rotating creel (3), which carries cones

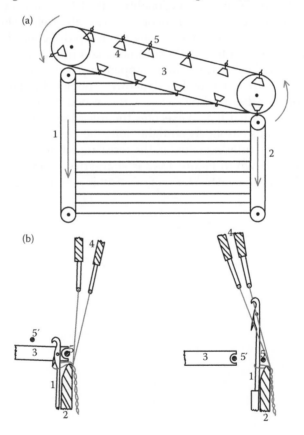

FIGURE 13.43
Magazine weft insertion on a Raschel. (a) A magazine. (b) Weft insertion process.

of yarn packages (4), and guides with clamps (5). Yarns of the required length drawn from each cone are gripped by clips carried on two rotating chains (1 and 2).

The creel rotates in a counterclockwise direction, and the two chains rotate in opposite directions. The tips of yarns from the cones are initially gripped by the chain on the left-hand side, which drags them away from the creel. Because of the continuous movement of the creel and the rotating chain, the corresponding yarn segments are simultaneously unwound from respective cones and moved closer to the knitting zone. When the unwinding cones reach the chain situated on the right-hand side, the corresponding yarn ends are gripped by clips of this chain, and a cutter snaps the yarn segments linking the cones and the clips. Thus, a series of yarn segments equal to the distance between the chains, which is set equal to the length of weft to be inserted, are cut off from each cone by this system and moved toward the knitting zone at the required speed. As a result, a magazine of pieces of weft yarns is continuously developed by this system.

These yarn segments are carried forward up to the front of the needle of the Raschel machine by a special sinker, shown in Figure 13.43b. Here, 1 is the needle, 2 the trick plate, 3 the special sinker, 4 the guide bars, and 5 (5′) the cross-sectional view of two consecutive weft yarn segments fed by the magazine. When the needle bar sinks, the yarn (5) is pushed bodily to the yarn sheet coming from the guide bars (4). Subsequently, as the needles rise, the sinkers withdraw, and the yarn segment gets trapped on the back side of the needles in the form of a continuous straight weft. Such a state is shown on the right-hand part of the figure. This weft yarn would finally occupy the core of the multilayered warp knit. The sinker withdraws and catches the next weft segment (5′) in its groove to be pushed into the knitting zone in the next cycle. The right-hand part of figure demonstrates this state.

Mis-lapped straight warp threads together with magazine-inserted wefts result in a strong orthotropic mesh. A clever juxtaposition of this concept with a modified inlaying approach leads to the isotropic multiaxial warp knit, described in Section 16.3.2.

13.5.1.4 Loop or Pile Structure

One or more of the following techniques, namely special points or other elements in the knitting machine, excess feeding of the pile yarn, and raising or brushing of the pile surface during finishing, is normally involved in production of loop or pile structure on warp-knitting machines.

Uncut piles can be produced on Raschel machines by employing an additional bed (2) of points (3), the principle of operation of which is illustrated in Figure 13.44. Yarn from the guide bar (1) (the back guide bar) wraps additionally around the point, which is subjected to an appropriate up and down motion. A fall plate (4) pushes the pile yarn below the latch of the needle, thus ensuring that these yarns are tucked into the structure. This step results

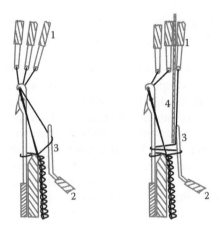

FIGURE 13.44
Raschel for knitting pile.

in a reduction of tension in the pile loops and growth in the volume of each pile. At the end of a cycle, the points are moved down, releasing the course of piles. Through selective sequencing of points and programmed lateral movements, ornamental piles can be developed on the technical back side.

13.5.2 Double-Bed Spacer and Cut Plush Fabrics

The concept of generating the 3-D spacer fabric and that of cut plush fabrics has been discussed in Section 13.3.3.

13.6 Comparison of Warp-Knitting Process Vis-à-Vis Other Yarn-to-Fabric Conversion Processes

The process of warp knitting a sheet of parallel yarns into a sheet of fabric is unique in the sense that all yarns are engaged at every instant of time in the fabric-formation process. In this sense, time utilization is maximized in the warp-knitting process.

In the commercial monophase weaving process, one observes that the processes of shedding, picking, and beating-up occur sequentially. In other words, all warp yarns are first split into a shed, after which the picking of weft takes place, during which shed lines do not change. When picking is over, beating up takes place. Insertion of a pick during the beating up process is inconceivable. This sequential nature of primary motions slows down the production process. To overcome this drawback, the concept of multiphase weaving was developed, in which all the three primary motions are operational at any given time, either across or along the warp sheet. However, multiphase weaving machines have not yet experienced commercial success.

The next best system is found in weft knitting and, more specifically, in multifeeder circular knitting machines. The actual process of conversion of feed yarns takes place simultaneously at multiple locations along the periphery of the knitting cylinder. Even then, a substantial fraction of needles, which convert yarns into loops, remain idle between adjacent feeders. This means that the process of addition of courses along the active fabric edge takes place in a fragmented manner. As many courses as there are number of feeders are formed simultaneously although only one loop gets added to each course at any given instant of time. For example, if the machine has 3000 needles and 120 feeders, then only 120 loops are under formation at any given instant of time and not 3000 loops.

In a warp-knitting machine, however, all the needles are forming loops all the time such that formation of all loops of an entire course takes place simultaneously. Accordingly, all the warp yarns are engaged in the formation of fabric continuously.

The other unique feature of the warp-knitting process is observed in the manner of effecting a change in construction.

In the process of weaving, different constructions are effected by altering the sequence in which warp yarns are split into the shed. The possible location of a yarn follows binary principles, and possible combinations of such binary locations under certain constraints result in different woven constructions. With modern electronic Jacquards, it has been possible to achieve a very high degree of individualization of warp yarns, resulting in a very high combination of the binary locations.

In the process of weft knitting, the object of selection is not the yarn but the needles. Each needle can occupy three different locations during the loop-formation process, and the possible combinations of needle locations under certain constraints result in different weft-knitted constructions. Of course, there are other methods of generating new constructions, such as through loop transfer, racking of beds, etc. These can be clubbed under the category of additional or supplementary tools and hence not taken into consideration here, just as extra weft-insertion or lappet and leno processes in woven fabric manufacturing are regarded as additional tools and kept out of purview. Just as in the process of woven fabric manufacture so in the process of weft knitting, electronic selection of needle positions has resulted in a very high degree of individualization insofar as control on the object is concerned.

In the warp-knitting process, however, this aspect of individualization is markedly restricted. The object of choice is neither the individual warp yarn nor the individual knitting needles but the guide bars. Each guide bar controls a set of warp yarns, and therefore, all warp yarns within a set behave similarly. In this sense, the selection principles can be compared to the dobby of a weaving machine, which decides upon the binary location of a heald that controls a set of warp yarns. However, the selection of location of a guide bar is subject to a much higher degree of freedom than the binary location of a heald, given by the extent and direction of underlap by which a guide bar can be made to move laterally. In this sense, the selection process enjoys a much higher degree of freedom than the weft-knitting or weaving process.

To sum up, it may be stated that the technology of warp knitting is more productive than weaving or weft knitting, and the highest degree of freedom in selecting the location of the object, central to determining the possible range of constructions in a given setup, is also highest in warp knitting. With respect to the degree of freedom in selecting the location of the object, central to determining the possible range of constructions in a given setup, weft knitting has an edge over weaving. However, with respect to the possible number of objects of selection in a given setup, both weaving and weft knitting enjoy a similar superiority over the process of warp knitting.

Further Readings

Paling D (1970). *Warp Knitting Technology*, Columbine Press, Buxton, Great Britain.
Raz S (1987). *Warp Knitting Production*, Verlag Melliand Textilberichte, Heidelberg.

14

Formation of Braids

14.1 Introduction

Braids are produced from single yarns as well as from a collection of parallel yarns, termed strands, or even from flat tapes. The products are invariably of narrow width and may come in the form of flat bands with two firm selvedges or as tubes, either hollow or stuffed with a core material. Braided products have been traditionally used as decorative trimmings and functional elastic components in apparel goods. They are also used in demanding technical applications, such as for shielding wires from electromagnetic interference or for absorbing very high impactful energy in the form of ropes, fishing lines, parachute cords, etc., and also for satisfying fairly modest and less demanding functions in household goods in the form of draw threads for curtains, wash lines, or even the ubiquitous shoelace (Brunschweiler, 1953, 1954a; Ko, 1987). In view of their structural peculiarities, braids can absorb the high energy of deformation and can take very complex contours without putting any high demand on the constituent yarns, strands, or tapes. Indeed, as opposed to knitting and weaving processes that impose very high and exacting standards on yarn quality primarily for withstanding the strains of the conversion process itself, the braiding process makes no such demands, and choice of yarn is governed entirely by the functional requirements of the end use.

14.2 Geometry of Tubular Braids

Going by the definition given in ASTM D123-49, a braided fabric is produced by interlacing several ends of yarns in such a manner that the yarn paths are not parallel to the braid axis. A tubular braided product meant for geotechnical end use is exhibited in Figure 14.1. It is a hollow braided tube, the top of which has been partially cut open to reveal 16 thick parallel plied threads encased within the tube. These parallel threads are not an integral part of the

Brecodrain—16
(54 ml/s)

FIGURE 14.1
Braided prefabricated vertical drain.

braided tube and are meant to promote the discharge capacity of this geotechnical drain. They can be ignored for the present while taking note of this
important facility of braiding technology that enables the braiding of yarns
into a sheath form around a core. This facility is made use of in manufacturing a wide range of braided products such as shoelaces or parachute cords or
shielded conducting elements, to name a few.

The braided tubular structure depicted in the figure has been formed, in
this instance, from a certain number of tape-like strands, each strand being,
in turn, made up of five thinner yarns. The surface of the tube reveals that
half the strands move from left to right, and the other half moves from right
to left. Without having a basic understanding of the structural features of
braided constructions, it is not possible to make out the number of strands
employed for generating this tube as part of the fabric surface is hidden
from view and the strands follow a helical path around the tube surface.
However, a clear pattern of structural repeat along and across the braid axis
is discernible. There are very clear zigzag lines with well-defined ridges
and furrows on the fabric surface. It is also observed that the crossing angle
between braiding strands in this particular case is greater than 90°. Indeed,
the crossing braiding strands can subtend a wide range of acute and obtuse
angles, and this variable constitutes an important parameter for designing
a braided product. The interlacement pattern among the strands resemble,
however, that of a woven two-up, two-down twill if the figure is rotated
through 45° and the obtuseness of the angle of interlacement between the
strands is ignored. There are apparently some similarities between braided
and woven fabrics.

Three symmetric weaves are shown in Figure 14.2. As one moves from a
1/1 plain weave to a 2/2 twill and then to a 3/3 twill, the imprint of bands

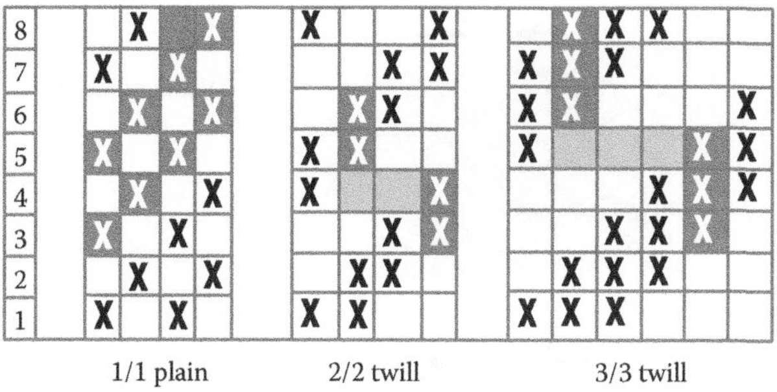

8		X		X		X			X			X	X	X		
7	X		X					X	X	X	X	X				
6		X		X			X	X	X	X						X
5	X		X			X	X		X					X	X	
4		X		X	X			X					X	X	X	
3	X		X				X	X			X	X	X			
2		X		X			X	X		X	X	X				
1	X		X		X		X		X	X	X					
	1/1 plain			2/2 twill				3/3 twill						

FIGURE 14.2
Three woven constructions.

created by floats of warp (dark strips) and weft yarns (gray strips) become more and more pronounced. If these weaves are mentally rotated through an angle of 45°, then the corresponding zigzag bands would become parallel to an imaginary X-axis. Similar bands along the width direction of the exhibited braided tube in Figure 14.1 can also be easily recognized. The nomenclature employed for some basic braids, which are essentially equivalent to corresponding weaves, are displayed in Figure 14.3 (Brunschweiler, 1953, 1954b). A particular note needs to be taken of the terms "line" and "plait" and the distances they represent in the corresponding braided interlacements. It is also observed that the braided tube exhibited in Figure 14.1 has the construction of a regular braid.

An important structural characteristic of a braid pertains to its repeating unit. A piece of braided fabric is made up of a certain number of lines and plaits, and its repeating unit exhibits a certain line width and a certain plait height. A line width is equivalent to the repeating distance of one zigzag band, and a plait height is a measure of the distance between two successive zigzag bands along the length direction of the braid or, more appropriately, along the braid axis (in this case the Y-axis). In this sense, the visible side of the braided fabric exhibited in Figure 14.1 is made up of four and a half lines along its width. Therefore, taking account of an equal number of lines along its width on the hidden surface, the entire tube width must have been made up of twice as many lines, namely nine lines.

The schematic diagram of a regular braid is shown in Figure 14.4. The angle of interlacement in this diagram has been kept at 90°, and the rectangle made up of eight strands in each direction conforms to the pattern of interlacement of the sample exhibited in Figure 14.1. Within this diagram is shown the rectangle ABCD containing one repeat of the 2/2 twill weave, and another rectangle JKLM outlines the corresponding appearance of the braided interlacement enclosing two lines and two plaits. The arm BA covers the space of four strands and is equal in length to the arm B'A', which, when projected on the imaginary X-axis, equals JK. Accordingly, one infers that the width JK of two lines is made up of the effective space occupied by one

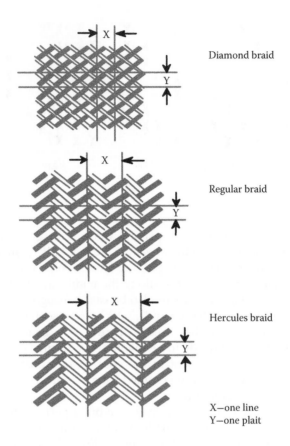

FIGURE 14.3
Three braided constructions. (From Brunschweiler, D., *Journal of Textile Institute,* 44, 666–686, 1953, CRC Press.)

strand interlacing with four crossing strands when the same is projected on the X-axis. Hence, the space of nine lines can be generated through interlacement of one strand with a total of 18 crossing strands. This would suggest that the tube exhibited in Figure 14.1 made up of nine lines was braided from 18 + 18 = 36 strands. This line of analysis leads to another simple observation that each strand interlaces with half the total number of strands during one complete helical path.

The corresponding plait height is given by half of the length of KL and represents the length added to the braided fabric in the course of one cycle of interlacement between the interlacing strands. The number of plaits produced in a regular braid in one complete helix traced by a braiding strand is evidently equal to half the total number of strands employed for producing the braid as every strand has to interlace successively with each strand moving in the opposite direction before it returns to its origin. In this particular example, then a total of 18 plaits are generated per helix.

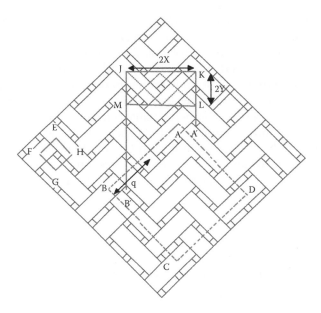

FIGURE 14.4
Simplified view of a regular braid.

Referring further to Figure 14.4, it is observed that a rectangle EFGH has been marked out at the left-hand corner for the purpose of relating it to the rectangles ABCD and JKLM. A blown-up view of this rectangle is shown in Figure 14.5. It resembles a trellis made up of four arms, each of modular length ℓ. This trellis covers the space occupied by one line and two plaits in

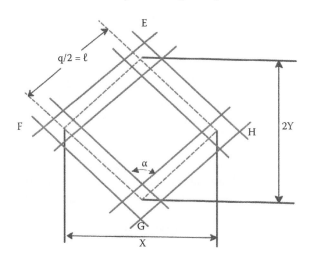

FIGURE 14.5
Trellis model of a braided cell.

a braided structure. Modular length ℓ (mm) of each arm of the trellis is being defined as the axial length of yarn between two neighboring interlacements.

Let n represent the total number of strands employed in producing a braid. Evidently, each yarn strand contributes a length ℓ within one plait, and as there are in all n number of strands taking part in the braiding process, it is concluded that an amount of $(n\ell)$ length of yarn is consumed for forming each plait. As a total of $(n/2)$ number of plaits is formed in one complete helix of the braiding strands, the total length of the yarn strand contained in the braided fabric within this helix unit must be $(n^2\ell/2)$. Knowing the count of the yarn strand (T tex) and the mass of fabric contained in a complete helix (w gm), it is possible to estimate the value of ℓ. Accordingly,

$$\ell = [2w/n^2T] \times 10^6$$

Let

d_b = diameter of braided tube

P = flattened width or part perimeter of braided tube

X = line width of braided fabric

Y = plait height of braided fabric, and

α = braid angle

It follows from Figure 14.5 that

$$X = 2\ell\sin(\alpha/2)$$

and

$$Y = \ell\cos(\alpha/2)$$

Hence,

$$\pi d_b = 2P = (n/4)X = (n/2)\ell\sin(\alpha/2)$$
$$= (n/2)[2w/n^2T]\sin(\alpha/2) \times 10^6$$
$$= [w/nT]\sin(\alpha/2) \times 10^6$$

If the length of braided fabric contained in one complete helix of braiding strand is termed the pitch and abbreviated as h, and the length of the strand in a helix be abbreviated as L, then it follows from Figure 14.6 that

$$L = \pi d_b/\sin(\alpha/2)$$

and

$$h = \pi d_b\tan(\alpha/2)$$

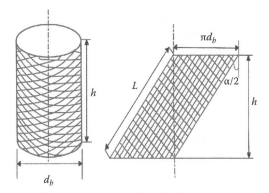

FIGURE 14.6
Geometry of a braided tube. (From Zhang, Q. et al., *Journal of Textile Institute*, 88, part 1, 1, 41–52, 1997.)

Accordingly,

$$L = [w/nT] \times 10^6$$

This relationship suggests that if a helix-length of strand within one braid pitch is increased, keeping the count of the strand and its number undisturbed, then the weight of the fabric produced in that length would increase proportionately.

The model of the braided structure that emerges from the consideration of a trellis unit forming the unit cell can be summarized as follows:

If a total number of n strands each of count T (tex) are braided into a symmetric structure with half of the strands moving in a direction opposite to the other half, then it follows that

- length (mm) of each strand within one braid pitch = $(n\ell/2)$ (14.1)
- total length (mm) of strands contained in one plait = $n\ell$ (14.2)
- weight (gm) of fabric in one plait = $(n\ell) T \times 10^{-6}$ (14.3)
- total length (mm) of strands contained in one pitch of braided fabric = $(n^2\ell/2)$ (14.4)
- line width (mm) = $2\ell\sin(\alpha/2)$ (14.5)
- plait height (mm) = $\ell\cos(\alpha/2)$ (14.6)
- width (mm) of opened out tube = $(n\ell/2)\sin(\alpha/2)$ (14.7)
- length (mm) of fabric per pitch of braid = $(n\ell/2)/\cos(\alpha/2)$ (14.8)
- weight (gm) of fabric per pitch of braid = $(n^2\ell/2)T \times 10^{-6}$ (14.9)
- diameter of braided tube = $(n\ell/2\pi)\sin(\alpha/2)$ (14.10)

Evidently, the two crucial parameters of a braid structure are modular length ℓ and braid angle α. Incidentally, for woven fabrics, the corresponding parameters are modular length ℓ and crimp angle θ.

14.3 Elements of a Tubular Braiding Machine

A braided fabric may be produced in two forms: flat and tubular. In both instances, the basic machine elements remain the same (Brunschweiler, 1954c). The strands to be braided are wound on flanged bobbins, which are then mounted on spindles. A spindle with its yarn-tensioning accessories along with its negative let-off system forms one complete carrier unit. On a typical braiding machine, the underside of the base of the spindle is equipped with projections and depressions, which lock in a close fit into corresponding grooves and projections cut on the upper surface of the base of the carrier unit (Figures 14.7, 14.8, and 14.9). In this manner, the spindle is

FIGURE 14.7
View of the underside of the base of a spindle.

FIGURE 14.8
Side view of the upper surface of a carrier.

FIGURE 14.9
Plan view of the upper surface of a carrier.

coupled positively to the carrier. The carrier base is, in turn, loosely mounted on an upright shaft, which is rigidly connected to a platform that rests on the machine bed plate. This platform has an extension in the shape of an oblong shaft that passes through a cam track–shaped slot cut in the base plate. This shaft is driven by a horn gear system, which therefore moves the entire carrier unit along a specific serpentine path around the periphery of machine base plate.

Ratchet teeth are cut along the exposed upper surface of the carrier base. A projecting finger from a block B_1 that slides along an upright spindle S_1 rests in the ratchet of the carrier base (Figure 14.10). This block B_1 is pushed down by a compression spring, keeping the carrier base in a locked condition. In

FIGURE 14.10
Locking device of a spindle carrier.

effect then, a finger presses against the ratchet teeth on the carrier base, preventing it from rotating and releasing the strand from the bobbin unless the locking finger is lifted away from the ratchet. The strand from the bobbin is taken around a set of pulleys (Figure 14.11) before exiting through a ceramic eye of the carrier unit. The lower set of pulleys of the carrier unit is mounted on block B_2 that slides along a spindle S_2 (Figure 14.12). This block is pressed down by a long compression spring K_2. When the strand from the bobbin is pulled out through the outlet eye, the block B_2 rises up against the force of compression spring K_2 and approaches the base of an extended loop formed out of the top of the tripping arm A (Figure 14.13), which, in turn, is connected to block B_1. Simultaneously, the tension in the yarn strand also rises to a critical value. Eventually, the block B_2 pushes the tripping arm upward, which, in turn, lifts up the block B_1 and thereby moves the locking finger away from the ratchet on the carrier base. At the moment of release of the locking finger, the tension in the yarn strand exerts a sufficiently high tangential force to the bobbin forcing it to rotate and release a certain length of strand. The resultant drop in strand tension allows the blocks B_1 and B_2 to slide down under individual spring pressures and activate the locking finger afresh. Such a negative let-off motion results in a sawtooth tension

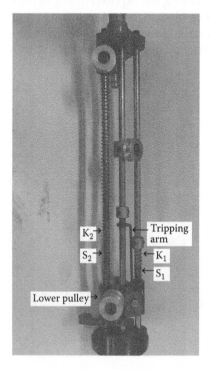

FIGURE 14.11
Pulley system for tensioning of a yarn strand.

FIGURE 14.12
Spring-loading system of a lower pulley block.

FIGURE 14.13
Functional arrangement of lower pulley block and tripping arm.

profile quite akin to the one described for weaver's beams in Figure 7.28. By suitably selecting the stiffness of compression spring K_2 in conjunction with properties of the braiding strand and adjusting its extent of compression, it is possible to maintain an average tension level in the strands to the lowest possible value that permits a hassle-free braiding. Indeed, it is possible to set tension values in keeping with the nature of the strands and also set different tension values on different carriers for causing a desired distribution of modular length among the braiding strands, inducing thereby an ornamental periodic distortion in the product.

As mentioned in the foregoing, carriers are driven by a horn gear system along a serpentine track, illustrated in Figure 14.14. For tubular braids, the movement is broadly along the periphery of a circle, and flat braids require a reciprocating motion. Discussion in this section would be restricted to rotary braiders meant for producing tubular braids only.

On a rotary braider (Figure 14.15), half the carriers move in a clockwise direction, and the other half moves counterclockwise. Two sets of eight carriers have been illustrated in the diagram, alternating between locations on an

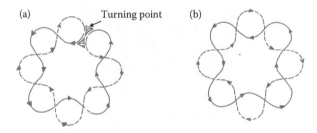

FIGURE 14.14
Types of braid. Flat (a) and tubular (b) braids.

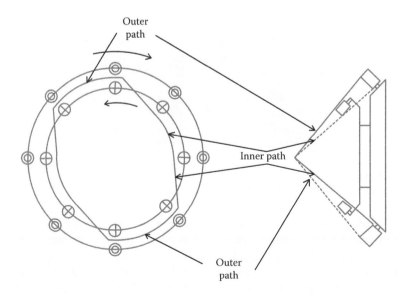

FIGURE 14.15
Set up of spindles on a rotary braider.

outer circle and an inner concentric circle while maintaining their overall motion in opposite directions. Braiding under such an arrangement can be carried out in two possible manners.

In the more commonly encountered solution, a carrier can be located in either of the two circles, dragging the strand it carries along a path indicated by the continuous serpentine line shown in Figure 14.14. This path would be traced by all carriers moving in the same direction and would be overlapped by another path with a 180° phase difference traced by the carriers moving in opposite directions. It is this crossing of two paths that leads to interlacement among strands, which, in the case illustrated in Figure 14.15, would lead to a 2/2 or a regular braid. If the frequency of interlacement among strands is doubled by forcing each carrier to cross over from one circular track to the other after moving past just one coming from the opposite direction, then a 1/1 or a diamond braid would be formed. This logic can be extended for other symmetric constructions, such as 3/3 or 4/4, etc. A reduction in interlacement among strands would tend to group them closer together leading to a drop in fabric diameter and porosity.

In the other solution, each carrier is permitted to continue moving in either of the two tracks. Thus, instead of a complicated combination of reciprocating and circular motions, carriers enjoy a continuity in direction of motion, enabling them to move faster. However, in such a situation, some additional intervention is necessary for interlacing the two layers of strands. To this end, strands from carriers mounted on the outer circle are periodically moved back and forth across the path of carriers moving along the inner circle. The resultant path of the corresponding strand carried by a carrier from the outer circle is represented by the zigzag track shown in Figure 14.15 along with the corresponding view of switching of strand position shown alongside. However, such a solution demands a special flexible mounting of spindles on the carriers occupying the inner circle. These spindles are mounted so as to be under control of the corresponding carrier at one of its ends only leaving an open space at its other end through which the strand from an outer carrier can pass through during its reciprocating motion. Subsequently, there would be a switch in the controlling end of the spindle by its carrier on the inner circle, and the strand from the outer carrier can come out through the corresponding gap during its return motion, encircling in this process the strand from the carrier on the inner circle. Evidently, such a continuous and precise switching of controlling ends of a large number of packages on their carriers calls for complicated and very precise engineering, making such machines fairly expensive.

For tubular braids, carriers are equally divided between two groups that move in opposite directions. The total number of carriers employed for braiding simple tubular fabrics is always a multiple of two. An uneven number can be employed for producing flat braids although it is possible to produce flat braids from an even number of carriers as well.

The principle employed in realizing the serpentine motion of carriers is illustrated by the Maypole braider shown in Figure 14.16. The term Maypole

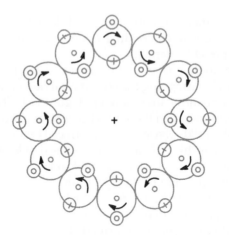

FIGURE 14.16
Nature of spindle movement on a Maypole braider.

is borrowed from the ancient European tradition of bedecking a very tall pole with flowers, colorful flags, and painted stripes while people dance around it on the first of May in celebration of the onset of summer. In the schematic view of the braider shown in Figure 14.16, a total of 12 large circles arranged around the center of a base plate house 24 carriers, represented by small circles, which are divided into two equal groups moving in opposite directions. Each of the 12 circles represents a horn gear. Horn gears have certain segments cut out in the form of notches and are set in such a way that an opposing pair of notches of neighboring horn gears always face each other. As neighboring horn gears always rotate in opposite directions, a carrier guided by a particular notch and moving clockwise along a semicircle when transported to a notch corresponding to a neighboring horn gear would move counterclockwise along the succeeding semicircle.

A simplified schematic diagram of the principal elements involved in a Maypole braider is shown in Figure 14.17. Strands from individual carriers converge to the braiding point close to a suitable take-up unit. An imaginary vertical column passing through the braiding point forms the Maypole, and the carriers moving along their serpentine tracks constitute the proverbial dancers. The individual segments of the serpentine track are not really circular but somewhat oblong, their width narrowing down near the crossover points, forcing the shaft below the carrier base to slide into the neighboring track (Figure 14.18). A cross-section of the carrier base shaft is matched suitably with track geometry for ensuring a smooth takeover by the notch of the succeeding horn gear at a crossover point. Each carrier base is equipped with a suitably designed platform that keeps the carrier axis stable while sliding along the track plate. The frictional resistance to sliding between the platform of the carrier base and the plate is overcome in a typical solution by providing for cam track guidance of an extension of the carrier base in an

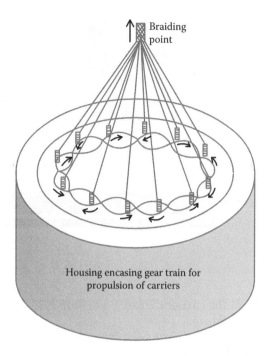

FIGURE 14.17
Principal elements of a Maypole braider.

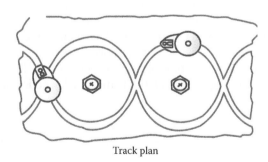

Track plan

FIGURE 14.18
Track plan on base plate. (From Brunschweiler, D., *Journal of Textile Institute*, 44, 666–686, 1953, CRC Press.)

enclosed oil bath (Turner, 1976). This results in a smoother motion and, hence, higher carrier speed in addition to preventing oil smudges on the strand as no external lubrication of mutually sliding surfaces is required anymore.

Horn gears are mounted on an intermeshed train of spur gears, which are mounted along a circular path on the machine base plate (Figure 14.19). The main motor is linked directly to one of these gears. The manner in which a

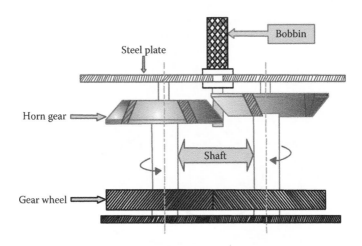

FIGURE 14.19
Drive to horn gears.

shaft extending from the carrier carrying a bobbin is guided by a horn can be observed from the figure.

The number of horns on a horn gear depends on the braid construction. Thus for 1/1, 2/2, and 3/3 braids, a gear, respectively, with two horns, four horns, and six horns would be required. A gear with four horns can also be employed to braid a 1/1 with a suitable arrangement of carriers. Thus, a machine with 18 horn gears each equipped with four horns can be employed to braid a 2/2 construction involving 36 carriers or even be employed to braid a 1/1 construction involving 18 carriers each carrying two strands.

Arrangement of carriers on five neighboring horn gears each equipped with two, four, and six horns for braiding, respectively, 1/1, 2/2, and 3/3 constructions is shown in Figures 14.20, 14.21, and 14.22. Carriers are represented

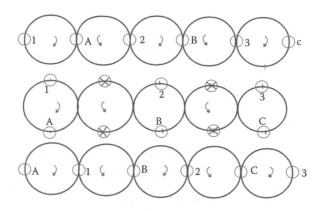

FIGURE 14.20
Arrangement and motion of carriers for a diamond braid.

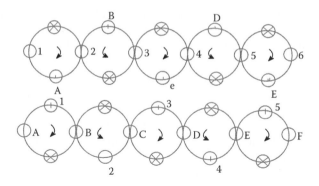

FIGURE 14.21
Arrangement and motion of carriers for a regular braid.

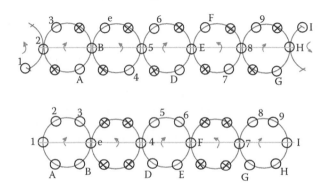

FIGURE 14.22
Arrangement and motion of carriers for a Hercules braid.

by small circles located along the periphery of horn gears, which, in turn, are represented by large circles. A horn that is free of any carrier is represented by a crossed-out small circle. The set of alphabetically designated carriers moves from right to left (or in a counterclockwise direction if these gears are assumed to be arranged along the arc of a circle), and the ones designated by numbers move in the opposite direction. When two horns of neighboring gears come face to face, there is always an exchange of carrier, the one bringing it to that point handing it over to the neighboring gear. A plait is formed when an exchange of carrier between neighboring horns takes place. In Figure 14.20, such an exchange occurs twice in one complete rotation of every horn gear, and in Figures 14.21 and 14.22, such exchanges occur four and six times, respectively. These numbers match precisely with the number of horns in the respective gears as well as with the number of plaits per repeat of braid, explaining the logic behind the choice of number of horns for the corresponding braid. One also observes that a half rotation of the horn of Figure 14.20 corresponds to one plait, and a one-fourth and a one-sixth rotation of the corresponding horns of the succeeding two figures represent one plait for the respective braids, which leads to the generalized statement that one complete rotation of a horn coincides with one complete braid repeat.

By looking at the arrangement of carriers on their horns and keeping in mind their directions of motion, one can find out the number of plaits required for any particular carrier to reach a position occupied by another carrier in its path ahead. For example, the carrier C would always be four plaits behind the carrier A in Figures 14.20 and 14.21 but would lag by only two plaits in Figure 14.22.

Strands of carriers designated by alphabets would interlace with those of carriers designated by numbers, and each carrier of any particular alphabet would interlace sequentially with strands of all numerically designated carriers. Obviously, the reverse also holds true. It is observed that in three stages depicted in Figure 14.20, that is, in a one-half rotation of each horn, strands of all carriers designated by alphabets move across and below the strands of the neighboring carriers that are numerically numbered. In another half a rotation, they move above and across the respective numerically numbered carriers, completing, in the process, one repeat. Switching over to Figure 14.21, it is observed that, in one step, the horn gears move over by a quarter of a rotation, completing one plait. Four such steps are required for one complete braid repeat. But in every single step, all alphabetically numbered carriers shift by one unit distance toward the left and the numerically numbered ones shift by one unit to the right. Moreover, considering the carrier D, for example, it is observed that, initially, it is located above carriers 4 and 5, and its movements suggest that, in the subsequent two plaits, it is going to move below and across carriers 2 and 3 while its neighbor E, which is initially positioned below carriers 5 and 6, is going to move above and across carriers 3 and 4. Similar mental observations may be carried out for all the carriers shown in the diagram by means of which a rational correspondence between the quasi-static carriers, and the corresponding braid can be formed. The layout of five complete horn gears each equipped with six horns with carriers located suitably for realizing a 3/3 braid has been illustrated in Figure 14.22 for two consecutive plaits and observations made in the foregoing for the other constructions apply here as well.

As opposed to its counterpart the weaving loom, in which changing over from a 1/1 construction to a 3/3 construction implies only a change in the shedding motion assembly associated with driving a larger number of healds with different drafting and lifting plans, a corresponding change on a braiding machine would imply changes in horn gear assembly and the corresponding gearing system, a feat difficult to execute on production floors. Moreover, an increase in the number of carriers on a horn gear demands bigger gear systems, and hence, if the carrier assembly has to remain unchanged, then the diameter of each horn gear and its driving gear along with the space between the centers of neighboring systems also have to increase, meaning thereby effectively a completely new machine. In this sense, a braiding machine has extremely limited scope as compared to a weaving loom.

The system of arrangement of horn gears along a circular track with the braiding point located at the center of the circle at a level considerably higher than that of the continuously moving carriers imposes a serious ergonomic limitation insofar as accessibility of individual strands as well as of braided fabric to the operator is concerned, especially when spindles are large in size and number of carriers is fairly large. Switching over from a vertical braider to a horizontal braider provides an elegant solution to this problem. As a result, the moving machine elements become more difficult to access, which should not be touched when in motion anyway, but the textile material is brought in convenient proximity to the operator for continuous monitoring. Such an arrangement becomes even more important when braiding is carried out on a mandrel with a specific contour. In such cases, the braided tube is meant to develop a specific shape quite different from that of a hollow circular cylinder, and the mandrel is moved in a programmed manner along the braiding axis through the braiding point while the machine is speeded up or slowed down correspondingly so as to maintain structural homogeneity. Clearly, a horizontal braider would be preferred for such tasks.

Let

$$r = \text{rpm of a horn gear (min}^{-1})$$

$$H = \text{number of horn gears on braiding machine}$$

$$a = \text{number of horns in each horn gear}$$

$$\Omega = \text{angular velocity of carriers about the braiding axis (rad/min)}$$

It follows then that if, at any given time, all horns are optimally occupied by spindles, then

$$n = aH/2 = \text{total number of carriers on machine} \qquad (14.11)$$

Each carrier moves through an extent of angular space occupied by a pair of horn gears around the braiding axis during the time taken by a horn gear to rotate once. Hence, in $(1/r)$ minutes a carrier covers an angular space equaling $[2(2\pi/H)]$ radians. Thus, the angular speed of the spindle around the braiding point can be expressed as

$$\Omega = 4\pi r/H \text{ rad/min} \qquad (14.12)$$

Substituting the expression of H by the expression of n, it follows that

$$\Omega = (4\pi r/H) = (2\pi ra/n) \text{ rad/min}$$

If a carrier completes m number of revolutions per minute around the braiding point of the machine, that is, it effectively covers $2\pi m$ radians per minute, then

$$2\pi m = \Omega = 2\pi r a/n$$

Hence,

$$m = ar/n = 2r/H \text{ (min}^{-1}) \tag{14.13}$$

This relationship shows that rpm or angular velocity of carriers around the braiding axis is inversely proportional to the number of horn gears employed.

The number of plaits generated during the period of one rotation of the carrier around the braiding axis being equal to $(n/2)$ and the number of such rotations in 1 min being given by m the number of plaits N produced in 1 min can be stated as

$$N = m(n/2) = ar/2 = nr/H \tag{14.14}$$

The weight of the tubular braid produced in 1 min can be expressed as W where

$$W = N(n\ell)T \times 10^{-6} = (2r/H)(n^2\ell/2)T \times 10^{-6} \text{ g/min} \tag{14.15}$$

If the braided fabric is taken up at the rate of V m/min, and in this duration, $(\Omega/2\pi)$ number of helices get formed, then it follows that in $(2\pi/\Omega)$ minutes one helix is formed during which time a length of $(2\pi V/\Omega)$ m of fabric of diameter d_b is produced (Zhang et al., 1997). Hence, the braid angle can conveniently be expressed by the expression

$$\tan(\alpha/2) = d_b\Omega/2V \tag{14.16}$$

Referring to Equation 14.10, it is recalled that

$$\text{diameter of braided tube} = d_b = (n\ell/2\pi)\sin(\alpha/2)$$

Consequently, it is possible to find out the value of ℓ for given values of d_b, Ω, V, and n if α is numerically eliminated from Equations 14.10 and 14.16.

As the rate of take up and the speed at which the yarn carriers move can be set on a machine, a predetermined diameter of braided tube, by the way of securely braiding around a circular core, should yield values of a host of dimensional parameters provided the modular length is controlled within acceptable limits. However, surficial and elastic properties of strands also play important roles in this conversion process, which are yet to be established properly.

14.4 Differences between Flat and Tubular Braid

Referring back to Figure 14.14, it is observed that horn gears for formation of both types of braid are mounted along a circular path. However, carriers on flat braid machines do not keep moving continuously along a circular path. At its turning point, the crossover element in the track plate is substituted by a returning element, which forces the carriers to reverse their direction of motion. Thus, carriers on flat braid machines carry out a reciprocating motion along a circular path spanning nearly 4π radians. In other words, a carrier on a tubular braiding machine reaches its starting point after executing an angular displacement of exactly 2π radians around the machine center while a corresponding carrier on a flat braid machine has to usually move by nearly twice that much before reaching its starting point. As a consequence, the length of the pitch of a flat-braid fabric equals the product of the number of carriers with plait height and can be expressed as

$$\text{length (mm) per pitch of flat braid fabric} = n\ell/\cos(\alpha/2) \qquad (14.17)$$

This results in a major structural difference between flat and tubular braids. Each strand in a tubular braid maintains its helical direction, be it clockwise or counterclockwise. Each strand in a flat braid, on the other hand, reverses its direction after striking the fabric edge. In terms of the rest of the dimensional quantities, a flat braid is fairly similar to its counterpart the tubular braid when the same is slit open and laid flat, with fabric strips of each pitch of opened out tubular braid joined along their length after reversing the surface each time.

The horn gears on two sides of the turning point of a flat braiding machine are designed differently from the rest of the horn gears of the system (Brunschweiler, 1954c,d). These two gears may have one additional horn and be of a proportionately higher diameter so that the arc distance between the centers of two neighboring carriers remains same while they move at the same linear speed as rest of the carriers. Hence, the number of teeth on the gear wheels of these horn gears is also proportionately higher. A typical arrangement for a 2/2 braid is illustrated in Figure 14.23. In this diagram, three horn gears each equipped with four horns are shown with carriers suitably located on respective horns. An additional reversing horn gear, which is equipped with five horns is located at the extreme right. The additional horn on the reversing horn gear is always empty in all the five stages during formation of four successive plaits. A close perusal of the diagram reveals the manner in which, for example, the carrier number 6 comes back to the last crossover point after reversing its direction of motion. This reversal in direction of motion of carriers gets reflected in the resultant fabric in which individual strands no longer follow a helical path but move in a zigzag manner along the principal axis of the braid. This reversal also results in a fabric

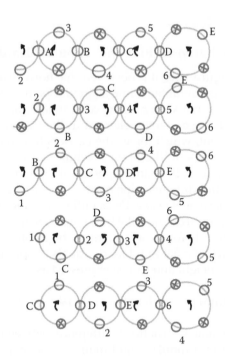

FIGURE 14.23
Arrangement and motion of carriers on a flat braider with a larger reversing horn gear.

selvedge, which is different from the rest of the fabric body (Brunschweiler, 1954c).

Reversal in the direction of motion of horn gears can also be carried out by a reversing horn gear having one horn less than the rest of the system. Such a solution has been illustrated in Figure 14.24. The carrier number 5 has clearly reversed its direction of motion in the course of the two plaits shown. The resultant selvedge would be quite different from that resulting out of Figure 14.23. A comparative blown-up view of the two respective selvedges is shown in Figure 14.25. The selvedge depicted on the left side of the diagram results from the arrangement of Figure 14.23, and that on the right side corresponds to Figure 14.24. It is observed that with an additional horn on the reversing gear the interlacement pattern among strands remains at 2/2 throughout whereas with the other option the interlacement pattern after reversal at the selvedge changes to 1/1 before shifting back to 2/2.

A flat-braiding machine equipped with two reversing horn gears each having one additional horn would run on an odd number of carriers as well as would a machine equipped with reversing horn gears each having one horn less than other gears in the system. Referring back to Figures 14.23 and 14.24, if two reversing gears are mentally placed on the extreme left-hand side of each of the two systems and the total number of carriers counted,

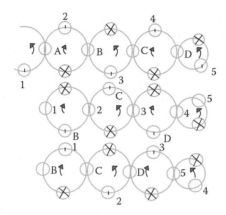

FIGURE 14.24
Arrangement and motion of carriers on a flat braider with a smaller reversing horn gear.

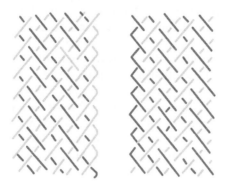

FIGURE 14.25
Selvedges on a flat braider.

then a total of 11 carriers are found to be required for system with reversing gears equipped with an additional horn whereas a total of nine carriers suffice for the other case. In other words, for a 2/2 braiding system employing five horn gears, one system requires $(5 \times 2 + 1)$ carriers whereas the other system needs $(5 \times 2 - 1)$ carriers. Logically then, if a system is designed with one reversing horn gear having an additional horn and the other reversing horn gear has one horn less than those of the rest of the horn gears, then one can employ an even number of carriers for braiding a flat fabric. Thus, a flat-braiding system can be operated with both an even number as well as an odd number of carriers. However, the flat braid produced on a machine with an odd number of carriers would yield similar selvedges on two edges, and dissimilar selvedges would result with an even number of carriers.

A notable difference in terms of association of a track to its carriers between flat-braiding and tubular-braiding systems relates to the fact that, in a flat-braiding system, each carrier has to travel through both tracks whereas in a tubular braid there is a clear division. Referring to Figures 14.20 through 14.22, it is observed that, for tubular-braiding systems, the carriers numbered alphabetically move along only that part of the track that takes them in a counterclockwise direction, and reference to Figures 14.23 and 14.24 reveals that all carriers must alternately switch tracks after reaching their reversal points.

14.5 Limitations of Braiding Systems

The compulsion of having to move packages along a nonlinear path constitutes a serious handicap insofar as speed of carriers and, consequently, the productivity of braiding machines are concerned. As yarn thickness rises and also as the number of carriers increases, packages have to become more and more massive so as to guarantee an acceptable downtime, which is governed by the necessity of replenishing exhausted packages. These factors impose considerable restrictions to product range.

The diagonal paths followed by strands held under tension with one half sloping positively and the other half sloping negatively force the net assembly of strands on a braiding machine to collapse toward the braiding axis. This feature of the braiding process limits the width of product.

A continuously shifting location of the carrier with respect to the braiding point results in a continuous fluctuation in the free length of strand between its point of exit from the carrier and its point of merging into the braiding zone, resulting in a continuous change in its tension. This fluctuation is aggravated by the intermittent and indeterminate negative let-off system of carriers. As a result, control of modular length and, hence, of product quality is primarily governed by the skill and experience of braider.

Thus, although braiding as a concept is intrinsically simpler than weaving or knitting processes and is also much less demanding on the quality of feedstock, inherent limitations outlined in the foregoing as well as a lack of understanding about the mechanics of the braiding zone has restricted its domain to a narrow range of low-width, relatively low-quality products manufactured on low-productive systems.

References

Brunschweiler D (1953). Braids and braiding, *Journal of Textile Institute*, 44, 666–686.

Brunschweiler D (1954a). The structure and tensile properties of braids, *Journal of Textile Institute*, 45, T55–T76.

Brunschweiler D (1954b). Braiding technology: Part 1, introduction and Maypole braider, *Skinner's Silk and Rayon Record*, *28*, 254–257.

Brunschweiler D (1954c). Braiding technology: Part 2, tubular braids: Methods of track and carrier construction, *Skinner's Silk and Rayon Record*, *28*, 380–382.

Brunschweiler D (1954d). Braiding technology: Part 3, haul-off methods and fancy braiders, *Skinner's Silk and Rayon Record*, *28*, 472–474.

Ko F K (1987). Braiding, In *Engineered Materials Handbook, Vol. 1, Composites*, 519–528, CRC Press, ASM International, Metals Park Ohio.

Turner J P (1976). The production and properties of narrow fabrics, *Textile Progress*, *8*.

Zhang Q, Beale D, Adanur S, Broughton R M, and Walker R P (1997). Structural analysis of a two dimensional braided fabric, *Journal of Textile Institute*, *88*, Part 1, No. 1, 41–52.

15

Formation of Nonwoven Fabrics

15.1 Introduction

Nonwoven fabrics are produced primarily from fibers, and woven, knitted, braided, laced, or tufted fabrics are produced primarily from yarns. The term nonwoven is a misnomer, and in spite of many attempts at precisely defining the exact nature and domain of this class of products, ambiguity continues to prevail. This is also partly fueled by the ever-expanding scope of the cluster of technologies broadly grouped under this class of fabric-formation system.

Textile products, such as felts or quilts fabricated by craftsmen primarily from fibers have been in existence over a very long time. However, it is only during the 1960s that nonwovens as an industrial product started coming into prominence with the development of needle-punching technology. Since then, many more nonwoven production processes have been introduced, which have proved to be commercially successful. According to the estimate made by EDANA and INDA, worldwide production of nonwovens was expected to reach 8.41 million tons by 2012 at a compound annual growth rate of around 7.9% (Holmes, 2009). Although the nonwoven industry in North America, Europe, and Japan has reached a mature stage, it still shows a compound annual growth rate of about 5% compared with Asia at about 14% (Leigh, 2010). Incidentally, the bulk of this production comes from developed economies, which is indicative of technological sophistication and high investment.

Modern nonwoven products are also distinguished by their end use–specific characteristics, and this has enabled a goal-oriented growth of specific sectors of this industry. This is aptly demonstrated by a symbiotic growth of the relatively new hydro-entanglement technology along with sharply rising demands of a modern hygiene and health care sector.

Nonwoven fabric formation is, in essence, a continuous two-step process. Raw material in the form of fibers or filaments is first converted to a web, which is then reinforced suitably. The first step of web formation from fibers is a severely truncated version of the conventional process of yarn formation from fibers, entailing thereby considerable all-around savings, and the second step of web reinforcement permits dispensing with the yarn preparation section altogether. Continuity of the two steps imparts certain compactness to the nonwoven fabric-formation process and should logically contribute to the economics of production. This aspect is reinforced further by coupling the formation of raw material itself through online spinning of fibers or filaments from polymer chips. The resulting scenario of polymer chips being fed into a continuous production line that delivers fabric at the other end could be regarded as the ultimate solution for mass production of textile fabrics. One would expect such products to be very cheap and therefore disposable. This expectation is reinforced by the much higher production rate of nonwoven fabric manufacturing systems as compared to modern high-production weaving and knitting systems.

The evidence provided by the global market during the past half century has, however, led to the realization that the strong point of the nonwoven fabric-formation process is not disposability of the resultant products per se—although some nonwoven products are indeed disposable—but the extra degree of freedom that permits a judicious choice of raw material, machine, and process parameters for a focused generation of functional properties in resultant products. For example, considerable strain on constituent yarns is imposed by a modern woven or knitted fabric-formation process, and hence, the yarns must possess certain properties, which are demanded by the fabric-formation process itself and not necessarily by the functional requirements of end use. Properties of constituent fibers and of the resultant web in the nonwoven fabric-formation process are chosen in keeping with the functional requirements of the resultant product. Similarly, the directional requirement of functional properties, such as complete isotropicity (for load bearing) or even a complete uni-axiality (for unidirectional fluid transmission), can be more easily built in nonwoven fabrics than in the more conventional forms. Such attributes and many others that nonwoven fabrics and nonwoven fabric-formation processes exhibit have led to their rapid commercial growth over the past half century. However, certain deficiencies in fabric properties have restricted application of nonwoven fabrics primarily to the sector of technical textiles. The most glaring shortcoming of this class of fabrics is its poor drapability. The rotational freedom at yarn-to-yarn contact points enjoyed by yarn segments of a fabric made purely from yarns leads to ease of shear deformation of the product, which, in turn, leads to its good drapability. A nonwoven product developed through the entanglement of fibers or filaments is usually deficient in this respect. Hence, wherever drapability is a fundamental functional requirement, such as in common forms of apparel goods or in certain classes of household textiles, nonwovens are not commonly employed. However, through judicious choice of fibers and the nonwoven fabric-formation process, very soft and highly drapeable nonwovens are also being specially engineered for specific applications in medical and protective textiles.

15.2 Classification

Nonwoven processes may be classified into two broad groups according to the manner of web formation, namely the dry method and the wet method. In the dry method, the fibers or filaments are dispersed in space by mechanical or aerodynamic forces before being reassembled in an orderly manner into a continuous web, and in the wet method, hydrodynamic forces and an aqueous medium are the key elements. The two broad methods of web formation are also further characterized by fiber lengths suited to the respective systems. Very short fibers are in the exclusive domain of the wet method, and short to moderately long staple fibers are converted to a web by a conventional mechanical carding process or by the aerodynamic dispersion and deposition method. Online spinning of filaments and fibers requires a modified form of dispersal and deposition aided by static charge and airflow. Hence, some authors (Massenaux, 2003) suggest a separate grouping for webs formed directly from polymers although it is also essentially a dry process.

Reinforcement of the web can be carried out by chemical, thermal, and mechanical means. Suitable binders glue the fibers together in the chemical method, and thermoplastic fibers can be bonded together by heat energy. Such gluing or bonding takes place at discrete points or zones in the fabric body. The mechanical method involves ramming the fibers into discrete entanglements so that frictional forces hold the fibrous assembly together. Fibers entangled mechanically enjoy a greater degree of freedom than chemically or thermally bonded ones.

A nonwoven fabric may be viewed as an assembly of a discretely bonded fiber mass wherein the nature of the bonds along with the nature and orientation of the fibers influence the properties of the resultant fabric. The functional properties of a nonwoven fabric can be skillfully developed by subjecting it to more than one mode of reinforcement so as to combine desirable attributes of the differing types of bonds. Thus, a combination of frictionally entangled bonding points in the body of a product coupled with very closely spaced thermally bonded points on its surface makes for a good dust filter fabric as the dust cake can be allowed to form on the thermally treated surface and be easily shaken off by a reverse pulse jet, and the relatively flexible deeper structure provides an ideal tortuous medium for trapping dust particles while permitting ease of airflow.

15.3 Fibers in Nonwoven Fabrics

While designing a nonwoven fabric for satisfying a specific combination of functional requirements, a great deal of attention needs to be paid to the physical properties of the fibers. Issues of cost-effectiveness as well as of demands made on the downstream processes also need to be taken care of.

The linear density of a fiber decides the specific surface area available for bonding as well as its stiffness while affecting the covering power and permeability of the fabric. Finer fibers provide more bonding sites for the same fabric areal density, and the fabric would be softer, more opaque, and less permeable. The shape factor of the fiber, given by its length-to-diameter ratio, plays an important role in the production process and the fabric properties. A higher shape factor of the fiber makes for more difficult processing while improving fabric properties, such as dimensional stability and flexural rigidity. Similarly, an altered cross-sectional profile of the fiber, such as a triangular cross-section or a multi-lobal cross-section as opposed to a circular section, drastically alters the fiber's surficial and mechanical properties in addition to affecting a void in the resultant fabric. Fibers with a built-in void result in lower fabric density and higher thermal resistance, moisture absorbency, and product stiffness. Fibers with a higher water-retention capacity disperse more easily in aqueous media and, hence, are preferred for the wet method of nonwoven fabric formation.

The fiber crimp, fiber coefficient of friction, and fiber length also play critical roles in the nonwoven fabric-formation process and in the properties of the resultant product. In addition to contributing to fabric bulk, a higher fiber crimp improves cohesion among fibers and, therefore, in the resultant web and fabric. On the other hand, fibers with higher crimp pose greater difficultly to individualization and contribute to ease of cluster formation while being dispersed in aqueous media. Fibers with a higher coefficient of friction contribute to improved bond strength while offering greater resistance to individualization.

Fiber length is critical in determining the route to web formation. As a rule of thumb, very short fibers in the range of 1.5 to 10 mm are converted to a web by hydrodynamic deposition (wet method), and aerodynamic web formation is resorted to for fibers in the range of 10 to 50 mm, and mechanical methods, such as carding, come into question for fibers in the range of 20 to 150 mm (dry methods). These ranges are only indicative and are by no means definitive.

Although all types of fiber, natural and synthetic or organic and inorganic, can be converted into nonwoven fabrics, manmade fibers are usually favored in view of their uniformity, cleanliness, elasticity, desired shape factor, and cross-sectional profile as well as resistance to mechanical, chemical, and thermal strains, which can be engineered as per requirement.

15.4 Web Formation from Fibers

15.4.1 Dry Method

15.4.1.1 Carding Process

On a classical carding machine, fibers that have been opened up to some extent as well as cleaned of major impurities are fed either by a lattice or by a chute to a fast-revolving licker-in roller. The surface of a licker-in (or taker-in) roller is covered with sawteeth, which are so directed that they strike against the fiber mass protruding into their path and thereby pluck off small tufts from the nip of the feeding system. During the onward rotation of the licker-in roller, these tufts of fibers are thrashed against a series of stationary bars for ridding them of residual impurities. Subsequently, as these cleaned tufts are taken around by the licker-in roller, they come against a faster-moving toothed surface of a revolving cylinder. Most of the fibers get picked up by cylinder wires, which, in turn, carry these fibers to an extended combing zone created either by slowly moving wires of a series of revolving flats or by a series of workers and strippers mounted atop the cylinder. The respective combing zones are formed by the coming together of fast-moving cylinder wires against the slow-moving flat wires or against the slow-moving worker wires. The two sets of wires move in the same direction, but they are inclined in opposite directions. The combing action around the cylinder is responsible for individualization of fibers as well as their orientation along the machine direction. Some very short fibers as well as tight clusters of fibers that cannot be opened get removed as waste. While being combed, the fibers remain anchored to the cylinder teeth. Removal of fibers from the cylinder surface in the form of a fine web is affected by a vibrating comb, which literally digs into the cylinder wires and removes the fibers bodily. Owing to cohesive forces between fibers, the entire sheet appears as a semitransparent gossamer-like sheet. The web so generated on a carding machine is extremely light (about 10 to 20 g/m^2) and somewhat uneven.

Two variations of carding machine are suited for short-fiber (revolving flat card) and long-fiber (roller and clearer card) systems. Revolving-flat carding machines perform better in respect to fiber opening and orientation but are less versatile than roller and clearer cards in respect to their ability to handle a range of fibers and fiber lengths. The flat strips of revolving-flat carding machines are more delicate than workers and strippers and tend to sag in the middle more easily. As a result, revolving-flat card machines are comparatively narrow in width, which affects production. Because of these two deficiencies, namely narrow width and less versatility, roller and clearer cards are encountered on nonwoven lines and not the revolving flat cards. Similarly, a limitation in the speed of a vibrating doffer comb as well as a switch over from the classical flexible clothing to metallic clothing has resulted in a pair of rollers, comprising one with card clothing and another with a grooved or smooth surface, carrying out the function of removing the web of fibers from the doffer in modern machines.

Webs removed either by doffer comb or by a pair of rollers have a preferential orientation of fibers along the length direction. For many technical applications, however, isotropicity is desirable and it requires a random orientation of fibers. A randomized web can be generated on conventional carding machines through employment of a randomizing roller after the doffer, which, however, results in a drop in production in terms of m/h.

15.4.1.2 Laying of a Carded Web

A carded web needs to pass through another operational step, namely web laying, so that the resultant ordered collection of fibers, termed a batt, could have the desired mass per unit area, distribution of fiber orientation, and width. The desired mass of batt is achieved by laying a certain number of layers of web over each other. This doubling process enhances uniformity of the product although some additional irregularities might also be introduced owing to extra handling.

The desired orientation of fibers in a batt may be of four types as shown in Figure 15.1. The longitudinally oriented batt, also termed a parallel-laid web, can be generated by a simple doubling process. By this process, however, the width of the batt remains the same as that of a carded web. The most popular commercial process of generating a batt from a carded web is by cross laying. Fibers in a cross-laid batt exhibit a distribution of orientation that is skewed in the machine (MD) and cross (CD) directions. A randomly oriented batt requires a carded web in which the fibers are randomly oriented. Such webs can be generated by using a randomizing device on the card and cross laying the same although the aerodynamic route, described in Section 15.4.1.3, is preferred commercially for such batts. The transversely oriented batt has hardly any commercial significance.

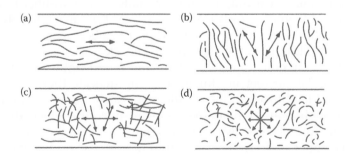

FIGURE 15.1
Types of fiber orientation in web. (a) Longitudinal orientation, (b) transverse orientation, (c) longitudinal and transverse orientation, and (d) random orientation.

The horizontal cross-lapper shown in Figure 15.2 shows three parallel lattices (belts) across the path of a conveyor belt. A carded web is guided in by a feed belt located at the top of the cross-lapper. The lattices of the feed belt execute a steady clockwise movement. The two belts, A and B, situated below the feed belt execute a simultaneous to-and-fro motion so as to spread the carded web in a desired manner across the width of the conveyor belt while the lower belt executes an additional rotary motion. The batt so formed on a conveyor belt is moved forward to the next station for reinforcement.

The speed at which the carded web is fed to the system (V_W), the speeds and directions of the transverse motion of belts A and B (V_{AT} and V_{BT}), and the surface speed of belt B (V_{BS}) are mutually related to ensure that tension in the web during its conversion to a batt does not fluctuate beyond a specified range. In other words, the speed of each point of the web beyond the feed belt has to be same until it lands on the conveyor belt. During the movement of belts A and B from the right edge to the left edge of the conveyor belt, the belt A takes up the amount $2V_{AT}$ from that supplied by the feed belt while belt B releases the amount $V_{BS} + V_{BT}$ to the conveyor belt. In order that the resultant belt movements yield a constant feed rate V_W of the web to the conveyor belt, the following condition would hold:

$$V_{BS} + V_{BT} - 2V_{AT} = V_W$$

On the other hand, the corresponding condition during the movement of belts A and B in the opposite direction can be expressed by the following form:

$$V_{BS} - V_{BT} + 2V_{AT} = V_W$$

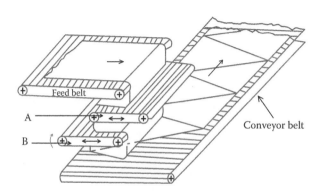

FIGURE 15.2
Horizontal cross-lapper.

Equating the two conditions stated above leads to the following relationships:

$$V_{BS} = V_W \tag{15.1}$$

$$V_{BT} = 2V_{AT} \tag{15.2}$$

Uniformity in batt thickness, number of layers of web across the batt, and orientation distribution of fibers in the batt would be affected by the interplay between the web speed V_W, the translational speed of the lower belt V_{BT}, the speed of conveyor belt V_C, and the respective widths w and b of the web and desired batt. If a web is laid at an angle θ to the length direction of the conveyor belt (Figure 15.3), then

$$V_{BT} = V_W \sin\theta$$

$$V_C = V_W \cos\theta$$

Hence,

$$V_{BT} = V_C \tan\theta \tag{15.3}$$

Evidently, if all fibers in a carded web are perfectly oriented along its length, then fibers in the batt would be oriented along angles $\pm\theta$ to the batt MD. Hence, for the general carded web exhibiting a certain orientation distribution of fibers around its length direction, the larger the value of the web laying angle θ, the stronger the skew of fiber distribution around the CD of

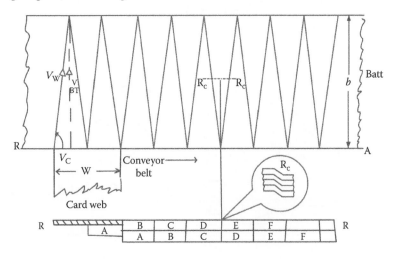

FIGURE 15.3
Geometry of cross-lapped web.

the resultant batt would be. A suitable choice of speed of the conveyor belt for given translational speeds of the oscillating belts is thus crucial for properties of the batt.

Having chosen the laying angle θ of a web of width w, one can then choose width b of the batt for fixing up the number of layers of web in the batt, according to following relationship.

$$\text{number of layers of web in batt} = n = (w/b) \sec\theta \qquad (15.4)$$

If mass per unit area of carded web is g, then the corresponding mass of batt would be given by the following expression:

$$\text{mass per unit area of batt} = g(w/b) \sec\theta$$

Expressing the web-laying angle in terms of web speed and belt speed, a modified form of mass per unit area of batt can be stated as

$$\text{mass per unit area of batt} = g(w/b) \sec[\sin^{-1}\{V_{BT}/V_W\}] \qquad (15.5)$$

If the speed setting of various elements of a cross-lapper is carried out strictly according to geometrical relationships with the given web dimensions and desired features of batt, then the batt may exhibit some nonuniformity in its final thickness caused by unavoidable deformation of the web during the laying process and also due to some elastic recovery later on. The viscoelastic nature of web and batt necessitates a certain degree of fine tuning in the speed of various elements of the cross-lapper so that a cross-section of the finished batt shows the same number of web layers all along its length as shown schematically in Figure 15.3. The view of the edge RR of the batt shows two layers of folded web, making up a total of four layers in its body whereby, when viewed closely, the sharp and short vertical line segments of web reversal during its laying would actually be made up of inclined members within the body of the batt. Irrespective of the actual configuration of webs at any location within the batt, the total number of layers must always remain the same.

The simple horizontal cross-lapper shown in Figure 15.2 suffers from three major drawbacks. A continuous to-and-fro motion of the belts, A and B, henceforth referred to as the carriage, limits the speed at which laying can be executed. This speed is, moreover, not constant when viewed in small intervals of time as a to-and-fro motion necessarily involves multiple decelerations and accelerations. At the two reversal points where the web folds back on itself, the carriage speed passes through a zero value, and at the center of the batt, the carriage speed attains its maximum. The web, however, continues to be delivered by the card at a constant speed. As a result, the batt exhibits more material at its folded edges accompanied by a progressive reduction toward its center. The third issue concerns the unguided movement of the web along belt surfaces compounded by two reversal points within the carriage and, more specifically, the frictional strain around the reversal zone on the upper belt within the carriage, which becomes critical for lighter webs.

The carriage of a modern cross-lapper is constructed differently from that shown in Figure 15.2. A typical example is the profile cross-lapper of NSC (France), which is schematically represented in Figure 15.4. It is made up of two segments, each segment being made up of very light rollers around which a continuous belt is mounted. These rollers can be moved independently, and they constitute the moving elements accounting for a much higher speed of such carriages. A carded web does not slide across belt surfaces but is positively guided throughout its journey in a sandwiched state between the pair of belts of opposing surfaces of two segments of the carriage. The web comes out of the belt grip only at the exit point located at the bottom. Through a controlled drive of the rollers, it is possible to store some quantity of the carded web within the carriage system during the reversal points of laying while the stored amount is suitably released during the translational phase. The thickening of the folded edges of the batt is countered in this manner. It is indeed even possible to control the roller movements and impart a specific profile to the batt along its thickness so much so that the batt would be thicker at the center and thinner at the edges. The thickness of such a fat-bellied batt is finally made uniform through a subsequent drawing process, which partly reorients the fibers along the MD to the extent required by the end use. Such systems can handle carded webs from 5 g/m² onward of widths up to 3.5 m and convert them into a batt of 1% mass CV in widths up to 16 m, consisting of as many as 120 layers at a laying speed of up to 295 m/min. These figures merely represent extreme values and are not meant to suggest a possible combination of variables.

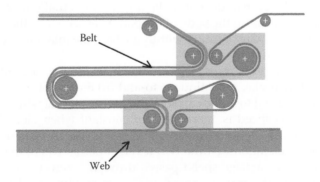

FIGURE 15.4
Profile cross-lapper of NSC France. (From Brydon, A.G. and Pourmohammadi, A., Fig. 2.38, Chapter 2, *Handbook of Nonwovens*, edited by S. J. Russel, 2007, CRC Press.)

15.4.1.3 Aerodynamic Process

Randomized webs can also be generated by dragging down the web from the doffer with the aid of a stream of air as is done in the K21 carding machine of Fehrer & Co. This machine not only generates a randomized web, but also has a very high production rate owing to employment of four cylinders in tandem (Jakob, 1990). There is a screen conveyor underneath the series of cylinders, and a force of air suction not only holds the fibers deposited on the screen firmly, but also assists in pulling the fiber mass away from the cylinders, aided by centrifugal forces acting on the fibers caused by rotation of the cylinders at high speed. The fiber mass is fed to the first cylinder on one extreme side, and most of it gets transferred to the second cylinder after being combed by the pair of workers and the stripper atop the first cylinder. Whatever does not get transferred is sucked down onto the screen below. This stage-wise process continues with the rest of the cylinders, and the final web is made up of four layers intermingled to a degree owing to the cross-current of the airstream. Such systems are capable of producing at the rate of 300 kg/m/h with 1.7 denier fibers as compared to 80 to 90 kg/m/h on single-cylinder carding machines employing a doffer comb. The attainable web weight with such a carding system is in the range of 10 to 100 g/m^2, and fibers in the resultant web attain a high degree of randomness. The aerodynamic doffing system, along with four tandem cards, provides an elegant solution for producing a random web, which, at the same time, yields much higher production in terms of mass as well as length per unit time. The quality of this web, in terms of residual stress, is also claimed to be superior to the mechanically produced ones.

A limitation of the mechanical carding process lies in achievable areal density (g/m^2) of the resultant web as well as in the thickness of the fibers that can be processed. This limitation is overcome by allowing mechanically opened fibers to be transported by an aerodynamic stream for deposition on a suction cage or on suction belts. A web mass in the range of 300 to 4000 g/m^2 can be achieved in this manner at one go, generating a web width of up to 5.4 m while processing fibers up to 300 denier. Evenness of the resultant webs is, however, not as good as those formed by mechanical process.

15.4.2 Wet Method

In this method, fibers of appropriate fineness and length are first formed into aqueous slurry. Subsequently, the slurry is pumped to a head box from which it is spread evenly onto a moving screen. Hydro-extracting systems located under the screen remove water and circulate it back to the slurry preparation systems.

The wet method is traditionally used for making paper from wood pulp. The length of fibers in the wood pulp varies between 0.5 and 2.5 mm, depending on the source and location of the wood. On the other hand, nonwoven fabrics need to be made of much longer fibers, in the range of 5 to 10 mm or at times even much longer. The longer the fiber, the more translucent and permeable is the product and the lower is the resistance to in-plane shear. When the slurry is required to be made from longer fibers, then the dilution of slurry, the velocity at which the slurry is fed to the filtering screen, and the rate at which water must be removed from the screen rise rapidly. This is explained in the following.

Let fibers, each of length 2ℓ, be made into slurry. In order that the fibers are dispersed thoroughly in an aqueous medium and not form any agglomeration, one may impose the extreme condition that no contact is made by any fiber with its neighboring fibers. In other words, each fiber is submerged in a sphere of water of volume equal to $[(4/3)\pi\ell^3]$. Hence, although a fiber may occupy any spatial configuration, the encasing sphere of water keeps it away from coming into contact with its neighboring fibers. If there are N fibers at any instant in the slurry, then one needs a total volume of water equal to $N[(4/3)\pi\ell^3]$. Let now another set of fibers of the same nature but of length $(2\ell/n)$ be formed into slurry. Each fiber would now require remaining submerged in $[(4/3)\pi(\ell/n)^3]$ volume of water to fulfill the condition mentioned earlier. As a result, water of volume of $(N/n^2)\,[(4/3)\pi\ell^3]$ would be required to form an equivalent slurry with the same fiber content as, in this case, the number of fibers in the slurry would be $N.n$. This simplistic approach demonstrates that, every other factor remaining the same, the volume of water required would go up by n^2 times if fiber length goes up by a factor of n.

Now, let two fabrics of the same areal density be made of the two perfect dispersions, one containing fibers of length 2ℓ and the other containing fibers of length $(2\ell/n)$. If each fiber of length 2ℓ weighs δ gram, and if the required areal density of the fabric is G g/m², then (G/δ) number of fibers needs to be deposited in 1 m length of fabric-forming screen if the width of the screen is assumed to be 1 m. In other words, a volume of $(G/\delta)\,[(4/3)\pi\ell^3]$ slurry needs to be fed to the screen while the screen moves by 1 m. For fibers of length $(2\ell/n)$, the equivalent volume would be n^2 times less. Hence, the n times longer fibers would need to be fed at n^2 times higher velocity to the filtering screen in order to form fabrics of equivalent areal density. By the same token, the water extraction rate from the filtering station would also need to be faster by n^2 times. These altered hydraulic conditions necessitate a fundamental change in machines designed for making nonwoven fabrics from those designed for making papers.

The schematic outline of a traditional straight wire paper-making system is illustrated in Figure 15.5. Fiber water slurry is pumped into a pond through an opening, termed traditionally as a slice, which, in turn, feeds the slurry onto a moving wire screen at hydraulic head h. The velocity v at which slurry emerges from the slice is given by

$$v = [2gh]^{0.5}$$

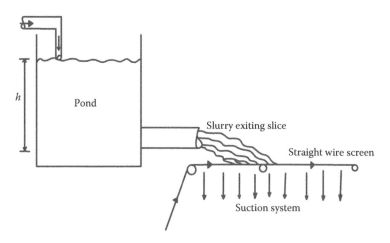

FIGURE 15.5
Schematic outline of a straight wire paper-making system.

In this expression, the factor *g* represents acceleration due to gravity. The velocity of slurry can be adjusted by altering the hydraulic head, evidently within narrow limits. A drastic change in velocity would not be possible with this arrangement. Moreover, if the slurry were to emerge at very high velocity on the screen, then a highly disturbed hydraulic condition would prevail around the fiber settling zone on the screen. This fiber settling zone develops over the interface of the emerging mass of slurry from the slice, the screen, and the corresponding water suction system arranged below the screen. The conditions over this zone should be congenial to allowing the dispersed fibers to move steadily toward the screen and not get scattered all over. Under equilibrium conditions, fibers from the emerging slurry should move toward and gradually settle down onto the screen, which should carry the collection of fibers in sheet form to the next stage of reinforcement and drying. The conceptual alteration in machine design for generation of high-velocity slurry and negotiation of high-energy slurry at the fiber settling zone is schematically outlined in Figure 15.6.

Two major changes in the altered design are given by the enclosed nature of the pond and inclined positioning of the wire screen. The enclosed pond permits slurry to be maintained and fed under pressure so that a pressure differential Δp between the inlet and outlet can come into effect. This differential, in turn, causes a rise in velocity (v) of slurry exiting the slice. Accordingly, if the density of slurry is represented by ρ, then (White, 2007)

$$v = [2(\Delta p/\rho) + gh]^{0.5}$$

The inclined wire screen effectively creates a wedge-like tapering of slurry exiting the slice and ensures continuity in the slurry–wire screen–suction system that stabilizes the flow in the fiber settling zone. As a result,

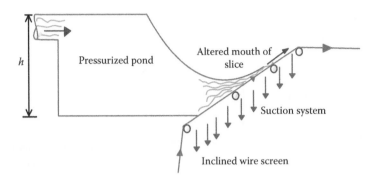

FIGURE 15.6
Schematic outline of an inclined wire wet-laying system.

a well-formed fiber sheet emerges from the narrow slit-like mouth of the extended slice. Machines constructed on such principles can deliver, for example, a web of width 5 m produced at the high speed of 600 m/min.

15.5 Web Formation from Polymer Chips

Two broad routes, namely spunbonding and meltblowing, are commercially exploited for formation of a web from polymer chips. Other methods, such as flashspinning and electrospinning, have had limited commercial applicability. Webs of spunbonded fabrics (S) are products of continuous filaments and those of meltblown fabrics (M) are made up of fibers. Moreover, the nature of the raw material, in terms of its melt flow rate (MFR), molecular weight, and molecular weight distribution (MWD), process details as well as product properties are quite different for the two routes. As a result, it has often been found to be advantageous to prepare sandwiched products of fabrics produced from these two routes, such as SMS or SMMS, and exploit the attributes of each in a synergistic manner.

While in the spunbonding route, filaments can be melt spun, dry spun, or wet spun from suitable polymers; the meltblown route is by and large restricted to thermoplastic resins. The MFR of resins for the meltblown route is much higher (up to 1500) than for the spunbonding route (less than 100), and their molecular weights are on the lower side with narrow MWD. Resins with higher molecular weight and higher MWD can be effectively processed though the spunbonding route. The meltblown route is advantageously adopted for producing webs made of microfibers (in the range of 1 to 5 μm in diameter equivalent to 0.007 to 0.17 dtex for PP) and creating a very uniform mesh of low areal density (as low as 5 g/m²), high surface area, and low porosity. The spunbonded filaments are usually much thicker (15 to

50 µm equivalent to 1.6 to 17 dtex for PP) with consequent effects on fabric properties. However, spunlaid webs produced from filaments, which can be either split by mechanical action or leached chemically/thermally, approach the domain of fineness of the meltblown route. Owing to process peculiarities, the filaments of the spunbonded route are much stronger than fibers generated through the meltblown route. Meltblown fabrics are, as a result, the much weaker of the two.

15.5.1 Web Formation by Spunlaid Route

A broad schematic outline of the spunbonding process is depicted in Figure 15.7. Dried polymer granules are, at first, passed through an extruder. Most spunbonded fabrics are made from PP and PET, which are meltspun. The technology of meltspinning in terms of the rheology of the polymer, the nature of the screw, the feed rate, the thermal gradient, the uniformity and distribution of melt within the extrusion chambers, etc. plays its role at this stage. The melt is filtered and subsequently pumped through an array of spinnerets arranged in the form of a beam along the machine width, resulting in a curtain of molten filaments. Airstreams cool, draw, and guide these filaments through a web-laying device toward a perforated apron. Air suction forces acting through an apron surface hold down the web securely, which is carried to the next station for thermal/mechanical/chemical reinforcement.

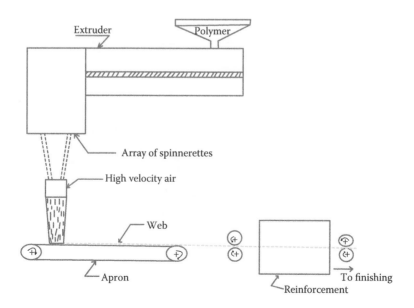

FIGURE 15.7
Schematic outline of a spunbonding process.

A closer view of the web-formation process is shown in Figure 15.8. Of major importance here are the two zones, namely of cooling and drawing the curtain of filaments by air and the controlled deflection of individual filaments onto the apron, for formation of a uniform web.

In its simplest form, air as hot as the emerging melt from spinnerets blows down on the molten filaments, stretching them, in the process, leading to improved orientation of molecular chains. This air mingles with the ambient air leading to gradual cooling off. The crystalline structure of the filaments depends on the intensity and speed of cooling, and this structure, in turn, affects its mechanical properties. Hence, in the course of further developments, a separate injector was introduced down the line, which employs the Venturi principle for quenching and drawing the filaments. This process change leads to finer filaments accompanied by improved orientation and a rise in production speed. Cooling and drawing by an additional airstream can be done separately for fine tuning the process in terms of desired filament fineness, orientation, and speed of production. Drawing the filament curtain additionally by mechanical means such as by Godet rollers leads to further improvement in orientation and crystallinity.

The choice of the exact process sequence depends on the type of polymer and the desired properties of the filaments. The mass throughput amount through the nozzles remaining the same, a higher filament speed results in finer filaments. Commercial filament production speed has gone up from around 1000 m/min to 5000 m/min with improvements in aerodynamic and

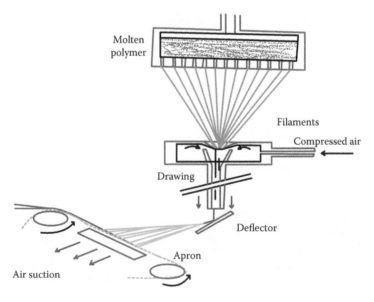

FIGURE 15.8
Web formation in spunbonding system.

mechanical drawing systems with controlled heat removal from the system. Accompanying improvement in filament properties in terms of orientation and crystallinity has paved the way for production of finer and stronger filaments, which then can be converted to a desired web varying between very light (about 10 g/m²) to fairly heavy (1000 g/m²) ones. Webs lighter than 10 g/m² are also produced via the spunbonding route by spinning multicomponent filaments, which are split up by mechanical action or leached chemically or thermally later on.

Transforming the so-formed filaments into a web of desired isotropicity, evenness, and areal density is the next crucial step in spunbonded fabric formation. Each filament can theoretically be viewed as being moved spirally over a perforated speeding apron by rotating deflector planes as shown in Figure 15.8. Orientation distribution of filaments in the resultant web would depend upon the relative speeds of the filament and apron as well as the nature of the looped path of the spiral. If the spiral executes a perfectly circular motion, the moving apron would distort the circles into ellipses with major diameter along the length direction of the web. The resultant web would have preferential orientation along the length direction. This ellipticity can be suppressed by increasing filament speed and introducing an ellipticity in the spiral itself. Evidently, an isotropic web would require filaments to be laid in circles on the apron. In addition to forming circular loops on the apron, the individual filaments also need to be disturbed from their linear path so that they can land randomly on the apron surface. This can be achieved by, for example, subjecting the filament curtain to an electrostatic field or by action of bounce blades prior to landing on the apron. Laying the filaments on the apron can also be executed by aerodynamic forces, which ensure intermingling as well as randomization of filaments in the final web. Evenness of the web becomes a critical issue for lightweight fabrics, especially when aerodynamic forces are employed for web laying. Spunbonded webs may be as wide as 5 m, and the lightest among them can be produced at a very high speeds of up to 600 m/min. In conjunction with moderate energy consumption, varying between 2.5 and 1.5 kWh/kg of polymer and a technically elegant and simple process for conversion of polymer granules into textile fabrics, the spunbonding process enjoys a unique appeal among all fabric-formation methods.

15.5.2 Web Formation by Meltblowing

In this process, a high MFR and low-viscosity polymer melt is extruded through the orifices of a linear die into the stream of a converging jet of hot air blowing at speeds varying between 6000 and 30,000 m/min.

A schematic view of the meltblowing process is illustrated in Figure 15.9. A close-up view of the core of the meltblowing zone is depicted in Figure 15.10. The molten filaments emerging from the orifices are rapidly attenuated into extremely thin fibers, which are then blown by another jet of cold air onto

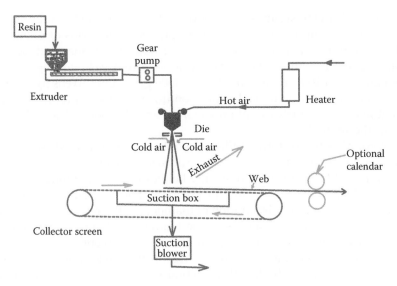

FIGURE 15.9
Schematic view of meltblowing process.

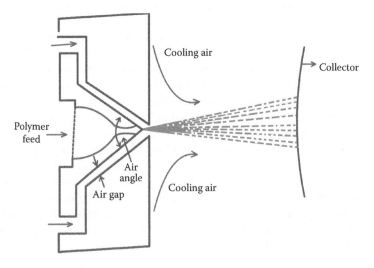

FIGURE 15.10
Close up view of meltblowing zone.

a collecting screen that rapidly carries away the entangled mass of the net-work of fibers in the form of a web to the next processing station. As opposed to filaments of spunbonded webs, no drawing of solidified fibers takes place in this process. The secondary stream of cold air only cools the attenuated collection of fibers and blows them toward the collector. Hence, the resultant fibers do not exhibit a high degree of orientation and crystallinity, which affects their mechanical properties. The tackiness of the fibers landing on the screen contributes to their bonding among themselves. This may be reinforced further by additional subsequent calendaring.

As compared to spunbonded filaments, meltblown fibers exhibit a lower degree of orientation, greater variability in diameter, and lower strength owing to existence of voids (Choi, 1988). Besides the MFR of the polymer melt, the other factors that play important roles in governing the properties of meltblown fibers and webs are the air angle, the volumetric flow rate of air, and the distance between the die and the collector. The strength properties of the web peak at an optimum value of MFR and decrease with a further rise in MFR (Jones, 1987). A smaller air angle yields a greater number of parallel fibers, a higher degree of attenuation, and less fiber breakages, and a steeper air angle results in a higher dispersion in fiber length and a greater degree of randomness (Milligan and Haynes, 1991). A volumetric flow rate of primary hot air and not its velocity is the primary factor affecting average fiber diameter. The air gap that affects air velocity plays a secondary role in the attenuation process. Air permeability and pore sizes of meltblown web increases with a rise in the die-to-collector distance (Lee and Wadsworth, 1990).

The areal density of meltblown webs typically varies between 8 and 350 g/m^2 (Bhat and Malkan, 2007). As compared to spunbonding, the meltblowing process consumes a much higher amount of energy. According to Malkan (1994), the meltblowing process consumes 7 kWh/kg of polymer as compared to 2.5 kWh/kg of polymer in the spunbonding process, which can be further reduced to 1.1 to 1.5 kWh/g of polymer with suitable modifications in the air suction system.

15.5.3 Web Formation by Flashspinning and Electrospinning

In a process patented by DuPont, high-density polyethylene is, at first, dissolved in hot solvent at very high pressure varying between 40 and 70 bars (1 bar = 1 atmospheric pressure = 100 kPa) after which the solvent is removed from the solution in a controlled manner leading to a network of fine fibers varying in thickness between 0.5 μm and 10 μm (typical human hair is 75 μm thick and spider silk has a thickness of 1 μm), which is then suitably collected and reinforced to the product Tyvek®. This lightweight material is permeable to water vapor but not to water and has a host of applications in the sector of protective clothing.

The drive to generate fibers having a diameter in fractions of micrometers (10^{-6} m) and approaching nanometers (10^{-9} m) is governed by the realization that properties of matter, be they electronic, mechanical, electrical, optical, or surficial, undergo drastic changes as one approaches the dimension of molecules. Noting that the diameter of a hydrogen atom is one fourth of a nanometer and that the width of molecules constructed out of diverse atoms would therefore be measurable on the nanometer scale, one can infer that a substance of width of, say, 50 nm would basically be a collection of very few molecules or, in some cases, probably be constituted by one molecule only! Just as properties of a brick are different from that of a wall constructed with the bricks, so would the properties of a macroscopic fiber of, say, a 10-μm PP (nearly equal to one denier) be very different from a 10 nm PP. A 10-μm PP fiber is a large-scale assemblage of a complex arrangement of many long-chain molecules having many amorphous and crystalline zones and cross-linkages among themselves. A very limited number of active end groups would be available on the outer surface of such a fiber for interacting with the immediate environment. Its properties would indeed be a result of properties of the molecules, their linkages among themselves, and the structure of their arrangements within the fiber body. A single molecule, on the other hand, would have all its own attributes available for responding to external stimuli. In the scientific world, the term nanomaterials refers to dimensions varying between 1 and 100 nm as the effects of the nanoscale start becoming apparent only at less than the 100 nm dimension. The electrospinning process has reportedly been employed to generate fibers in the range of 40 to 500 nm (Kittelmann and Blechschmidt, 2003).

The basic concept of electrospinning is outlined in Figure 15.11. A polymer solution or melt is pushed through a narrow orifice and then is pulled by electrostatic forces toward the surface of a conducting collector, the potential difference between the orifice and collector being in the range of 5 to 30 kV or, in some cases, even higher. Deformation of the liquid droplet on the orifice tip into a Taylor cone and its subsequent attenuation into an ultrathin fiber or filament is generated by interaction between very high electrostatic forces, the surface tension of the fluid, and the molecular forces of the polymer. At the outset, a critically high concentration of the electrical charge on the droplet results in a sufficiently high electrostatic force that accelerates it into a jet stream. Away from the source, the electrical attenuating force weakens

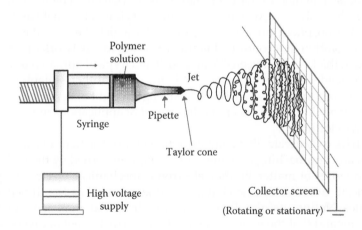

FIGURE 15.11
Concept of electrospinning process.

rapidly, and the jet has to overcome air resistance and gravitational forces. This causes the stream to deviate from a linear path and start bending away. Simultaneously, the concentration of like charges on the jet stream gives rise to strong forces of repulsion within itself, which makes the jet attenuate further. The extremely high exposed surface in the attenuated stream permits rapid cooling or rapid evaporation of solvent leading to a rapid solidification of the stream (Ren et al., 2010). Thus, the thin stream of polymer emanating from the orifice tip rapidly develops into a gossamer-like fine funnel of filament before landing on the collector surface, usually another flexible material, such as a textile fabric.

A wide range of biopolymers, electrically conducting polymers, and liquid crystalline polymers, both synthetic and natural, such as collagen, silk fibroin, polyurethane, polystyrene, polylactic acid, etc., have been electrospun to fibers as fine as 10 nm and upward. Such fibers are of special interest for filtration, tissue-engineering, drug-delivery, and wound-dressing systems (Bharadwaj and Kundu, 2010). The material variables playing important roles in the electrospinning process are concentration, viscosity, molecular weight, surface tension, conductivity of the melt or solution while applied voltage, tip-to-collector distance, type of collector, and ambient conditions, such as temperature and humidity of the immediate surroundings, constitute relevant process variables. Basu et al. (2013) observed that the diameter of fibers in a single-syringe electrospinning process depends primarily upon dope properties, and Cramariuc et al. (2013) opine that applied voltage and polymer rheology play decisive roles in determining fiber diameter.

The rate of production from a single syringe (needle), which, depending on material and process variables, can vary between 0.01 and 0.1 g/hr (Varabhas et al., 2008) or from 0.1 to 1g/hr (Forward and Rutledge, 2012) can hardly be of any commercial significance. The logical step of electrospinning with an array of needles for raising the production rate is subject to the effect of strong charge repulsion between adjacent needles and jets, which calls for wider spacing of needles, limiting thereby the production rate per unit area of collector. Multiple needles have been substituted advantageously by porous tubes with multiple holes (Varabhas et al., 2008) to suppress some of the associated problems achieving, in the process, 250 times higher production rate than that of a single needle (Dosunmu et al., 2006). Mutual repulsion among jets due to electrical field interference also tends to disturb onset of secondary bending in individual streams of jet, thereby affecting the morphology of the product (Theron et al., 2005). The nozzles of the needles or pore openings on the tubes are also prone to clogging.

The drawbacks of electrospinning with the help of needles and tubes as mentioned in the foregoing are elegantly addressed in what is termed as free-surface electrospinning, wherein the propensity of a liquid surface at its interface with another suitable medium—air, for example—to break up into multiple waves and finally into multiple jets under the action of favorable electrostatic field is made use of. Such controlled rupture of the liquid surface is facilitated by an external agent, such as wires, cylinders, disks, or even gas bubbles, which penetrate the highly charged liquid surface and come out of it carrying thereby entrained liquid on their surfaces (Figure 15.12). This film of entrained liquid around the external agent undergoes subsequently a dewetting process and breaks up into a multiple of charged droplets, which then individually get deformed into Taylor cones and finally accelerate into jet streams. These three stages have been schematically represented by a different shape of liquid matters on the three wires positioned above the liquid surface in the front view of Figure 15.12. The lowest wire just having come out of the liquid exhibits the collected fluid in the process of separation into individual entities, owing to the dewetting process followed by the intermediate wire carrying these entities in the form of clearly defined droplets spaced widely apart, and on the third wire, the droplets have been transformed into Taylor cones. Thus, as a suitable array of external agents regularly dip into charged liquid and come out of it, a large number of jet streams keep on emanating from the system and streaming out in the usual manner toward a suitable collector.

A system for commercial production based on the concept of free surface electrospinning is marketed under the trade name "Nanospider" (Jirsak et al., 2005) with a production rate of 1.5 g/min per meter length of the surface. A considerable amount of scientific investigation is currently under progress to understand, characterize, and optimize productivity as a function of equipment geometry, fluid properties, and applied voltage in free surface electrospinning (Forward and Rutledge, 2012). It is understood, however, that the desired morphology and size of fiber may be difficult to achieve though the free-surface electrospinning process (Zhou et al., 2009).

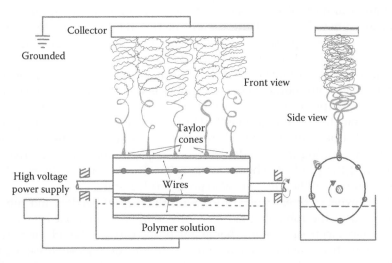

FIGURE 15.12
Concept of free surface electrospinning.

New innovations in electrospinning include flat spinneret electrospinning (Zhou et al., 2009); coaxial electrospinning for generating porous core-sheath fibers aimed at controlled drug release in wound-healing patches (Nguyena et al., 2012); and also for producing hollow and porous nanofilaments, which can be employed in fields as diverse as ultrafiltration, fuel cells, membranes, tissue engineering, etc.; two-pole air-gap electrospinning for generating 3-D cylindrical construction aimed at providing multiple channels to direct axon growth (Jha et al., 2011); and blow-assisted electrospinning for increasing the production rate of filaments (8 to 10 g/hr per jet (Chase et al., 2011), etc. It is evident from the current global attention focused on this method of web formation that in the coming decade many more exciting developments both in terms of technology of production and fields of application can be expected.

15.6 Reinforcement of Web

After the desired web or batt has been prepared in one of the various ways described in the foregoing, it is reinforced suitably with a view to achieving functional properties in the end product. Reinforcement can be carried out by either of any of the processes described in the following or even by a combination of more than one. There are four routes for supplying the required reinforcing energy through mechanical, chemical, thermal, and sonic means.

15.6.1 Mechanical Method

Two methods based on a similar principle of creating entanglements within a fibrous assembly and thereby consolidating the same are commercially practiced widely. They, however, impart very different properties to the product. The needle-punching method, which involves repeated thrusting of rough-edged metallic needles into the fibrous assembly, results in a very high degree of consolidation and a higher amount of residual stress in the resultant product as compared to the hydro-entanglement (spinlacing) method in which very fine and powerful water jets pierce the assembly at multiple points causing entanglements among the fibers, leaving behind very low residual stress. The hydro-entangled fibrous mass is accordingly very soft and highly absorbent while a needle-punched product is stiffer and denser. It important to note that the structure of reinforced units within the respective products is also very different.

Stitchbonding a web is the third mechanical method of web reinforcement. This method is a hybrid of the warp-knitting and stitching processes and comes under the broad banner of MALIMO technology. The word MALIMO is an acronym of three German words: MA stands for Mauersberger, the inventor of this process; LI stands for Limbach-Oberfrohna, the name of an industrial town in which this process was mechanized; and MO stands for Molton, a German word for plain weave (Ploch et al., 1978). The basic process involves employing a stitching principle to reinforce a fibrous web or join together diverse textile systems, such as yarns and yarns, yarns and a fibrous web, or even yarns and fabric, the stitches being created by warp-knitted loops to form products primarily for household and technical applications.

15.6.1.1 Needle Punching

Central to this system are barbed needles, two very basic types of which are depicted in Figure 15.13. A needle has at its base a crank, followed by the stem or shank, both crank and shank being circular in cross-section. Subsequently, there may be one or, in some cases, two reductions in needle diameter, leading to the working part, namely the blade, which is usually triangular in section, and equipped with barbs along its three edges. Blades having other types of section, such as a quadrilateral or a star or that of a water drop with one sharp edge only, are also in vogue for special purposes. Double reduction of the stem leads to a thinner blade and retains the needle strength around its base. It is the needle blade aided by barbs that is responsible for transporting fibers from their original position within a fibrous assembly, resulting in entangled plugs or pegs of fibers. Beyond its working part, the needle tapers gradually to a tip. Needle tips as well as needle barbs come in various shapes and dimensions.

A simplistic view of a barb is shown in Figure 15.14. This element may be viewed as having been created by the slashing edge of a needle blade by a sharp tool. As a result, a barbed throat is created having a certain length, width, and depth and therefore a certain volume of enclosed space. This

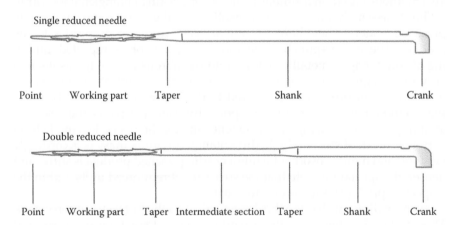

FIGURE 15.13
Two basic types of barbed needles.

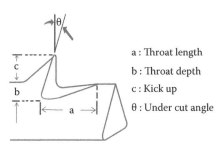

a : Throat length

b : Throat depth

c : Kick up

θ : Under cut angle

FIGURE 15.14
Close up view of a barb.

throat is responsible for catching fibers that happen to come in its way during the punching motion of the needle and transporting them across a thickness of fibrous mass.

The number of fibers transported by a barb throat depends on the fiber thickness as well as on the location of the barb on the needle. Thicker fibers occupy a larger throat space, causing a lesser number of fibers to be transported by a barb. Clearly then, thicker fibers require larger throats, which, in turn, can only be created from thicker needles. This has led to some empirical relationships relating needle gauge with fiber linear density.

Barbs located nearest to the needle tip transport the largest number of fibers, and the furthest barb usually does the least work. There are two reasons for this phenomenon. The barb closest to the needle tip is the first one to enter the fibrous mass and hence enjoys a maximum chance of catching fibers. Only those fibers that escape this barb would be available for the succeeding ones that follow an identical path. Logically then, the furthest barb can be expected to encounter the least number of fibers and therefore would have to do the least work. Moreover, the inward thrust of a needle into the fibrous mass tends to push fibers away, thus increasing the distance between barbs that follow and potential fibers that might be caught in their throats. At the same time, if resistance to the slippage of fibers among themselves is high enough, then the fibers dragged by a barb would tend to pull in the surrounding fiber mass closer to the succeeding barb. These two opposing tendencies decidedly affect load distribution of the barbs located along the same edge as the needle.

The actual workload of individual barbs is affected by a complex interaction between variables, such as thickness of fibrous mass at the instant of punching (t), its distance from the needle tip (ℓ), and the distance that the needle tip travels beneath the lower surface of the punched material (p). The last named variable, that is, p, is commonly termed as depth of penetration. The workload of a barb goes up for higher values of t and p and a smaller value of ℓ. A barb remains idle if $\ell \geq p + t$, and for $\ell \leq p$, its workload is the highest (Figure 15.15). This workload can be quantified by the number of fibers transported and the resistance that fibers offer to slippage among

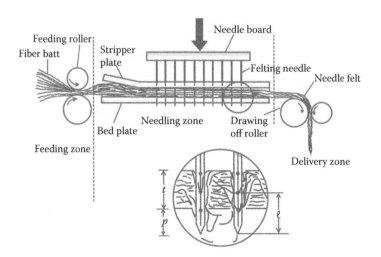

FIGURE 15.15
Action of needle punching.

themselves. Thus, a partly consolidated fibrous mass would impose a greater workload on the first barb nearest to the needle tip as opposed to a fiber mass yet to be punched. Such workloads impose a strain on the barb and affect their wear and tear. Moreover, the net resistive force acting on a barb arising out of this workload gets translated to a bending moment around the base of the shank of the needle. An equitable distribution of barbs around the needle axis along a spiral path goes some way toward alleviating this situation although, due to an unequal workload among different barbs on a needle, there is always a net bending moment that eventually causes cyclic strains on the needle. Hence, needles need to have a required rigidity commensurate with the expected workload. Accordingly, needles employed for punching partly consolidated batts need to be, in principle, stronger than those employed for punching fresh ones.

The barb shown in Figure 15.14 has a pronounced kick up, which is harsh on fibers. Such a treatment is acceptable for fibers, such as coir or wool, that have high breaking elongation and surfaces not prone to easy damage, but fibers that are more brittle, such as glass or ceramic or whose surface properties are critical to their performance as many synthetic fibers are, a much gentler treatment is imperative. Innovations in barbed needle manufacturing have resulted in different barb geometries that subject fibers to gentler treatment with similar if not enhanced performance of fiber transport. An example of such an improvement is shown in Figure 15.16.

Irrespective of the nature of barbs, most three-edged needles sport three barbs per edge spaced at equal distance. Regular (R), medium (M), close (C), and fine (F) barb spacing are illustrated in Figure 15.17. A needle with only one barb along each edge and spaced extremely close together come in a

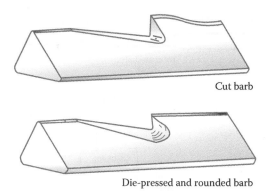

Cut barb

Die-pressed and rounded barb

FIGURE 15.16
Barbs with improved geometries.

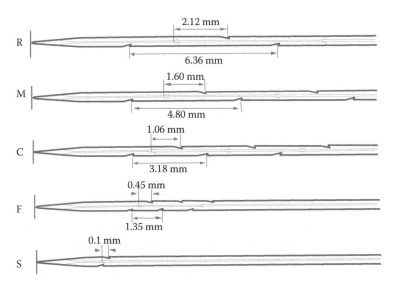

FIGURE 15.17
Barbs with different spacing.

class (S) of its own. Evidently, the throw of each of these needles differs considerably, and each type serves a different function.

The regular barb needles are invariably employed for batts that are punched for the first time. Such batts are lofty with large spacing between fibers in the direction of batt thickness. A typical description of such a needle is illustrated in Figure 15.18. The number of needle wires of a circular section occupying a 1-inch space when placed side by side represents its nominal gauge. Thus, a wire of gauge 15 measures 1/15th of an inch across its diameter. A needle is specified by the nominal gauge of its blade, and the illustration is of a 36-gauge regular barbed (RB) needle with double reduction. This

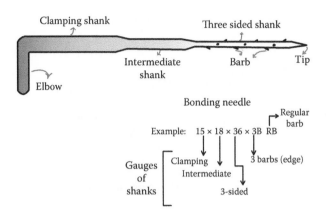

FIGURE 15.18
Characterization of a needle.

needle has been fabricated from a very thick wire of gauge 15, indicating that it is expected to withstand considerable strain although, being of effective gauge 36, it is actually meant to transport reasonably fine fibers in the range of 20 deniers and thereabouts. Needle gauge may vary between a very coarse one of 12 gauge to extremely fine ones of 42 gauge. A 42-gauge needle is expected to handle microdenier fibers, and very coarse fiber such as coir (up to 600 dtex) can be handled by the coarsest needles. The blade of a RB needle is the longest of the lot, and its frequency of operation is therefore lowest and not suited for fast-running machines. Shorter blade lengths are more suited for higher operational frequency. As a result, they would punch a batt more vigorously unless the blades are made finer with reduced throat space in the barbs. Finer blades also occupy less space, and if made from single reduction wires, they can be grouped together more tightly, leading to a much larger punching density on the batt. During the initial course of formation of pegs, the fibrous mass undergoes a pronounced reduction in thickness and some amount of lateral spreading. The accompanying lateral spreading depends on the instantaneous tensile resistance of fibrous mass, and therefore, generation of a certain number of pegs in a totally unconsolidated batt would cause a greater amount of lateral deformation than when the same is carried out in a partly consolidated batt. One can thus visualize a batt being punched initially by RB needles so that the fibrous assembly gains some initial integrity with limited areal deformation, which can subsequently be subjected to a more vigorous punching regimen employing different types of needles (MB or CB needles) to higher punching density for improving the extent of consolidation through formation of a very large number of smaller pegs spaced close together. The larger the density of such pegs in a fibrous mass, the greater would be its consolidation. However, every needle entry leaves a mark on the surface of the fibrous mass, and thicker needles leave larger marks. The goal is to punch the fibrous mass sufficiently without leaving behind any specific pattern of these pockmarks. Finer needles make finer perforations and leave smaller imprints on the surface of the product. Indeed, if such imprints are extremely closely spaced, as in pixels on a monitor screen, they may even appear to the eye as continuous.

Reduced barb spacing manifests in higher frequency, a necessity, for example, with spunbonded webs that move much faster than a cross-laid batt made from staple fibers. Similarly, a continuous movement of the batt during the punching process demands that the dwell time of a needle within a batt be as short as possible, which is also facilitated by very close barb spacing. Indeed, for very fast moving batt, such as a spunbonded web, the needles need to be moved along an elliptical path instead of a simple liner to and fro (or up and down) motion so as to avoid undesirable damage during the punching process.

The special S needle, also known as a crown needle, has the specific purpose of pushing out an umbrella of fibers through the lower side of the batt with a view to creating a surface effect. Similarly, a needle with a fork-like extremity instead of a tip as well as devoid of any barb has also the function of creating surface texture. Such special needles are not meant for reinforcement of the web and hence will not be gone into.

Besides its barbs, the other functionally important element of the needle is its tip. Needle tips can be very sharp or can have a smoother and rounded surface like that of a ball. Different needle tip variations differing in the extent of roundness of tip are illustrated in Figure 15.19. They come into question when the batt carries materials, such as yarns, that may be damaged by the sharpness of the punching agent. Yarns may be present along one direction in the batt or may be present in more than one direction in the form of a reinforcing fabric, and yarns may be fine or may be coarse, and inter-yarn spacing also might differ from one type to another. A properly chosen roundness of tip ensures that the yarns are pushed away laterally without being pierced through. Even then, the barbs that follow the needle tip may also damage yarns. This problem is negotiated by barbs without any kick up or/and barbs that are suitably oriented vis-à-vis the needle crank or even needles having barbs on one edge only.

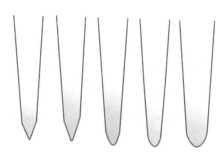

FIGURE 15.19
Different types of barb tips.

Needles are mounted on boards with holes distributed in a matrix (Figure 15.20). Needles are inserted through these holes and held in the desired position by clamping their cranks suitably, keeping in mind that in their standard execution the three apices of a triangular blade are displaced by 60°, 180°, and 300° to the crank axis. This spatial positioning of barbs has a bearing on the extent of fiber transport vis-à-vis the orientation of fibers in a fibrous assembly and also on potential damage that may be caused to the reinforcing yarns. The diagram shows a diagonal needle set out, which is sure to result in a discernible pattern, termed as tracking, on the punched batt. The principle of randomness is normally resorted to while creating holes on a needle board by ensuring that distance between neighboring rows and columns of needles follows a random pattern. Such a step suppresses the problem of tracking considerably (Young, 1970). A multiple of needle boards, each of a specific length and width and carrying the desired number of needles, are mounted on the beam of a needle-punching machine. This beam is imparted in a reciprocating motion, and thus all the boards carry out a reciprocating motion in unison. A side view of such a mounted system is shown in Figure 15.20, revealing the width view of one needle board equipped with 10 rows of needles. The length dimension of the assembled boards coincides with the width dimension of the punching machine and is directed normally to the plane of the depicted figure. Two stationary perforated plates, namely the stripper plate and the bed plate, are mounted on the machine in such a way that needles pass through matching holes in these plates while executing a reciprocating motion. The depth of penetration (p) of the needle, defined as the distance between the upper surface of bed plate, that is, the lower surface of a batt being punched, and the farthest point of the needle tip below the bed plate (Figure 15.16) is adjusted by moving the bed plate. A greater value of p is associated with more intense punching and is regarded as a critical process parameter for adjusting product quality. A positive value of p larger than the distance between the tip and the first barb of the needle,

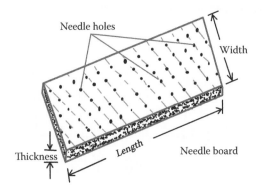

FIGURE 15.20
View of a needle board.

which is usually 6.4 mm, is associated with the emergence of fiber loops through the lower surface of the batt being punched. Otherwise, fibers from the lower surface of the batt being punched would be pushed into the holes of bed plate for a positive value of p. As the needles retract from the holes of the bed plate, the batt has a tendency to rise up along with but restrained by the stripper plate. Fibers left behind in the holes of the bed plate would be pulled out and brought flush against the lower surface of the batt during its subsequent forward motion. In the subsequent cycles of punching, these protruding fibers may chance to get interlocked with another group of fibers punched afresh into the holes of the bed plate. Hence, under favorable circumstances, reinforcement of a batt may occur at two levels; the pegs or plugs within the body of the batt and an interlocked or intermatted layer of fibers on its lower surface.

The setup shown in Figure 15.15 serves the purpose of pre-needling or tacking with a view to generating a degree of cohesion in the fibrous assembly. The resultant product would have very different properties about its two surfaces and would need further controlled and intense punching for satisfying demanding end uses. A standard follow up procedure is a double-sided punching. A double-sided punching can be carried out in two successive stations although a single-station double-sided punching may also suffice certain end uses. A single-station double-sided punching is carried out by employing two needle beams face to face arranged on the two sides of perforated plates. The bed plate for one beam functions as the stripper plate for the other. Such an arrangement not only ensures an intensive punching at one station but also results in a product that is balanced with respect to properties about its two surfaces.

Double-sided punching may be simultaneous or alternate. In the case of alternate punching, when one bed moves in to punch, the other board retracts, and in simultaneous punching, both boards move in to punch at the same time. For the same punching density and the same machine frequency, the rate of production in the case of alternate punching would theoretically be twice as much as that with simultaneous punching, and a great amount of saving in needle inventory can be achieved with the latter because of offset arrangement of needles between the upper and lower needle boards. There would, however, be a considerable difference in internal structure between the products of the two processes.

The internal structure of a punched fibrous assembly is altered when needles move at an acute angle to the plane of the batt. Corresponding pegs occupy a longer path across the batt thickness and create a more intense entanglement among fibers resulting in a higher degree of compactness for equivalent punching density. Such a structure can be realized by passing the batt through similarly curved stripper and bed plates, which forces the batt to assume the common curved path while the needles move along a linear path. An extreme scenario is obtained in circular needling for manufacture of needled tubes, varying in diameter between 10 and 500 mm, wherein fibrous assembly wrapped around adjustable rollers are punched from the top and bottom. Alternately, the fibrous mass may be kept flat and punched from both sides while the needles are made to follow a curvilinear path.

Punching density P is a critical factor affecting the extent of compaction of the batt and is expressed by the number of punches per square centimeter (cm^{-2}). Factors that determine the value of P are punching frequency n of the machine (min^{-1}), speed s of the batt (m/min), and the number of needles N per linear meter width on the needle board.

For every punching cycle, there would be N punches across 1 m width of the batt. If the number of rows of needles on the needle board is r and if d (m) is the distance between successive rows of needles, then a strip of batt having area $(d \times 1)$ m^2 receives a row of (N/r) punches as it enters the punching zone. This strip would have to travel a distance $d \times r$ (m) before it moves beyond the punching zone. The batt advances by s (m) in 1 min, during which n punches are executed by the machine. For travelling $d \times r$ (m), the number of punching cycles are $(d \times r/s) \times n$. Hence, the strip of batt having area d (m^2) receives $[(N/r) \times (d \times r/s) \times n]$ punches before it moves ahead of the punching zone. Therefore, $[N \times (d/s) \times n]$ punches land on an area of $(d \times 1)$ m^2 of batt. Hence, the punching density is given by the expression

$$P \ (cm^{-2}) = [(N \times n)/s] \times 10^{-4} \tag{15.6}$$

Modern needle-punching machines operate with a value of $1000 \leq N \geq 3000$ at the pre-needling stage for $10 \leq P \geq 75$, and these values, depending on the product, may go up to as high as 30,000 and 1000, respectively, in the finishing stages. These machines are characterized by a punching frequency of 1000 to 3000 per minute, producing fabrics at a speed of up to 60 m/min (Brunnschweiler and Swarsbrick, 2007). Felts for a paper-making machine of up to 15 m width are produced on specially designed needle-punching systems.

15.6.1.2 Hydro-Entanglement

The role of a needle board carrying a matrix of barbed needles is taken up in this process by an injector carrying rows of nozzles. Powerful jets of water emerging from the injector pierce a moving fibrous assembly, which is supported from below by a porous entity of suitable topology and opening. This support may be flat in the form of a moving belt of woven wire mesh (Figure 15.21), or it may be a curved surface of a perforated metallic sheet mounted on the skeleton of a hollow cylinder with large openings (Figure 15.22). Usually, a hydro-entanglement line is made up of a series of such injectors, which are supplied by pressurized clean water from individual pumps. Four such injectors are depicted in Figure 15.21. Jets from these injectors consolidate the web in stages but penetrate the same side of the web repeatedly. In many cases, both sides are alternately subjected to the action of the jets (Figure 15.23). Water passing through the porous belt or hollow

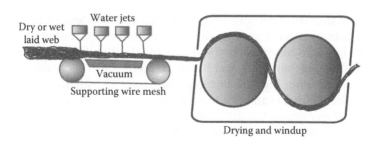

FIGURE 15.21
Schematic view of a flat hydro-entangling station.

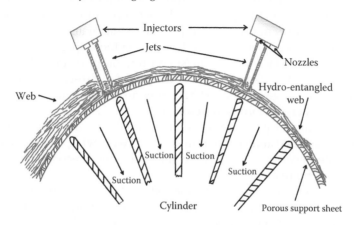

FIGURE 15.22
Schematic view of a curved hydro-entangling station.

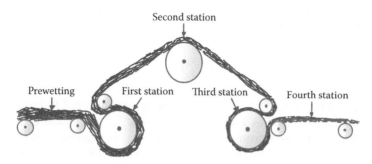

FIGURE 15.23
Hydro-entangling station with double-sided punching.

cylinder is subsequently sucked away, cleaned, and recirculated. The wet fabric is then hydro-extracted and dried.

The impinging water jets rearrange the fibers of the web, on occasions even splitting them up into multiple and very fine entities. By way of illustration, one may refer to the Evolon® range of products of M/S Freudenberg of Germany that involves splitting by the hydro-entanglement process hollow filaments of about 2 dtex made up of multiple segmented-pie PA and PET materials into microfilaments of 0.09 to 0.13 dtex fineness for production of lightweight, very soft, and highly drapable fabrics suitable for active sportswear and other high-performance end uses. During hygral treatment of a web, a considerable amount of residual stresses in the fibrous assembly along with many impurities get removed while hydrogen bonds are formed among fibers with suitable end groups. As opposed to intermittently punching needles, these jets flow continuously, and if the fibrous assembly is not continuously moved away from the jetting zone, it can even get damaged. The pressure at which these jets of around 100 µm diameter operate can be as high as 60 MPa (600 bar), and similar jets operating at five to 10 times higher pressure are industrially employed to cut materials ranging from paper to granite. Water jets employed as cutting tools operate in the Mach 2 and higher range (Mach 1 = 343 m/s) and rely on principles of supersonic erosion (Leach and Walker, 1966), and the jets of hydro-entangling systems operating in the range of 10 to 60 MPa pressure have much lower velocity and therefore energy. Fibrous systems punched by high-pressure water jets operating at a higher pressure range of up to 60 MPa are moved at speeds ranging between 300 and 600 m/min suited to wet laying and spunbonding lines, and the lower pressure jets are suited for line speeds in the range of 75 m/min and thereabouts.

In addition to the issue of the compatibility of jet energy with web speed, there is also the critical issue of spending an optimum amount of energy for maximizing an achievable strength level expressed by a specific energy coefficient (Moschler, 1997). Under a steady state of flow and assuming that density ρ does not change, the velocity v_w of a water jet is proportional to the square root of pressure and the force F with which it strikes a surface of area A. It can be expressed as

$$F = A\rho v_w^2$$

Thus, the force of impact would be directly proportional to the pressure applied, and a part of this force would account for the extent of displacement of fibers in the web from their original location, the rest going to waste. Energy E consumed by the unit mass of the web in this work is affected by web variables, such as its width, areal density, and speed, and jet variables, such as its number and the force with which each jet strikes the web. The specific energy coefficient is a ratio of this energy E to the resultant strength of the reinforced web. By varying the web and jet variables suitably, one can strike an optimum value of this ratio for economical operation. It is important to note that the density of liquid also plays a critical role in this process, and suitable additives in water can enhance the effectiveness of the jet without adding to the energy cost. Improvements in the jetting system have, over time, resulted in a steady reduction of energy consumption per kilogram of fiber processed. Typical values quoted in current literature vary between a low of 0.1 to a high of 1.3 kWh/kg of processed material (Mustermann et al., 2003).

As mentioned in the beginning, the web is supported by a suitable carrier material, which might be a woven wire mesh or a porous metallic sheet. A jet impinging on the web may, after passing through a fibrous barrier of web, encounter either an opening or a rigid obstruction. In the former case, it would simply pass through the opening and get removed by the drainage system. It has, in this case, simply pushed some fibers laterally apart along the web plane and even imparted a displacement in a direction normal to the web plane. The net displacement to affected fibers is tantamount to a partial rotation. Hence, a through-and-through passing water jet would result in a localized compaction accompanied by some twisting of the affected fiber bundle in a web mass. On the other hand, if a jet encounters a rigid obstruction after piercing the fiber mass, then it would get splintered and rebound onto the back side of the web, provided it is not damped by a captive water layer between the web and the carrier material. Thus, in the presence of an effective drainage system, the rebounding jet particles would do more work of further entangling, twisting, and consolidating the affected fiber mass. These partially twisted and compacted bundles of fibers account for reinforcement of a spunlaced web.

If a web is placed on another web, and the resultant sandwich is subjected to the process of hydro-entanglement, then one can visualize the formation of twisted and compacted bundles of fibers from both webs, thus, in effect, joining the two. Hydro-entangling is an elegant way of forming composites of spunbonded–melt bonded–spunbonded (SMS) or spunbonded–air laid pulp–spunbonded, and many other desirable combinations.

The topology of the carrier material can be exploited to impart surface texture, and the nature and distribution of its openings account for pores of suitable size, shape, and overall porosity in spunlaced material.

The surface texture of a woven wire mesh is determined by the thickness of interlacing wires, the nature of interlacement and density of the wires in two principal directions. The surface of a metallic sheet can have imparted a suitable topology by selective deposition of particles by the process of electrolysis, or its surface may be engraved. Indeed, a fibrous web may be made to look like a knitted or a woven fabric though use of a suitably engraved carrier followed by stabilization of the resultant texture through application of compatible polymeric binders.

Fibers suitable for hydro-entanglement should have low flexural and torsional rigidity, should not be brittle, and should have low surface tension with respect to water. To improve the wettability of the web, it is thoroughly wetted out prior to hydro-entanglement, somewhat akin to pre-needling in the needle-punching process although having a completely different purpose. The ability of constituent fibers to form hydrogen bonds among themselves contributes to improved strength of the reinforced web. Staple fibers as well as continuous filaments from a wide range of polymeric materials can be hydro-entangled whereby a staple length of around 50 mm and fiber fineness around 1.1 to 3.3 dtex are commonly hydro-entangled. Fibers not belonging to this preferred range, such as air laid wood pulp, can, however, be trapped in a sandwich mode between two layers of webs made from the suggested range of fibers for imparting additional functional properties to the product. Fabrics in the range of 15 to 450 g/m^2 to a maximum width of 5.4 m are commercially produced with this technology.

15.6.1.3 Stitch Bonding

The principle of web reinforcement employed in this process is best exemplified by going into the MALIMO method of fabric production. The term MALI is used here as a generic name as processes employing similar technologies but branded differently (such as the ARA series or the ATS series, etc.). A side view of the MALIMO fabric formation zone is depicted in Figure 15.24. In this process, a conventional warp sheet (16) and a layer of weft yarns (15) are made to converge to the fabric-formation zone. The continuous layer of weft yarns, generated in a manner similar to that described in Section 13.5.1.3, is made to lie over the warp sheet in an ordered manner, and the cross-over points are stitched together by employing pillar or tricot construction. Stitching of the two sheets is carried out with the help of compound needles (1 and 2), one guide bar that feeds stitching yarns (17) to the needles, and a knocking over sinker bar (4) that helps in casting off of the loops. Additionally, there is a bar (5) carrying a set of retaining pins, the tips of which are set flush against a supporting rail (6). The retaining pins primarily form a reed-like wall restraining the weft sheet from moving away

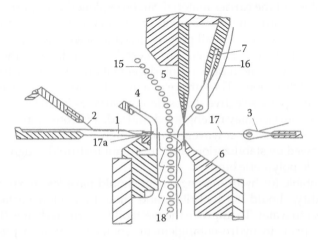

FIGURE 15.24
MALIMO formation zone. (From Ploch, S. et al., 1978. Permission obtained from Karl Mayer.)

beyond a certain plane during the piercing action by a series of sharp needle crowns while permitting the needle tips to pass through the dent-like gaps. In this process, the support rail absorbs the thrust through flexible contact with the tips of the pins. The retaining pins also assist in spacing out and precise positioning of the warp yarns through the gaps of which needles move in and out.

Two stitch-bonding systems, namely MALIWATT and MALIVLIES, are employed for reinforcement of a fibrous web. In the MALIWATT system, a cross-lapped web replaces the weft sheet, and warp yarns are totally absent. Stitching threads bind the web together by a tricot or pillar stitch following the same method employed in the MALIMO process to bind together warp and weft sheets. MALIVLIES is, however, a unique approach of reinforcing a web by creating binding elements from fibers of the web itself, and no yarn whatsoever is employed in this product.

In a MALIVLIES machine, the retaining pins are replaced by laying in sinkers. These are essentially more robust than retaining pins. These sinkers push the web back with sufficient resistance when the needles pierce through it, resulting in a degree of compaction. A MALIVLIES web has to be prepared from fibers of certain length, fineness, and inter-fiber cohesion. Moreover, they should be preferentially oriented along the width direction on the surface of the web facing the laying in sinkers so that the needle throat can pick up a sufficiently large number of fibers from this surface during its return journey and pull them securely through the web into a loop form, and the concerned fibers remain strongly gripped by their neighbors and do not slip out of their grip. Web compaction at this stage contributes to improved cohesion among fibers. The needle wire starts moving forward much earlier than in the MALIMO system and actually penetrates the web to close the needle throat space while the needle is about one third of its way back within the web. The reinforced web exhibits stitched loops on the technical front side only, and no underlaps form on the back side.

Both the stitch bonding systems come in machine widths in excess of 6 m and can process webs as fine as 60 g/m² (MALIWATT) on the lighter side to the fairly heavy ones of about 2500 g/m² (MALIVLIES) with production speed as high as 400 m/h for the light webs (Schreiber et al., 2003).

15.6.2 Thermal Bonding

Infusion of heat energy can cause thermoplastic polymeric materials to change from a solid to a liquid state, and removal of heat reverses this process. This property of thermoplasticity is a result of weak interactions between molecular chains of such materials. These weak interactive bonds gradually break down with increasing energy levels without affecting the long-chain backbone structure of the polymer. These bonds form afresh when the additional energy is removed. Such materials are characterized by the glass-transition temperature, softening point, and melting point.

The glass-transition temperature characterizes a boundary between a brittle and a rubbery state, and at the softening point, a material becomes tacky. The polymer loses its solid state at its melting point and becomes liquid, as a result of which it starts flowing. These three temperatures vary considerably from one type of polymer to another. By way of an example, typical values for PP are –(12 to 20)°C, 150°C to 155°C and 175°C, and the corresponding values for PET are 70°C to 80°C, 230°C to 250°C, and 250°C to 260°C.

If, then, a web consisting of a mixture of PP and PET fibers is heated to a temperature of 175°C, then the PP component would melt while the PET component retains its fibrous identity. The numerous capillary channels formed by the PET fibers of the web would permit the molten PP component to flow and spread out, obstructed only by the many crossing points among the PET fibers themselves. Once the flowing PP material has stabilized across the web, a sharp reduction in temperature would cause its solidification, holding the nodes among the PET fibers firmly into solid bonds. Such bonds between nodes of crossing PET fibers and thin layers of binder in the form of solidified PP material are primarily adhesive in nature, and the strength of these nodes would therefore be dictated by the properties of interface between the PET fiber and PP binder. If, in this process, all nodes of the web get bonded rigidly, the resultant product would become highly resistant to shear deformation, becoming unsuitable for textile application. Thus, from the point of view of end use, a judicious choice of share of binder component in the blend, resulting in immobilization of a limited number of nodes, is crucial in such a process.

If, instead of blending PP and PET fibers into a fibrous assembly, one employs core sheath fibers made of PP in the sheath and PET in the core and converts such fibers into a web, then such a web only needs to be heated to the softening point of PP for all fibers to achieve a degree of tackiness of their surfaces. On rapid cooling, all these fibers would get bonded at points wherever they are in mutual contact. Such bonds are between PP fibers themselves and are therefore cohesive in nature. This same mechanism would evidently be in place for webs made purely from one type of thermoplastic fibers only.

It is inferred from above that fibrous assemblies made of thermoplastic fibers can be reinforced by adhesive and cohesive bonds when heat energy is suitably supplied and removed. Evidently, webs made from a combination of thermoplastic fibers with those that do not possess this property, such as cotton or jute, can also be reinforced similarly.

The internal structure of oriented polymeric fibers is dynamic in nature, and it exists in a given state of thermodynamic equilibrium at a given point of time, owing to motions of chain molecules, which, in turn, depend on temperature. Above the glass-transition temperature, there is a segmental mobility of backbone chains of its amorphous zones, and above the melting point, the crystallites go into a fluidic state. Thus, the internal organization of an oriented polymeric fiber is bound to get partly changed when it is exposed to heat even when the same is not softened or melted. After cooling, the heated fiber has therefore a tendency to take up a different dimension unless measures are taken to counter the same. Hence, thermobonding may be associated with undesirable thermal shrinkage of the web. Permitting such shrinkages may, on the other hand, contribute to increasing bulk of the product.

Thermobonding systems are designed to supply a required amount of heat energy to the web to be reinforced and cool down the heated web suitably for effecting the desired degree of reinforcement. As these systems are needed to operate commercially, the rate of heating and rate of cooling or, in other words, heat flow rate forms a crucial aspect. A certain degree of control over the web during these transitions is also important so that a product of desired dimensions (density, thickness, and width) can be maintained within acceptable limits on a long-term basis.

A material may be heated in various ways. It may be brought in physical contact with the heat source, in which case heat Q (kcal or Joule = watt second) flows from the surface of body at a higher temperature (T_h K) to a location within the body at lower temperature (T_ℓ K) over a distance X (m) and across a surface of area A (m²) as per the heat-flow equation.

$$(dQ/dt) = AC_T[d(T_h - T_\ell)/dX]$$

The factor C_T is thermal conductivity of the material receiving heat, and its unit is W m⁻¹ K⁻¹. It is inferred that the shorter the distance between the hot surface and the location in the object as well as the larger the area of contact between the two materials, the higher the rate of heat flow would be. There would, accordingly, be a temperature gradient in the object being heated.

One can visualize a web composed of multiple layers of fibers brought into physical contact with a hot plate (Figure 15.25). The path X between the hot plate and the unexposed side of the web is occupied by a number of fibers,

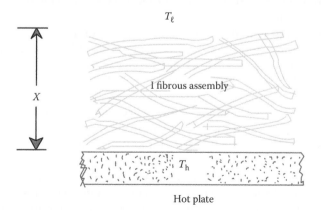

FIGURE 15.25
Model of conductive heat flow.

which may have different values of C_T as well as pockets of air in between. Typical values of thermal conductivity of PP, for example, is 0.1 to 0.3, and that of hot air is 0.04. Hence, a fibrous assembly of PP fibers would exhibit an effective value of $C_T \ll 0.1$. If such a web is held on the hot plate for a very short time, then one can expect the web surface in immediate contact with the plate to get heated to a temperature that would be much higher than that of the surface at distance X. However, this gradient can be reduced if the contact is maintained for a sufficiently long time. In order that all binder fibers across the entire web are raised to the same temperature, one then needs to dwell for a fairly long time. This time can be reduced by providing another source of heat on the unexposed side of the web, in which case the distance between the source and the object is reduced by a factor of two. Further, if the pockets of air are also gotten rid of by sufficiently compressing the web between the two hot plates, then not only would the overall thermal conductivity of the object get raised considerably, but there would be a further reduction in X and a rise in A. This would, however, be accompanied by a rise in the softening and melting points of fibers, partly offsetting the gains. During such a process, the reinforced web would also become fairly dense and thin. It is thus inferred that thinner and lighter webs meant for relatively thin products are suited for thermobonding by a conduction method employing contact between hot rollers. Usually, the surface of such rollers is engraved so that the resultant product is bonded only at selective locations with desired ornamental patterns.

The low effective contact time of milliseconds between the web and hot rollers even for moderate production speed varying in the range of 100 to 300 m/min is realized through high pressure (30 to 260 N/mm) and temperature (up to 400°C) for effecting the desired temperature rise in the binder elements. At high pressure and high working width, rollers tend to undergo bending deformation, which, in turn, causes an uneven gap along the nip. Similarly, raising the roller temperature to very high levels requires highly capital-intensive technology.

An elegant solution to this issue involves employing a flexible belt for providing the required nip. In a typical layout, a flexible silicone rubber continuous belt is pressed against a heated roller with a very large wrap angle (Figure 15.26). Silicone rubber is a polymer containing an Si–O–Si backbone with functional organic side groups. In addition to exhibiting very high thermal tolerance, this material also shows very high recoverable deformation because of the strong and long bond lengths and also the large bond angles in the backbone. The fabric that is fed into the nip between such a belt and a heated roller is forced to remain in contact with the heat source for a considerable length of time before it can emerge out of the enclosed space. This arrangement results in a very high contact time with the heat source, obviating, to a degree, the necessity of high pressure and high temperature.

In a continuous production system, two or, at times, even three hot cylinders are employed as a heat source, and the fibrous web is squeezed through

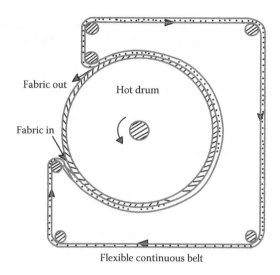

FIGURE 15.26
Principle of extended heating zone.

the narrow gap between corresponding rollers. The contact zone between cylinder surface and web extends over an area—and not over a line—due to compressibility and resiliency of a fibrous web. The web becomes gradually thinner as it gets closer to the narrow opening between the rollers and is at its thinnest along the hypothetical nip line. As it emerges from this narrowest space, the web recovers to a degree and swells up partially. This has been illustrated in Figure 15.27. Based on this understanding, it is possible

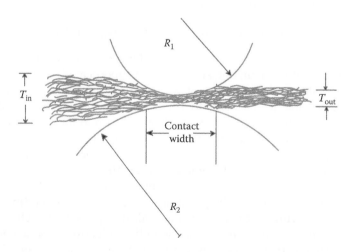

FIGURE 15.27
Model of contact zone in a nip.

to estimate the duration of contact between the web and the heat source for given roller dimensions, roller gap, web speed, and thickness and optimize the roller temperature.

duration of contact between rollers and web = [contact width/web speed]

contact width = [contact area/web width]

The contact width can be estimated from the material properties and dimension of rollers (Watzl, 2003).

$$\text{contact width} = 3.04 \ [PR/EL]^{0.5}$$

$$R = [R_1 R_2 / (R_1 + R_2)]$$

In this expression,

$$R_1, R_2 = \text{roller radius}$$

$$P = \text{applied pressure}$$

$$E = \text{elastic modulus of roller material}$$

$$L = \text{width of web}$$

Another method of estimating contact width is through measurement of web thickness at the input (T_{in}) and output (T_{out}) stages for given value of gap (a) between rollers (Pourmohammadi, 2007).

$$\text{contact width} = R^{0.5} \ [(T_{in} - a)^{0.5} + (T_{out} - a)^{0.5}]$$

Binder elements in a fibrous web can also be supplied heat energy carried by hot fluid, usually hot air, but, at times, even superheated steam. With such convective systems, the necessity of web compression does not arise. It is, as a result, possible to thermobond very thick—up to 200 mm—and fairly heavy webs—up to 4000 g/m²—at a high speed—up to 1000 m/min—in the process retaining or even improving the bulk of the product.

Heat energy can also be radiated from a suitable source. It is well known that electromagnetic waves of suitable wavelength can excite materials to a higher energy level. Microwaves, for example, in the range of 1.8 to 3.5 μm get absorbed by water, which gets quickly heated up. Depending on the absorbance spectrum of the binder material, it is possible to select a radiating source for the purpose of thermobonding. Electromagnetic energy in part of the IR spectrum is employed to induce molecular motions in polymers on the surface of thick nonwovens with a view to generating a glazing effect (Batra and Pourdeyhimi, 2012). Similarly, binder dots on nonwoven interlinings can also be melted by this method. The penetrative capacity of such waves deep within a nonwoven material is generally weak, and hence, to date, its application has primarily been restricted to the surface of nonwovens only. A major advantage of this method is the absence of any physical contact with the web, permitting a very relaxed processing.

15.6.3 Adhesive Bonding

15.6.3.1 Physics and Chemistry of Adhesive Binders

An adhesive is typically a liquid or semiliquid material that adheres to an adherent and may be used to join two similar or dissimilar adherents. A liquid adhesive solidifies subsequent to the evaporation of the solvent or emulsifying medium. Bond strength between the adhesive and adherent is usually due to van der Waals forces caused by the proximity of the molecules of the two materials, which, in turn, is affected by the ability of the adhesive to flow freely on the surface of the adherent. Electrostatic forces and, on occasion, even chemical bonds may hold the solidified adhesive and adherent together. Adhesives can be derived from a natural source, such as rubber, animal glue, or starch, but, more often than not, they are synthesized for developing desired properties.

Reinforcement of a fibrous mass with the help of adhesive binders is aimed at immobilizing all or a part of the large number of fiber intersection zones, leaving the rest of the body of individual fibers free of any adhesive matter. Such selective immobilization prevents or restricts inter-fiber slippage while permitting the fibrous body all three modes of deformation—tensile, bending, and shear—that are crucial for textile functions. The key to adhesive bonding lies therefore in selecting the most suitable binder for the given fibrous mass so that the resultant matrix of fiber and adhesive at the targeted intersection zones develops the desired rigidity and in ensuring that the applied liquid adhesive material quickly spreads among the targeted fiber intersection zones. In the context of the present treatment, binder fibers and binder powders are kept out of purview.

The rigidity of the fiber–adhesive matrix depends, on one hand, on the strength of the interface between the fiber and the adhesive—the adhesive strength—and the strength of the adhesive itself, that is, its cohesive strength. These adhesive and cohesive strengths at the fiber-intersection zones need to be commensurate to the strength of the fibers as well. A harmonious association of these three factors results in optimum reinforcement of the fibrous web. Of these three factors, the adhesive strength plays the most crucial role.

The ability of a liquid to flow along the surfaces of the fibers, which are arranged randomly in three-dimensional space and collect preferentially at fiber-intersection zones, depends on the surface tension or energy of the two materials as well as capillary action of the numerous inter-fiber channels.

Surface tension of liquid is governed by cohesive forces among the liquid molecules. Molecules within the body of a liquid are pulled equally in all directions by neighboring molecules, resulting in equilibrium of forces. The molecules that are located on the liquid surface are, however, subject to unbalanced cohesive forces and hence are pulled inward. This phenomenon causes freely floating drops of liquid in a vacuum to assume spherical shapes that satisfy the condition of minimum surface tension. Molecules on the surface of a solid are also subject to similar unbalanced forces caused by the disruption of intermolecular bonds. This imbalance in intermolecular forces manifests itself in the form of free surface energy. The unit of surface energy density is J/m^2, and that of surface tension is N/m. This is a convenient way of expressing the same phenomenon for two states of matter. Pure water has, for example, surface energy of $0.072 \, J/m^2$, and the magnitude of its surface tension is $0.072 \, N/m$. When a liquid is brought to bear on the surface of a solid, the imbalance in their respective surface energies would dictate the behavior of the interface. If the imbalance is in favor of the solid, then the liquid surface would get stretched on the solid surface manifesting in its wetting. Surfactants lower the value of surface tension of liquid and hence improve their ability to wet solid surfaces. The same phenomenon allows peeling off of liquid layers into bubbles as commonly experienced with soaps. Compositions of liquid adhesive binders usually exhibit suitable surfactants to ensure their rapid spreading out within the body of fibrous mass. Such rapid flow of adhesive liquid within a mesh of fibers is aided by capillary forces and is influenced by effective surface tension between concerned liquid and solid as well as the viscosity of liquid and diameter of capillaries. Maintaining as low a value of viscosity as possible of the binder formulation forms therefore another key guideline. Capillary flows are disturbed at fiber-intersection zones around which the flowing liquid eventually gathers.

The basic adhesive binder is a polymer whose building blocks are usually vinyl monomers. The backbone of vinyl monomers exhibit (C=C) groups or what is commonly known as alkene groups. Joining such groups end to end results in an extended alkene chain (...C–C–C–C...), which forms the backbone of vinyl polymers. The basic structure of some vinyl monomers is shown in Figure 15.28. As one of the functional end groups is changed from (H) to (CH_3) and then to (C_2H_3), the monomer changes from ethylene to propene and then to butadiene with the boiling point steadily rising from $-103.7°C$ to $-47.6°C$ and then to $-4.4°C$. The liquid monomer styrene having the boiling point of $145°C$, density of $0.909 \, g/cc$, and viscosity of $0.762 \, cP$ exhibits the aromatic benzene functional end group (C_6H_5). Such a monomer can be easily emulsified in water, and the free-flowing emulsion, when applied to a nonwoven, can quickly spread through the entire body of fibrous mass. Similarly, if the functional end group is (CN), which is commonly known as the nitrile group, then a colorless liquid monomer, acrylonitrile, is formed exhibiting a boiling point of $77°C$ and a density of $0.81 \, g/cc$. However, the polymers of styrene or of acrylonitrile, namely polystyrene and

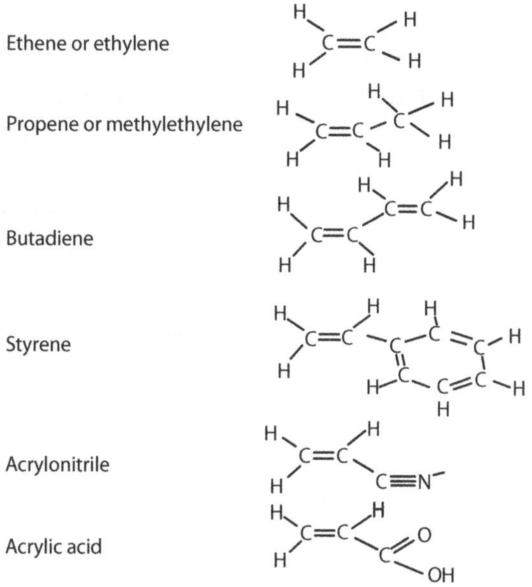

Ethene or ethylene

Propene or methylethylene

Butadiene

Styrene

Acrylonitrile

Acrylic acid

FIGURE 15.28
Structure of some vinyl monomers.

polyacrylonitrile, are fairly hard and stiff and are, as such not employed as adhesives for bonding fibrous systems. On the other hand, if the monomers of styrene and butadiene are polymerized together, forming a styrene butadiene copolymer, then the resultant material exhibits properties favorable for many commercial applications. Similarly, copolymers of styrene and acrylic acid monomers or those of acrylonitrile and butadiene monomers yield very useful adhesive binders, which are also widely employed. Typical characteristics of some of these binders are listed in Table 15.1. The exact process of polymerization adopted as well as the percentage content of the monomers and additives affect properties of the corresponding products.

If a (COOH) group constitutes the functional end group of the vinyl monomer backbone, then the colorless acrylic acid liquid with a boiling point of 141°C and a density of 1.051 g/cc is formed. This organic acid reacts with alcohols to form esters, which are commonly known as acrylates. Thus, reactions of acrylic acid with methyl, ethyl, or butyl alcohols yield methyl acrylates, ethyl acrylates, and butyl acrylates, respectively, binders with superior properties to the vinyl copolymers listed in the foregoing but substantially more expensive.

Instead of vinyl-based polymers, adhesives based on epoxy can also be used in bonding nonwovens. Epoxy is a term commonly employed for the epoxide functional group. An epoxide is formed by the oxidation of an alkene, such as oxidation of ethylene to ethylene oxide, and can be polymerized to

TABLE 15.1

Characteristic Properties of Some Commercial Binders

S. No.	Type of Binder	Characteristic Properties
1	Styrene butadiene (copolymer) Rubber	Self cross-linking leads to high water resistance; good solvent resistance; high elasticity and toughness; stiffness and hardness goes up with styrene content
2	Styrene acrylic copolymers	Forms soft film of low extensibility; exhibits good resistance to a wide range of chemicals and good aging properties.
3	Nitrile butadiene (copolymer) rubber	High elasticity and toughness; hardness goes up with acryonitrile content; high abrasion resistance and low thermoplasticity; acryonitrile component improves resistance to solvents, oil and moisture
4	Acrylic binders, such as ethyl acrylates, butyl acrylates, methyl methacrylates, etc.	Resistant to sterilization; soft and hydrophobic; self cross-linking; good resistance to dry cleaning; resistant to moisture and solvents
5	Starch, polyvinyl alcohol, polyethylene oxide, polyacrylates	Water soluble polymers
6	Polyester polyurethane copolymers	Excellent adhesion and film-forming properties; soft, elastic with good resistance to hydrolysis and light
7	Phenolic binders	High abrasion and temperature resistance
8	Waterborne epoxy resins	Excellent adhesion and high mechanical strength with excellent chemical, thermal, and electrical resistance

form a polyepoxide or, in common terms, to an epoxy resin, which is basically a polyether resin. Ether is formed when two similar or dissimilar alkyl groups are linked to the two valence bonds of an oxygen atom. The common anesthetic diethyl ether (C_2H_5–O–C_2H_5) is a well-known example of this class of compounds. Alkyl groups are members of the alkane family that exhibit single bonds between neighboring carbon atoms and are termed as saturated hydrocarbons. The generic formula for alkanes is C_nH_{2n+2} with methane (CH_4) being the simplest member followed by ethane (C_2H_6), propane (C_3H_8) and so on. As functional end groups to other elements, members of the alkane family are referred to as alkyl groups. Epoxides contain more than one epoxy groups, and they can be homopolymerized during curing to yield a thermosetting material. A typical epoxy resin, namely the bisphenol A diglycidyl ether, is a transparent, colorless liquid at room temperature and can be UV cured on homopolymerization. Ehrler (2003) underlines the suggestion made by Young (1996) that employing bisphenol-A–based epoxy resin is a method of reducing the free formaldehyde load of the bonded product. Epoxy resins are, however, more common as adhesive for composites.

Other types of binder systems based on polyurethanes or on natural resources also find application as adhesives for nonwovens, whereby the versatility, product cost, and ease of application of the binder form the bottom line for commercial exploitation.

15.6.3.2 Method of Application of Binders to Nonwovens

The most commonly adopted route of binder application to a fibrous assembly involves the emulsification of binder material along with relevant additives in an aqueous medium. After having been suitably applied, the emulsion is polymerized. The so-treated fibrous assembly is subsequently dried to remove the aqueous medium.

An emulsion is a mixture of two or more immiscible fluids. The dispersed fluid forms the dispersed phase, and the fluid in which the dispersion is carried out is termed as the continuous phase. Some energy is required to disperse a fluid in a continuous phase, and the stability of the many interfaces between the phases affects the eventual stability of an emulsion. Owing to differences in light refraction of the individual phases, a liquid emulsion is never transparent and may even assume a characteristic color. A liquid emulsion is inherently unstable, and over a period of time, the phases eventually separate out. Addition of emulsifiers or surfactants aids in improving the stability of the interfaces. Otherwise, the attraction between the dispersed droplets pulls them together, either into flocks (flocculation) of close bunches of droplets or merges the droplets into larger-sized drops or even leads to complete separation into separate phases.

An emulsion in which each member of the dispersed phase contains a number of macromolecules is termed latex. Thus, latex emulsion of, say, styrene butadiene rubber would contain emulsions of latex particles, each being made up of a number of SBR polymers. During removal of the continuous phase, the latex particles come close together, and depending on their nature, may form discrete particles or a continuous film. Additional cross-linking agents only improve the cohesive strength and other properties of the binder.

Latex particles are formed through emulsion polymerization, a process in which monomers, along with an emulsifier and an initiator, are mixed in an aqueous medium under a controlled condition. Most of the monomers are kept dispersed in separate droplets by emulsifiers, and the individual loose ones react with the initiators to start the polymerization process. Initiators are radicals that have one or more covalent bonds free to react with other substances, and they are, by nature, highly reactive. Free radicals are created in a number of ways, including breaking up large stable molecules through heating. Thus, an emulsion, when heated to a certain temperature, may cause formation of initiators through the breaking up of an otherwise stable substance and which then starts the polymerization process among the monomers. Ingredients of the emulsion also contain agents that control the degree of polymerization to the desired extent as well as plasticizers and cross-linking agents that control the softness and strength of binding agent.

A latex emulsion can be applied to a fibrous assembly through saturation, spraying, foaming, or even by printing or coating (Chapman, 2007). In the saturation process, the fibrous assembly is passed through the emulsion, and excess liquid is suitably removed through squeezing. Fibrous mass that has been pre-bonded by some other methods can withstand the strain associated with such a process. Such a process also reduces voids within fibrous assembly and makes it more compact, an effect that may be undesirable for voluminous products, such as wadding, insulation material, elements of sanitary material, etc. In the foam-application process, an applicator evenly spreads the foam across the nip zone of a padding system into which the fibrous assembly enters. This process essentially involves a similar strain on the fibrous material as in saturation bonding but is less demanding on the subsequent drying process as an appreciably smaller quantity of liquid needs to be evaporated. In spray bonding, a fine mist of droplets is applied to either one or both surfaces of a free-moving fibrous assembly. Flow properties of the droplets affect their penetration into the interior of fibrous material, and often, an even distribution of binder across the cross-section of the bonded fabric is difficult to achieve. Unevenness of distribution of droplets, even on the surface of material, as well as considerable wastage associated with spraying are other drawbacks of this process. However, the structural features of the fibrous assembly are left undisturbed by spraying. The surface of a fibrous assembly can be either printed or coated by a paste of the binder material. When printed either by rotary screen or by engraved rollers, the binder is applied only at certain locations on the surface of the fibrous assembly, leaving the rest of the material free. This approach is aimed at building in certain textile properties of the product, such as its drape, fluid transport, strength, etc. Coating with controlled penetration of the binder paste, on the other hand, affects the surface properties decidedly.

The last stage of the bonding process involves drying the treated fibrous assembly. During this process, the dispersing medium is evaporated, cross-links are formed, and the fabric assumes its final shape and form. The three methods, namely conduction, convection, and radiation, are resorted to for serving different purposes. Convection drying involves forcing hot air across the bonded fabric if it is sufficiently permeable; otherwise, hot air is blown along its surface against its direction of movement. The original structure of fabric remains undisturbed in this process. A multiple of hot cylinders with a desired thermal gradient are employed in conduction drying, which is cheaper and more suited for thinner nonwovens requiring moisture removal at a fast rate. The radiation process is slow and expensive but can form a segment of the drying section serving specific purpose, such as pre-drying or cross-linking only.

References

Basu S, Jassal M, and Agrawal A K (2013). Concept of MEV in electrospinning of PAN-DMF system, *Journal of Textile Institute*, *104*, No. 2, 158–163.

Batra S, and Pourdeyhimi B (2012). *Introduction to Nonwovens Technology*, DEStech Publications, Inc., Pennsylvania, USA.

Bharadwaj N, and Kundu S (2010). Electrospinning: A fascinating fibre fabrication technique, *Biotechnology Advances*, *28*, 325–347.

Bhat G S, and Malkan S R (2007). Polymer led web formation, In *Handbook of Nonwovens*, Edited by S J Russel, Woodhead Publishing, Ltd., Cambridge, UK, ISBN 13: 978-1-85573-603-0.

Brunnschweiler G, and Swarsbrick G (2007). Needle punching technology, In *Handbook of Nonwovens*, Edited by S J Russel, Woodhead Publishing, Ltd., Cambridge, UK, ISBN 13: 978-1-85573-603-0.

Brydon A G, and Pourmohammadi A (2007). Chapter 2, In *Handbook of Nonwovens*, Edited by S J Russel, Woodhead Publishing Ltd., Cambridge, UK, ISBN 13: 978-1-85573-603-0.

Chapman R A (2007). Chemical bonding, In *Handbook of Nonwovens*, Edited by S J Russel, Woodhead Publishing, Ltd., Cambridge, UK, ISBN 13: 978-1-85573-603-0.

Chase G G, Varabhas J S, and Reneker D H (2011). New methods to electrospin nano-fibers, *Journal of Engineered Fibers and Fabrics*, *6*, No. 3, 32–38.

Choi K J (1988). Strength properties of melt blown nonwoven webs, *Polymer Engineering and Science*, *28*, No. 2, 81–89.

Cramariuc B, Cramariuc R, Scarlet R, Manea L R, Lupu I G, and Cramariuc O (2013). Fibre diameter in electrospinning process, *Journal of Electrostatics*, *30*, 1–10.

Dosunmu O O, Chase G G, Kataphinan W, and Reneker D H (2006). Electrospinning of polymer nanofibers from multiple jets on a porous tubular surface, *Nanotechnology*, *17*, 1123–1127.

Ehrler P (2003). Binders, In *Nonwoven Fabrics*, Edited by W Albrecht, H Fuchs, and W Kittelmann, Wiley-VCH Verlag GmbH & Co. KGaA, Weinheim.

Forward K M, and Rutledge G C (2012). Free surface electrospinning from a wire electrode, *Chemical Engineering Journal*, *183*, 492–503.

Holmes R (2009). Italian trade mission, Available at http://www.thenonwovens institute.com/ncrc/italian_showcase/presentations/speaker/rory_non wovens_markets_and_opportunities.pdf; accessed on 07.08.2011.

Jakob H (1990). Aerodynamic web forming of light weight nonwovens, *Textile Technology*, *7*, 138–141.

Jha B S, Colello R J, Bowman J R, Sell S A, Lee K D, Bigbee J W, Bowlin G L, Chow W N, Mathern B E, and Simpson D G (2011). Two pole air gap electrospinning: Fabrication of highly aligned, three-dimensional scaffolds for nerve reconstruction, *Acta Biomaterialia*, *7*, 203–215.

Jirsak O, Sanetrnik F, Lukas D, Kotek V, Martinova L, and Chaluopek J (2005). A method of nanofibers production from a polymer solution using electrostatic spinning and a device for carrying out the method, International Patent, WO 2005/024101.

Jones A M (1987). A study of the melt flow rate and polydispersity effects on the mechanical properties of meltblown polypropylene webs, Book of Papers, *4th International Conference on Polypropylene Fibres and Textiles*, Nottingham, UK, September 23–25, 47.1–47.10.

Kittelmann W, and Blechschmidt D (2003). Extrusion nonwovens, In *Nonwoven Fabrics*, Edited by W Albrecht, H Fuchs, and W Kittelmann, Wiley-VCH Verlag GmbH & Co. KGaA, Weinheim.

Leach S J, and Walker G L (1966). The application of high speed liquid jets to cutting, *Philosophical Transactions of the Royal Society of London, Series A: Mathematical and Physical Sciences*, 260, No. 1110, 295–310.

Lee Y, and Wadsworth L C (1990). Structure and filtration properties of melt blown polypropylene webs, *Polymer Engineering and Science*, 30, No. 22, 1413–1419.

Leigh R (2010). Double-digit Asian growth to drive global nonwovens market to 2015, Available at http://www.pira-international.com/double-digit-asian-growth-to-drive-global-nonwovens-market-to-2015.aspx; accessed on 07.08.2011.

Malkan S R (1994). An overview of spunbonding and meltblowing technologies, *Proceedings Conference on Nonwovens*, Orlando, FL, 31–37.

Massenaux G (2003). Introduction to nonwovens, In *Nonwoven Fabrics*, Edited by W Albrecht, H Fuchs, and W Kittelmann, Wiley-VCH Verlag GmbH & Co. KGaA, Weinheim.

Milligan M W, and Haynes B D (1991). Air drag on monofilament fibres: Meltblowing application, *American Society of Mechanical Engineers*, 54, 47–50.

Moschler W (1997). Technical factors influencing hydroentanglement of fine fibrous assemblies, *15th International Nonwovens Colloquium*, Brno.

Mustermann U, Moschler W, and Watzl A (2003). Hydroentanglement process, In *Nonwoven Fabrics*, Edited by W Albrecht, H Fuchs, and W Kittelmann, Wiley-VCH Verlag GmbH & Co. KGaA, Weinheim.

Nguyena T T T, Ghosh C, Hwang S G, Chanunpanichc N, and Park J S (2012). Porous core/sheath composite nanofibers fabricated by coaxial electrospinning as a potential mat for drug release system, *International Journal of Pharmaceutics*, 439, 296–306.

Ploch S, Boettcher P, and Scharch D (1978). *MALIMO*, VEB Fachbuchverlag, Leipzig.

Pourmohammadi A (2007). Thermal bonding, In *Handbook of Nonwovens*, Edited by S J Russel, Woodhead Publishing, Ltd., Cambridge, UK, ISBN 13: 978-1-85573-603-0.

Ren Z F, Liu G Q, Wang Y F, and Cao C (2010). Theoretical analysis of electrostatic field in electrospinning systems, *Journal of Textile Institute*, 101, No. 4, 369–372.

Schreiber J, Wegner A, and Zaeh W (2003). Loop formation process, In *Nonwoven Fabrics*, Edited by W Albrecht, H Fuchs, and W Kittelmann, Wiley-VCH Verlag GmbH & Co. KGaA, Weinheim.

Theron S A, Yarin A L, Zussman E, and Kroll E (2005). Multiple jets in electrospinning: Experiment and modeling, *Polymer*, 46, 2889–2899.

Varabhas J S, Chase G G, and Reneker D H (2008). Electrospun nanofibers from a porous hollow tube, *Polymer*, 49, 4226–4229.

Watzl A (2003). Thermal processes, In *Nonwoven Fabrics*, Edited by W Albrecht, H Fuchs, and W Kittelmann, Wiley-VCH Verlag GmbH & Co. KGaA, Weinheim.

White C (2007). Wet laid web formation, In *Handbook of Nonwovens*, Edited by S J Russel, Woodhead Publishing, Ltd., Cambridge, UK, ISBN 13: 978-1-85573-603-0.

Young G (1970). The problems of tracking, *Textile Month*, February Issue, 44–47.

Young G C (1996). Waterborne epoxy resin systems for the use as binders in nonwovens and textiles, *TAPPI Proceedings*, Charlotte, NC.

Zhou F L, Gong R H, and Porat I (2009). Three-jet electrospinning using a flat spinneret, *Journal of Material Science*, 44, 5501–5508.

Further Readings

Kittelmann W, and Bernhardt S (2003). Production of fibrous webs by carding, In *Nonwoven Fabrics*, Edited by W Albrecht, H Fuchs, and W Kittelmann, Wiley-VCH Verlag GmbH & Co. KGaA, Weinheim.

Krcma R, and Lennox-Kerr P (1971). *Manual of Nonwovens*, Textile Trade Press, Manchester, UK.

Lunnenschloss J (1972). The effect of fibre length, fibre fineness, crimp and matting on the needling process and the properties of needled nonwovens, *Melliand Textilberichte (English Edition)*, 1, No. 2, 128–134.

16

Formation of Triaxial and Multiaxial 2-D and 3-D Fabrics

16.1 Introduction

Textile fabrics designed for composites that are meant for complex load-bearing functions, such as in aerospace and marine applications, can be required to exhibit functional elements in multiple directions. In this sense and restricting to behavior in a two-dimensional plane, a biaxial woven fabric being orthotropic exhibits poor functionality in bias directions. The same is true for a Raschel-knitted fabric incorporating warp and weft inlays. A braided fabric, on the other hand, fares better in the bias direction but is ill equipped to handle strains in principal directions. On the other hand, many nonwovens made up of randomly oriented constituent elements enjoy a degree of isotropicity. However, conventional products of all four fabric-formation systems are invariably planar sheets, and in spite of having, on occasion, perceptible thickness, none exhibits a functional element in the third direction. And indeed even if such planar sheets are joined one on top of the other, as in the lamination process, the functional elements in the third direction would still be missing.

In order to achieve an improved planar isotropicity in woven, braided, and knitted fabrics and thereby approach planar isotropicity to a degree, yarns can be introduced in respective structures in different directions. This approach results in triaxial woven and triaxial braided fabrics as well as multiaxial woven and knitted fabrics. Introduction of load-bearing elements in the third direction in woven, knitted, and braided structures has led to the three-dimensional or 3-D fabric-formation systems. A large number of conceptual solutions to 3-D fabric-formation systems have been reported in literature, some of which will be gone into in the following.

The 3-D fabrics being targeted specifically in the composites sector have attracted the attention of scientists and engineers from a wide range of disciplines that have contributed and continue to contribute to the advancement of this segment of textile engineering. If textile materials, on account of grossly superior strength-to-mass ratio as compared to conventional

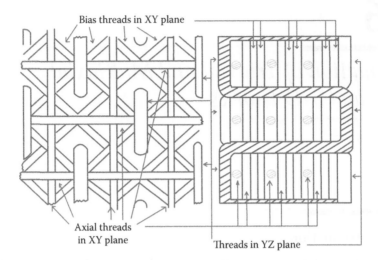

Bias threads in XY plane

Axial threads
in XY plane

Threads in YZ plane

FIGURE 16.1
Hypothetical 3-D fabric construction.

engineering materials, are to provide a viable alternative, then textile fabrics have to mimic comparable directional functional homogeneity. The fabric construction shown in Figure 16.1 is made up of 12 layers of yarn, each layer being arranged along the XY-plane. Every four consecutive layers taken together constitute an isotropic sheet. Joining together a number of such sheets by interlacing yarns oriented along the YZ-plane results in a 3-D fabric with yarns oriented along all three principal directions and a degree of isotropicity along the XY-plane owing to the respective bias yarns. However, bias yarns being absent along the YZ- and XZ-planes detracts from the bulk isotropicity of such a material. This example illustrates the conceptual hurdle in designing a textile fabric that closely mimics conventional engineering materials.

16.2 Triaxial Fabric

16.2.1 Woven Triaxial Fabric

The concept of triaxial fabric was patented in 1969 (Dow, 1969) although hand-woven triaxial fabrics had been in vogue for a long time for weaving canes found in furniture. This patent was, however, followed up by a detailed description of a hand-operated complex loom (Dow and Tranfield, 1970) for continuous production of a variety of triaxial weaves described in the patent literature. For over a decade or so, this technology and the ensuing products received attention from researchers (Scardino and Ko, 1981), after which attention shifted to 3-D fabric-formation systems. Some basic concepts of this fabric-formation process will be gone into in the following.

The basic weave of a series of 15 weaves patented by Dow, who termed these as Dow weaves, is illustrated in Figure 16.2. It is made up of two sets of warp and one set of weft yarns. One set of warp yarns, represented by letters, is inclined to the right, and the other, represented by numbers, is inclined to the left. Taking one member from each warp set, namely b and 1, it is observed that, at the bottom of the diagram, the two yarns cross each other. During one cycle of picking, the yarn b moves to the right by an amount, $0.5w$, and the yarn 1 moves to the left by an equal amount so that, at the end of one cycle, they are spaced apart by the amount w. This process continues in subsequent cycles. Hence, in this fabric-formation system, warp yarns are required to keep on shifting either to the right or to the left. This motion has been termed as transverse motion in the literature. It is also observed from the diagram that alphabetically numbered yarns are always located above weft yarns but under numerically numbered ones, and the reverse holds true for numerically numbered ones. This is possible due to a two-stage shedding process. During the first stage, the two sets of warps cross over among themselves with the numerically numbered ones located above those that are alphabetically numbered. This crossing over is accompanied with a lateral displacement of each set by $0.25w$. During the next stage and prior to pick insertion, there is a further lateral shifting with the numerically numbered ones occupying, on this occasion, the lower shed line. Each shedding stage is thus associated with a matching transverse motion.

Such a scheme of operation of warp yarns, namely repeated up and down motion accompanied by a steady lateral motion, calls for an innovative and

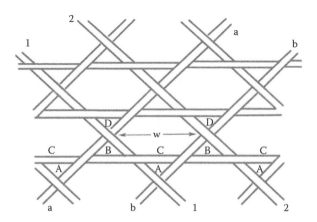

FIGURE 16.2
Pattern of yarn interlacement in a basic triaxial weave, US Patent 3446251 by Dow.

fairly complex warp guidance system. Evidently, conventional healds, which can only impart up and down motions to the warp, do not come into question here. Additionally, the warp guidance system must be equipped to allow yarns to change direction of shift once they reach the outermost edges of the fabric. Thus, after one cycle of pick, one member from the alphabetically numbered group changes over to the numerically numbered group at one fabric edge, and after the next picking cycle, the reverse happens at the other fabric edge. Such a measure, working with an odd number of warp yarns, ensures similarity of fabric edges. One may visualize the warp supply setup designed as an oblong-shaped clockwise-moving carousel, whereby the upper tier is occupied by alphabetically numbered yarns and the lower tier by the numerically numbered ones (Figure 16.3). Yarns in each tier move laterally during every pick by the amount $0.5w$ so that spacing of neighboring yarns of the warp layer grows by the amount w after each pick. Figure 16.3 shows how, over one pick cycle, yarns of the lower tier move to the left and those of the upper tier move to the right by $0.5w$ while one yarn from the extreme left of the lower tier moves to the top so that an overall 1/1 shed geometry is maintained. As indicated in the lower part of the figure, a yarn from the extreme right side of the upper tier would move to the lower tier during the subsequent cycle, reestablishing in the process the starting condition.

A conventional reed that spaces warp yarns, controls warp width, and beats up the inserted weft on a conventional loom is inadequate in the given scenario and gives way to a more flexible system that is made to execute functions of reed while being withdrawn from critical zones of the warp system during relevant periods of the fabric-formation cycle. The picking system alone bears some resemblance to conventional hand-operated ones whereby the guidance provided to a flying shuttle by a conventional sley is substituted by combs of flexible reed.

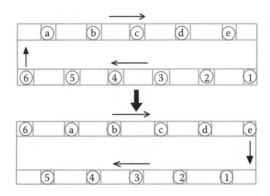

FIGURE 16.3
Carousel setup of warp supply to a triaxial weaving system.

16.2.1.1 Warp Guidance System

The warp guidance system constitutes a warp traversing system, a warp advancing system, and a warp shedding system. The roles of traversing and shedding systems have been outlined in the foregoing. The warp advancing system is synchronized with the beating up and shedding systems for providing sufficient lateral space between neighboring yarns of two warp layers for insertion of respective shedding and beating up elements.

Shedding involves pulling yarns from the top tier of the carousel to the lower one and vice versa. For this purpose, hooks can be employed. In order to insert the hooks through the two layers of warp sheet, it is necessary to align yarns of the top layer exactly on top of the yarns of the bottom layer so that sufficient space is created between each pair of yarns of the two layers. After insertion of the hooks, the warp layers need to be advanced again suitably so that yarns are aligned along the path of the hooks in order that they are caught securely and pulled to respective shed lines. Similarly, insertion of combs into the warp sheet for beat up also calls for creation of space between neighboring warp yarns. The warp advancing system executes this continuous and fairly involved lateral to-and-fro motion of warp yarns.

16.2.1.2 Beating Up System

This process is carried out in four steps by a series of cantilevers that carry combs at their free ends and that enter the warp sheet either from above or from beneath the warp layers. These elements have been termed as beaters in the literature (Dow, 1969). At the first stage, the beater A (Figure 16.2) is inserted from below after the warp yarns have been suitably advanced and necessary gaps created between them. Combs of this beater hold the fabric fell stable during the subsequent transverse motion of the warp yarns. After the second stage of shedding and prior to picking two more sets of beaters, namely B and C, are inserted one after the other. Beater B is inserted from above while C is inserted from beneath the warp layers. A considerable amount of advancing of warp yarns is required for accomplishing these insertions. A reference to Figure 16.2 illustrates how the respective combs of beaters A, B, and C are lined up vis-à-vis the two sides of a pick that gets inserted subsequently. Clearly, the beater A would hinder pushing of the last pick to the cloth fell if this arrangement is maintained. In practice, beater A is withdrawn after beater B has been inserted, and hence, the pick gets beaten up by the combs of B and C only. To provide a closer packing of yarns, an additional beater D is inserted next from above the warp line after withdrawing beater B. The combs of beater D are meant to push against the cross-over points of the two sets of oppositely inclined warp yarns. Clearly, beating up of warp yarns is a unique feature of the triaxial weaving process. As the beater D pushes on toward the cloth fell, the beater C is withdrawn. A close perusal of the sequence of insertion and withdrawal of beaters shows that, at no time, are two beaters from the same side of the warp sheet active, and therefore, one beater from beneath and one from above can be logically programmed to carry out the sequence of operations.

16.2.2 Braided Triaxial Fabric

The structure of a triaxial braided fabric is depicted in Figure 16.4. Strands oriented in bias directions are braided in the usual manner while an additional yarn or strand is inserted along the length direction of the product. This additional yarn, which can be termed as axial yarn, does not interlace with yarns of bias directions but simply gets trapped between the two sets. The main purpose of axial yarn is to impart a degree of lengthwise rigidity to the braided product. Such yarns are inserted into the braiding zone through tubes inserted into the base plate of the braiding machine along the centers of the circular tracks (Figure 14.17). These tubes pass through the centers of the respective horn gears and open up on the underside of the machine frame. Yarns from stationary spools suitably mounted on a side creel can be guided into these tube openings and easily pulled up into the braiding zone.

As hardly any strain is imposed on axial yarns during the braiding process, it is quite possible to exploit this facility and insert cheap and weak flat materials to the braiding zone for imparting body and width to the braided product. Such a method was adopted for converting cheap jute slivers into sapling bags held together by a very small number of braiding jute yarns (Figure 16.5).

16.2.3 Comparison of Woven and Braided Triaxial Fabrics

Both woven and braided triaxial fabrics illustrated in the foregoing exhibit yarns in two bias directions. The woven ones carry an additional axial yarn along the width direction of the product while the braided variety carries it along the length direction. In this sense, neither of the two is isotropic in the fabric plane as an axial yarn is lacking in the length direction for the woven

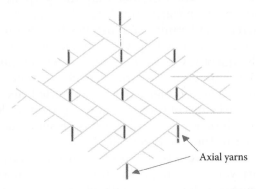

FIGURE 16.4
Construction of a triaxial braid.

FIGURE 16.5
View of a triaxially braided fabric.

product whereas an axial yarn is missing in the width direction from the braided one. The basic woven triaxial fabric is much firmer than its braided counterpart as all constituent threads securely interlace with each other. Hence, the woven variety would exhibit superior planar shear resistance. However, the production process of woven triaxial fabric is so complicated that one hardly encounters it in practice, and the simplicity of the production process of braided triaxial fabric accounts for many of its successful commercial applications.

16.3 Multiaxial Fabrics

16.3.1 Quartaxial Weaving

Iida et al. (1995) developed a process for weaving triaxial fabric with bias yarns α, α' whereby additional straight warp yarns β, β' (Figure 16.6) can also be interlaced with weft yarns Y, resulting in a quartaxial fabric. A close observation of the interlacement pattern of straight warp yarns reveals that yarns β' lie below the crossing bias yarns, and the yarns β lie above. The straight warp yarns never individually interlace with bias warp yarns but get trapped in the resultant fabric by interlacing with weft yarns while remaining on either side of the sheet of bias yarns. The warp yarns thus form a sandwich of three layers, the outer two being created by straight warp yarns while the interwoven bias yarns form the core.

The weaving process of quartaxial fabric described in the patent is broadly similar to the one described in Section 16.2.1. However, an additional layer of straight warp yarns β is provided above the bias yarn layers while another layer composed of straight warp yarns β' is arranged underneath. During the picking process, warp yarns β are always moved down while warp yarns β' are moved up for interlacing with weft yarns and subsequently brought back

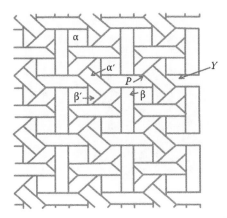

FIGURE 16.6
Pattern of yarn interlacement in a basic Quartaxial weave, US Patent 5,472,020 of Iida et al.

to their idle positions. Changes in the sequence of movement of straight warp yarns vis-à-vis bias and weft yarns lead to variations in structures. Thus, for example, β and β′ may be moved in odd picks while the interlacement of weft with bias yarns takes place in every even pick, or β and β′ may be moved in alternate odd picks while bias yarns interlace in even picks and so on and so forth. Seven different combinations are listed in the patent document.

16.3.2 Multiaxial Warp Knitting

A standard view of a multiaxial warp-knit fabric, incorporating a web of fleece along with four sets of inlay yarns and only one set of very thin knitting yarns, which knit pillar stitch and holds in the process the inlay yarns and the fleece of web together, is demonstrated in Figure 16.7. The figure on the left shows the view of the technical back side of the fabric, and the one on the right shows the sectional view. Evidently, the pillar stitching knitting yarns have to be drawn through the front guide bar, incidentally, the guide bar number 1 of a Raschel machine. There are two sets of inlay yarns, one along the width of the fabric and another along its length. Inlays along the width are inserted by a magazine weft-insertion system, and the ones along the length result out of mis-laps (0 0//). In Figure 16.7, the mis-lapped yarns occupy a position closest to the loops knitted by the front bar, which are visible on the technical front side of the fabric. Hence, these yarns must have been drawn through guide bar number 4. The inlays along two bias directions are accordingly controlled by guide bars 2 and 3.

The inlays along the bias directions in the figure move through four needle spaces over four knitted courses. Hence, the extent of lateral shogging is over one needle space for every course. However, the two sets seem to keep moving continuously in opposite directions without reversing after a certain number of courses, as is the case with the usual inlays. This means

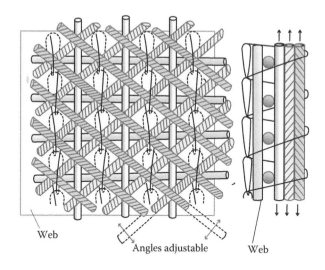

FIGURE 16.7
Schematic view of yarn assembly in a multiaxial warp-knitted fabric.

that while guide bar 2 keeps moving from left to right, the bar 3 moves from right to left. Such a situation is realized by arranging the two sets of bars on a loop-shaped track. Viewed as a whole, the two bars move in a clockwise direction although, viewed in isolation, bar 2 moves from right to left, and bar 3 moves in the opposite direction. The point to ponder is that these bars are working in conjunction with other bars that are mounted on the main machine frame. Hence, if these two bars execute a clockwise motion and the rest of the machine elements keep on working in the usual manner then a situation would arise that can be best demonstrated by Figures 16.8 to 16.10.

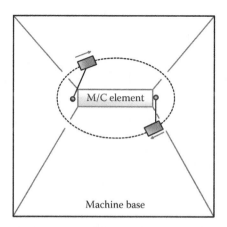

FIGURE 16.8
Functional model of a warp-knitting machine.

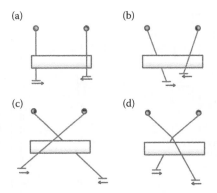

FIGURE 16.9
Hypothetical twisting together of warp inlay yarns. (a) Inlay yarns parallel, (b) inlay yarns approaching each other, (c) crossing of inlay yarns, and (d) twisting of inlay yarns.

A functional model of the machine is shown in Figure 16.8. It represents a plan view of the system, comprising machine elements anchored to the machine base and two rectangular blocks, representing two inlay guide bars that execute a clockwise motion. The line segments represent yarns joining the two bars to two rigid locations on the machine frame. Figure 16.9a to 16.9d provides a front view of the essential components of the model and demonstrates the eventual twisting of inlay yarns around each other. This can be avoided if the machine element represented by the rectangular block also rotates in a suitable direction. In Figure 16.10a–16.10c, the machine

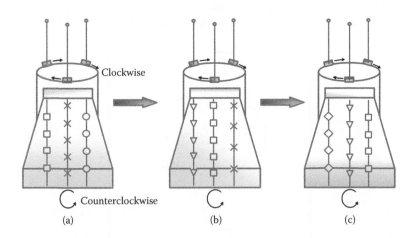

FIGURE 16.10
Principle of rotating machine components in multiaxial warp knitting. (a) Chain of squares in extreme left position, (b) chain of squares displaced to centre, and (c) chain of squares in extreme right position.

element represented by two rectangular blocks joined together by a truncated cone is shown to be rotating in a counterclockwise direction. The lateral displacement of the three vertical lines on the cone with three different symbols helps in visualizing this rotation. One observes that the line with small squares moves from left to right, and the three blocks on the elliptical track appear to be relatively stationary as the track itself moves in counterclockwise directions, being rigidly linked to the machine frame. Thus, the functional model demonstrates that, if the machine frame carrying relevant elements moves in an opposite direction to the inlay bars and if these opposing motions are synchronized properly, then yarns in two bias directions can be continuously inlaid in the fabric. The speed at which these elements rotate would obviously decide the bias angle; a higher speed would result in a larger bias angle.

An elegant alternative to this complex process of laying-in bias warp yarns was developed by Wunner (1989) who employed a magazine insertion system to carry all inlay yarns, along both the warp and weft as well as along the bias directions as shown in Figure 16.11. Carriages 8 and 9, equipped with a band of yarns, lay bias yarns at desired angles, and a similar carriage 14 lays the band of weft yarns across a conveyor system, which carries the assembled yarn sheet to the knitting zone of a single-bed Raschel machine. The warp inlay yarns are also preassembled on this sheet so that the knitting process simply requires one guide bar binding together yarns in the assembled sheet by means of a chain or tricot stitch. A fibrous web can also be fed suitably to the knitting zone for getting a product similar to that shown in Figure 16.7.

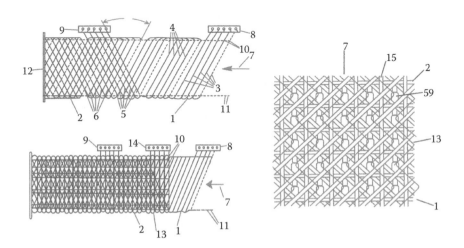

FIGURE 16.11
Multiaxial warp knit construction by modified magazine insertion principle, US Patent 4,872,323 of Wunner.

16.4 Three-Dimensional Fabrics

A wide range of textile fabrics having a pronounced third dimension are clubbed under this category, and some authors have devoted considerable attention to grouping and classifying them systematically (Khokar, 2002a; Bilisik, 2012, 2013). Such exercises have far-reaching consequences, both academic and commercial. In the context of this present chapter, attention would be restricted solely to such systems that generate products exhibiting load-bearing yarn elements or components thereof in at least three principal directions while the product is still on the production system. Thus honeycomb, sandwich, and shell fabrics, which assume a 3-D form subsequently, are kept out of purview although they are very important and interesting members of the family of 3-D textile fabrics.

16.4.1 Three-Dimensional Knitted Products

The arrangement of knitting elements on a double-bed Raschel machine for production of spacer fabric is depicted in Figure 16.12. Two sets of trick plates (6) and needle beds (7) are arranged back to back, and the resultant fabric is pulled down in the gap between the two sets. In this figure, five guide bars are arranged for feeding yarns to the needles. However, the guide bars 1 and 2 feed one set of needles, which may be termed the front bed, while the guides 4 and 5 feed the other set of needles on the back bed. Hence, two fabrics are formed separately on these two beds. The central guide bar (3) feeds yarn on both beds and links up the two fabric layers. Evidently, the distance

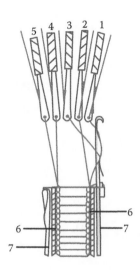

FIGURE 16.12
Side view of a double-bed Raschel.

between the two sets of needle beds and trick plates affect the length of yarn segments connecting the two layers of fabric.

If these yarn segments are made of stiff yarns, then the resultant fabric would have a pronounced third dimension with a high compressional resiliency. Such fabrics, known as spacer fabrics, are finding widespread use in diverse applications, such as compression bandages, body armor, boot soles, padding material in seats, geotextiles, etc. By employing more than one guide bar common to both beds as well as varying their underlaps and the distance between adjacent beds, which can be as high as 60 mm, it is possible to produce a wide range of such fabrics to suit specific applications.

Spacer fabrics can also be generated on certain types of flat-bed weft-knitting machines (Abounaim et al., 2009; Cherif et al., 2012). However, as the specifics of flat-bed weft-kniting technologies have not been discussed in this book in relevant sections on weft knitting, this issue is not taken up in detail.

16.4.2 Three-Dimensional Braided Products

A Maypole braider can be employed to generate a biaxial or a triaxial fabric along the periphery of a right circular cylinder. A class of 3-D fabric termed a multilayer interlocked braid was developed by Brookstein (Mouritz et al., 1999) by employing the classical Maypole braider. It involves braiding multiple concentric layers of triaxial braids that are linked together by serpentine movements of braiding spindles, which execute, in addition to circumferential motion, some through-the-thickness motion. For example, four concentric layers of triaxial braids can be produced in one piece by ensuring that each serpentine movement of a spindle covers a pair of horn gears in the radial direction, thus binding two neighboring layers together (Figure 16.13). Thus, the first layer is bound to the second, the second to the third and first, and the third to the second and the fourth, ensuring in the process that each layer retains its identity. This principle can be extended for selectively

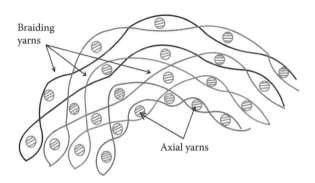

FIGURE 16.13
Principle of 3-D triaxial braiding on Maypole braider.

choosing a through-the-thickness path of individual spindles for generating 3-D circular braids with varying sections and thicknesses (Uozumi, 1995).

If the hollow of a braided cylinder is filled up with yarns laid along its diameters or along the chords of its thin sections, then the resulting product would have not only a body with pronounced third dimension, but would also carry segments of straight load-bearing yarns along various directions across the fabric thickness. This goal can be achieved by making braiding spindles crisscross over the surface of a circle instead of moving along its periphery and repeatedly touch its periphery at different points. The multiple linear yarn paths in the resultant product would, however, be traced by spindles moving along serpentine paths around and across other spindles. As a result, the braiding yarns would no more interlace but twist around each other, that is, they would get intertwined. A reference to Figure 1.8 suggests that suitably designed movement of braiding spindles along similar lines could also yield planar lace-like sheets if only two diametrically opposite points on the periphery of a circle are allowed to be touched by the spindles.

Applying this scheme of spindle locomotion, one can set up the horn gears in diverse geometric formats to develop 3-D braided products of varying profiles. One such example for developing a 3-D braid is illustrated in Figure 16.14 in which a 4 × 4 square matrix of horn gears move a total of 16 spindles along two diagonals and four line segments joining centers of four sides of the virtual square. After four revolutions of a horn gear, each of the four spindles of type "A" completes one return motion along the positively sloped diagonal, and each one of type D completes one return motion along the negatively sloped diagonal. Each member of the two groups of B and C types of spindle moves along the four short arms joining the centers of four sides of the imaginary square but in opposite directions. Thus, effectively, one structural repeat is completed in one revolution of horn gear. The

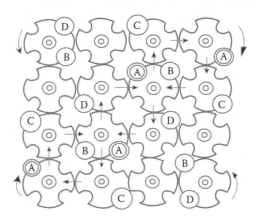

FIGURE 16.14
Principle of horn gear matrix for generation of solid 3-D braid. (From Ko, F. K., *Engineered Materials Handbook*, Vol. 1, Composites, pp. 519–528, 1987, ASM International.)

constituent yarns in such a product would be inclined in multiple directions with respect to the main axis of the assemblage and would not be able to bear any axial load. However, axial yarns can be introduced through the center of each horn gear for stiffening the product along its axis.

Propulsion of spindles by conventional horn gears is a low productive proposition. On one hand, a large number of horns remain unoccupied, and on the other hand, the guiding cam tracks cause considerable energy loss. In the example cited of a 4×4 matrix of horn gears, only one fourth of the 16 horns are, at any time, occupied by spindles. In such a matrix arrangement of horn gears, a particular member of the matrix is usually surrounded by multiple members, giving rise to the possibility of selection of transfer of spindles from one horn gear to any one of its neighbors. Evidently, electronic Jacquard selection techniques can be employed on cam elements, creating in the process flexible tracks that lead to structural variations in resultant products. Even then, drawbacks associated with track guidance would continue to persist.

An elegant solution to this issue has been reported by Mungalov and Bogdanovich (2004) the principle of which is illustrated in Figures 16.15 and 16.16. A pair of horn gears, H_1 and H_2, each equipped with four horns, is shown in the upper half of Figure 16.15. At four locations around each gear are stationed four rotary gripping forks (RGF), shown by shaded sections. A RGF can be rotated through 180° around its axis by a pneumatic actuator. In Figure 16.15, the RGF common to the pair of horn gears and marked as XY has been rotated by 180° as can be seen by comparing its alignment against the upper and lower halves of the figure. The two side walls of a RGF are carefully machined to perfectly match the curvature of side walls of spindle carriers 1 and 2, which are gripped by respective horn gears. Moreover, the two side walls of each spindle carrier also match the curvature of the horn walls perfectly. Each carrier is, in addition, flanged suitably along its upper

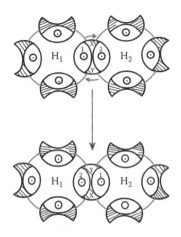

FIGURE 16.15
Horn gears with rotary gripping forks.

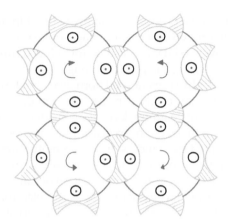

FIGURE 16.16
Principle of modular construction of 3-D braiding system employing rotary gripping forks.

and lower surfaces to enable a secure sliding fit with the exposed upper and lower surfaces of a horn. As a result, a rotating horn securely carries the gripped carrier along the arc of a circle without necessitating any additional guidance of track plates. Switching of the spindle carrier from one horn gear to its neighbor is effected by rotation of the corresponding RGF, which slides out a carrier from one horn and slides it into the neighboring one. In fact, each rotation of a RGF results in the transfer of a pair of spindle carriers to their respective neighboring horns. If, however, the RGF is not rotated, then the spindle carriers continue to dwell on their respective horns. In Figure 16.15, two horn gears support eight spindle carriers—indicated by small circles—and are neighbored by seven RGFs, one of them being common to the pair of horn gears, H_1 and H_2. The six free RGFs are meant for coupling additional horn gears in rows and columns for building up a matrix that may serve as one module for the system.

The smallest 2×2 matrix of horn gears along with 16 spindle carriers and 12 RGFs is shown in Figure 16.16. The four central RGFs are common to neighboring horn gears, and they form the active group while the other eight RGFs are inactive. An electronic selection device operates on the actuators of the four active RGFs for effecting desired transfer of respective spindle carriers. In case all active actuators are selected to remain idle, then the resultant product would be four separate and parallel twines, each composed of four yarns on account of the rotation of horn gears on their axes. If the actuators of the central RGFs are kept inactive while the actuators of the lower and the upper RGF are activated systematically, then two separate braids would ensue. If the actuators of all the four active RGFs are operated systematically, then a 3-D braided product would be developed.

The operating sequence of such a system involves systematic pauses after each quarter of a revolution of horn gears. During these pauses, the actuators are selected, resulting in transfer or no transfer of spindle carriers to their neighboring horns.

A typical module of such a braider may be a 4 × 4 or a 9 × 9 matrix of horn gears. Such modules may, in turn, be conveniently assembled or rearranged according to the desired section of the product in view. An assembly of 20 such modules for generating an X-section product is shown in Figure 16.17.

Similar to Maypole braiding machines, axial yarns that run along the length of product may be introduced through tubes inserted at the center of each horn gear. Even the free space enclosed within four neighboring horns may be used for insertion of such axial yarns.

A scrutiny of Figure 16.14 indicates that serpentine movement of spindles by the horn gear system results in a much longer path than if the displacement of spindles were executed in finite discrete linear steps along Cartesian coordinates. Such a path for the carrier "A" is traced out by the series of discrete arrows. A switch over from rotational systems employing horn gears to one of discrete linear movements of spindles along Cartesian coordinates has led to the concept of multistep braiding systems. Such a system has the additional advantage of dispensing altogether with the intermediary horns and applying motion directly to spindles.

Florentine (1982) patented the four-step braiding system, the schematic setup of which is shown in Figure 16.18. It must be noted, however, that Florentine characterized the process as weaving, probably as a generic term for conversion of yarns into fabrics. In his invention, a 10 × 5 matrix of carriers (12) support spindles (13) holding yarn packages (14). This core matrix is surrounded on four sides by additional rows and columns of alternately arranged carriers. Solenoids (15) with their plungers (16)—indicated by discontinuous lines—surround the carrier matrix (11). As a consequence of the given carrier configuration, only alternate plungers are in contact with respective exposed carrier walls both along rows as well as along columns. All four side walls of each carrier are equipped with permanent magnet

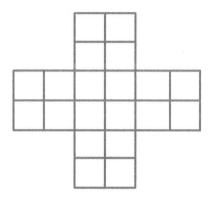

FIGURE 16.17
View of assembly of modules for braiding a profiled section product.

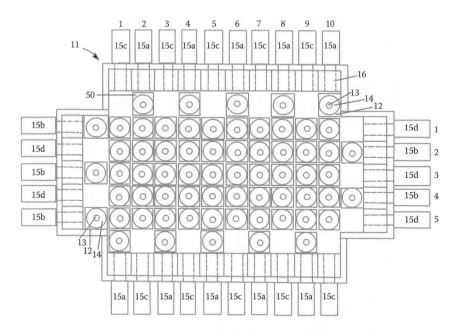

FIGURE 16.18
Principle of four-step braiding, US Patent 4,312,261 of Florentine.

arranged in the manner depicted in Figure 16.19. As a result, each carrier wall remains attracted to the wall of its neighboring carrier and is thus kept steady in its idle state.

The braiding process involves sequential linear displacement of rows and columns of the carriers by activating respective plungers. One such sequence is illustrated in Figure 16.20. In the first step, the even columns of a 10 × 5 core matrix arranged according to Figure 16.18 are pushed down while the odd columns are pushed up by respective solenoid plungers. The second

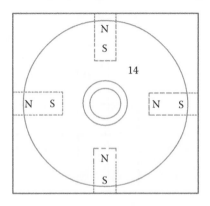

FIGURE 16.19
View of a carrier with magnets, US Patent 4,312,261 of Florentine.

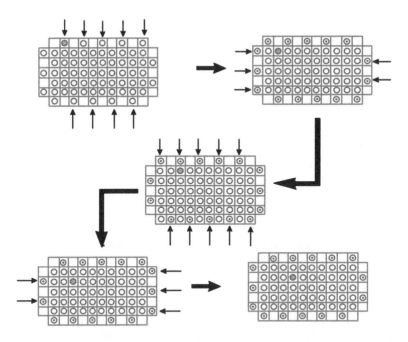

FIGURE 16.20
Principle of sequential displacement of carriers in four-step braiding system.

step involves displacement of odd rows to the right and of even rows to the left. This sequence is reversed in the subsequent two steps so that even columns are pushed up and odd columns pushed down during the third step whereas odd rows are pushed to the left and even rows to the right in the fourth step. This sequence of four steps forms one complete braiding cycle from the point of view of machine operation, and for the given setup, 11 such cycles are required for a particular carrier to return to its starting position. This is diagrammatically shown in Figure 16.21.

Retaining the essential character of the four-step braiding process intact, one can introduce variations in the sequence of displacement of columns and rows for effecting changes in the manner of intertwining of yarns from various carriers. Thus, instead of pushing both odd and even rows or columns in each step of a cycle, one can choose to push only the odd ones in the first cycle and the even ones in the second cycle. Alternately, one can mix up the odds and evens in every cycle, such as only odd rows and even columns are in one cycle and the even rows and odd columns are operated in the next. Such possibilities can be fruitfully explored through computer simulation as reported by Kostar and Chou (1994).

Florentine (1982) also worked out a four-step braiding process for developing braided hollow 3-D cylindrical fabric. The layout of corresponding elements is shown in Figure 16.22. Carriers (1) carry spindles (2), which hold spools of yarn (3) in place. These carriers are arranged along radial rows

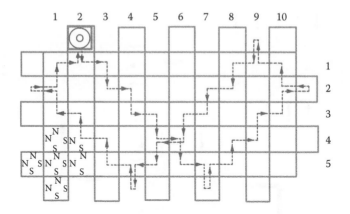

FIGURE 16.21
Schematic view of a typical path of carrier in four-step braiding, US Patent 4,312,261 of Florentine.

(4) along the surface of an annular ring only a segment of which is shown in the figure. As a consequence, these carriers automatically form concentric rings whose members are spaced apart by a finite distance. In such an arrangement, the carriers are not equipped with magnetic walls but are controlled by a fairly elaborate mechanical system, which will not be gone into here. Each radial row of carriers contains one vacant slot, which is alternately on the outermost and innermost rings from one radial row to the next. The radial rows are interspersed with partial radial rows (5) in which carriers are arranged in a fashion similar to those of (4) with the difference that only parts of the radii adjacent to the peripheral ring are equipped with carriers.

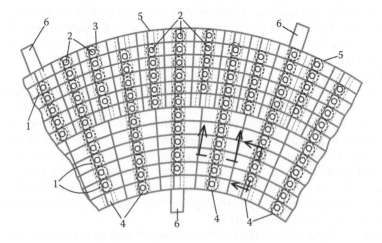

FIGURE 16.22
Setup of four-step braiding system for a hollow cylindrical product, US Patent 4,312,261 of Florentine.

These additional rows of partial carriers are meant to offset the progressively increasing circumferential gap between adjacent radial rows with a rise in diameter thereby counterbalancing a gradual reduction in yarn packing near the outer layers of the resultant cylindrical fabric. Solenoids (6) are employed for radial displacement while concentric rings are rotated in steps—also through solenoid actions—for circumferential displacement of carriers.

A typical displacement profile of one carrier on a system equipped with five concentric rings is shown in Figure 16.23. Out of the five rings, A to E, only the inner three, namely B, C, and D, need to be rotated in discreet steps, and the intermediate radial displacements are accomplished by suitable displacement of radial rows.

One limitation of the four-step braiding process relates to the absence of any provision for axial yarns. As all carriers must move and as there cannot be any unoccupied space within the matrix, it is not possible to insert axial yarns. This drawback is overcome in a two-step braiding system developed by McConnel and Popper (1988), a schematic diagram of which is shown in Figure 16.24. Vertical tubes (1) are arranged on a grid (2) in a manner corresponding to a sectional profile of the desired product, an X-section in this case. Stationary yarn supplies (3) feed axial yarns to respective tubes (1). Each tube is located in the center of a raised square, and the spaces between the squares constitute segments of a XY track for movement of braiding yarn carriers (4), indicated by T-shaped elements in the diagram. A carrier, at any

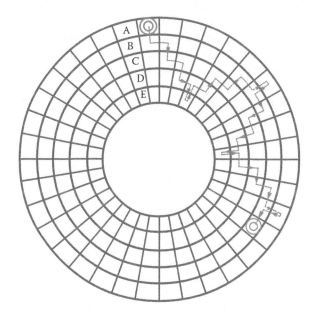

FIGURE 16.23
Path of a carrier in a four-step braiding system for a hollow cylindrical product, US Patent 4,312,261 of Florentine.

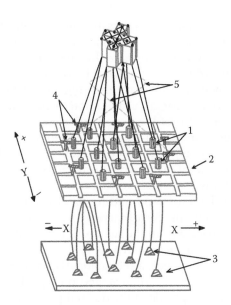

FIGURE 16.24
Setup of a two-step braiding system.

time, can move either in the X- or in the Y-direction, whereby the sense can be either positive or negative. The typical path traced by braiding yarns between carriers and the resultant product is shown by the lines (5). The methodology followed for locating tubes and braiding spindles on the grid as well as the principle adopted in moving the spindles for producing a 3-D braided product of X-section is explained in Figure 16.25.

The outline of an X-section is projected on the grid in such a manner that its boundary lines occupy diagonals of the grid squares. Tubes (1) are located on the centers of the enclosed squares so that the axial yarns coming out

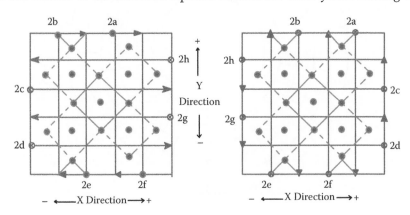

FIGURE 16.25
Principle of sequential displacement of carriers in two-step braiding system.

of these tubes get properly stationed along the periphery as well as on the inside of the projected X-section. All free corner points of the grid squares associated with the boundary lines of the X-profile are then identified as reversal points, and braiding yarn spindles are placed at each alternate reversal point. In the given example, a total of 16 reversal points come in question, and therefore, eight spindles designated as (2) are placed at alternate reversal points starting with the lowest XY point being referred to as the pivotal point. These eight spindles have been designated as 2a to 2h. At the first step, each of these spindles is moved in the X-direction, some in the negative sense and others in a positive one. As alternate reversal points had been kept free, each spindle can be moved until it reaches the nearest free reversal point in the X-direction on the opposite side of the section. Thus, in the first step, all braiding spindles cross right through the X-section from one boundary edge to the other along the X-direction. In effect then, each axial yarn is crossed over by a pair of braiding yarns moving in opposite directions. Subsequently, spindles in all reversal points are moved in the Y-direction until the opposite free reversal points are reached. As a result, the two remaining free sides of each axial yarn also get crossed over by a pair of braiding yarns moving in opposite directions. Hence, at the end of two steps of the braiding process, each axial yarn is enclosed from all four sides by segments of braiding yarns from four different spindles. These two steps may individually appear to be akin to insertion of weaving picks in the X and Y directions through sheds created between layers of rigidly held axial yarns. However, if the path of a particular spindle over multiple sequences of two-step units are followed, then it is observed that each individual yarn executes a tortuous path within the total assemblage of braiding and axial yarns. An impression of the internal structure of a two-step process 3-D braided assembly is provided by the model in Figure 16.26. It shows 12 pillars representing the boundary axial yarns for the X-profiled product and path of two braiding yarns, one covering a larger distance along the X-axis, such as that of the spindle (2h) in

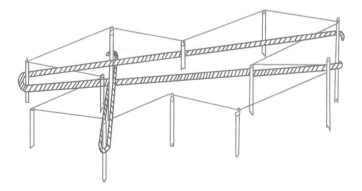

FIGURE 16.26
Schematic view of path of braiding yarns in a two-step braided product.

Figure 16.25, and the other covers the larger distance along the Y-axis, such as that by the spindle (2b).

Bilisik (2013) refers to a six-step braiding system comprising two sequences of two-step processes, during which axial yarns get intertwined with braiding yarns, interspersed with two individual steps of insertion of filling yarns along the transverse and thickness directions of the product, evidently through the many clear gaps created between the stationary axial yarns. This product is equipped with straight yarns in the X-, Y-, and Z-directions that are bound together by braiding yarns having a 3-D configuration in space. Such a product is classified as multiaxis 3-D braided material.

Considering the many different types of braiding systems that employ linear displacement of braiding spindles in finite discrete steps, Kostar and Chou (1994) proposed the broad umbrella of multistep braiding processes, the two- or four-step braiding processes being individual members of this generalized system. Such an approach permits computer simulation of the general process and working out an optimum solution for the specific product under development.

16.4.3 Three-Dimensional Woven Products

16.4.3.1 Formation of Three-Dimensional Flat Woven Products

The basic principle of assembling a collection of straight yarn segments that are oriented along three mutually perpendicular planes by employing techniques of weaving is illustrated in Figure 16.27. Fukuta et al. (1974) designed this system in which warp yarns aligned along the Y-direction are drawn through a matrix of tubes mounted on a flat board held along the XZ-plane. This board replaces the conventional reed of a weaving loom. A reed has many parallel dents each oriented along the Z-direction. In principle then, if each dent of a reed is split up into a multiple of tubes stacked on top of each other, then a reed gets converted to a flat board equipped with a matrix of tubes. The resultant warp ahead of such a flat board assumes the form of a matrix as opposed to a sheet form ahead of a conventional reed. This conceptual transformation of a reed into a flat board equipped with a matrix of tubes forms an essential element of 3-D fabric-formation systems based on principles of weaving.

The warp ahead of such a flat board can be viewed as being made up of a large number of parallel yarn sheets whereby in an $m \times n$ warp matrix, there would be m rows and n columns of parallel yarn sheets. As a result, the matrix of warp is presented to the yarn interlacement zone in the form of $(m - 1)$ sheds in the X-direction and $(n - 1)$ sheds in the Z-direction. Such sheds are, however, stationary and cannot be changed from pick to pick. By means of suitably designed picking systems, it is possible to insert an array of parallel picks, namely $(m - 1)$ picks along the X-direction and $(n - 1)$ picks along the Z-direction in a sequence ensuring that weft insertion devices do

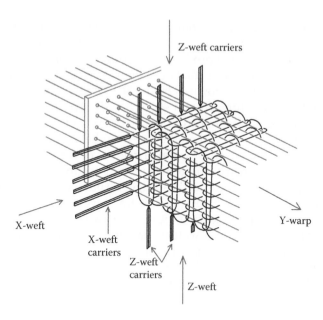

FIGURE 16.27
Principle of 3-D weaving after Fukuta et al., US Patent 3,834,424 of Fukuta.

not collide with each other. Such picks would literally fill up gaps between individual shed lines forming thereby a mesh, the openings of which are occupied by individual warp yarns. This mesh can subsequently be pushed bodily by suitable beating up devices toward the fabric take-up zone to which the warp matrix converges.

From the foregoing description, it is evident that the three sets of yarns, namely the Y-warp, X-weft, and Z-weft, do not interlace with each other within the body of the assembly. However, with the help of additional binder threads, the two sets of weft bind the assembly together along four boundary surfaces of the resultant rectangular parallelepiped during their return journeys across the warp matrix. As X-wefts and Z-wefts are inserted alternately, they occupy parallel and separate XZ-planes within the assembly. Tension in individual weft yarns would tend to pull in the entire warp yarn assembly toward the central axis of the rectangular parallelepiped thereby making the product more narrow and compact. This could simultaneously result in a raggedness of the four exposed surfaces, which is mitigated by employing binder threads for holding weft yarns securely at the extremes of their looped configurations.

A lack of interlacement of warp with weft yarns within the yarn assembly is caused by an absence of change of sheds although some kind of interlacement takes place on the four exposed surfaces of the parallelepiped. This phenomenon prompted Khokar (2002a) to coin the term Noobing, an acronym for Non-interlacing, Orthogonally Orientating, and Binding, for differentiating such processes from weaving. Absence of any interlacement between warp and weft yarns is, on the other hand, expected to improve the load-bearing behavior of the product and reduce the chances of damage to those very high-modulus yarns that have poor transaxial properties.

A 7 × 7 matrix of warp is shown in Figure 16.27 being assembled into a 3-D fabric by weft yarns coming from a set of X-weft carriers stationed on one side of the weaving device and from two other sets of Z-weft carriers, one set operating from above and the other set from below the warp matrix. In this particular case, the X-weft carriers carry loops of yarns though openings between rows of warp matrix and return to their original position while tips of the weft loops are held at the farthest edges by binder threads. Each X-weft is inserted as a double pick. Subsequently, the X-weft yarns are pushed forward to the fabric fell, thus clearing the space between the flat board and the fabric fell for easy passage of Z-weft carriers through the gaps of $(n - 1)$ columns. The geometry of the fabric-formation zone after passage of the Z-weft carriers is shown in Figure 16.28. The crossed Z-wefts form a clear passageway for the X-weft carriers, which are inserted again and retracted to their original positions, leaving behind another row of double picks that are held

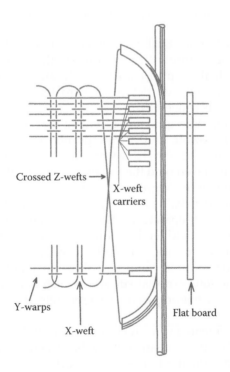

FIGURE 16.28
A schematic close up view of 3-D weaving zone after Fukuta et al., US Patent 3,834,424 of Fukuta.

at their free ends by binder threads. The Z-weft carriers are then retracted to their starting positions, enclosing in the process the newly inserted X-weft yarns. Beating-up elements are then activated to push the mesh of X- and Z-weft yarns firmly forward to the fabric fell.

The manner of movement of Z-weft yarns in the system designed by Fukuta et al. was elegantly duplicated by a set of shed-forming warp yarns in the modified design of a conventional loom developed by Greenwood (1974) for formation of 3-D fabric. As shown in Figure 16.29, a Y-warp sheet is supplied to the weaving zone from a weaver's beam through dents of reed without, however, being drawn through any heald eyes. This sheet is split into many layers and kept separated by means of pairs of bars. Raising or lowering a pair of bars lifts or lowers the corresponding Y-warp layer vis-à-vis the rest of the Y-warp sheet. By suitably operating these bars, it is possible to split the Y-warp sheet into successive rows through which X-weft yarns can be sequentially picked in a conventional manner by a reciprocating shuttle. Once the picks of weft have been propelled through all the rows, the Z-warp yarns, which are supplied from an additional beam and which are drawn normally through heald eyes and reed dents, undergo a shed line change, thereby trapping the stack of X-weft yarns. Subsequently, an entire stack of X-weft is beaten up to the cloth fell. The Z-warp and Y-warp yarns are drawn through the same reed although through different dents and can be distributed across the warp sheet in the desired manner.

Products of the two pioneering inventions described in the foregoing—both incidentally being patented in the year 1974—come in specific solid rectangular sections and a fixed width. It is not possible to cut such a product to a desired shape as a lack of inter-yarn interlacement would cause disintegration of the entire assembly. These drawbacks of 3-D woven (or noobed) fabric were overcome in the inventions of Mohamed and Zhang (1992) and of Khokar (2002b).

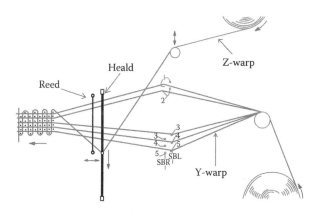

FIGURE 16.29
Principle of 3-D weaving after Greenwood, US Patent 3,818,951 of Greenwood.

The system invented by Mohamed and Zhang draws upon conceptual elements of inventions of both Fukuta et al. and of Greenwood while adding a new dimension for achieving the specific goal of producing 3-D fabrics with a desired cross-sectional shape. The essential conceptual development is illustrated in Figure 16.30 with the example of a 3-D fabric having the cross-sectional shape of an inverted T. The first difference relates to the manner of supplying Y- and Z-warps to the fabric-formation zone. The Y-warps are fed through dents of reed in a matrix form arranged according to the desired cross-sectional shape of the product. They are drawn from a creel and maintained under suitable tension so that relative spatial disposition of individual yarns is maintained throughout the fabric-formation process. Thus, both the perforated flat board of Fukuta et al. and the weaver's beam with separating bars as devised by Greenwood are dispensed with, and a direct feeding of a matrix of Y-warp yarns in the form of the desired cross-section was adopted by Mohamed and Zhang. Second, the Z-warp yarns are split up into groups along the fabric width, depending on the extent of movement required. The central ribbed portion in the illustrated example of the product has a larger number of rows than the two wings on the two sides. Hence, the corresponding Z-warp yarns (Z_1 and Z_2) are operated by separate healds, having much larger throw than the Z-warp yarns (Z_3 and Z_4) covering the two wings. The third development concerned the necessity to insert different lengths of X-weft yarns along different rows of the matrix. This was resolved by engaging picking elements, similar to that of the system of Fukuta et al., from both sides of the machine and adjusting the binding cord elements close to the corresponding fabric edges. Thus, for the central ribbed portion, having a smaller number of columns, the picking elements of the X-weft were inserted from the left with the binder cord stitching element moved close to the corresponding edge. However, for the lower portion spanning the entire width

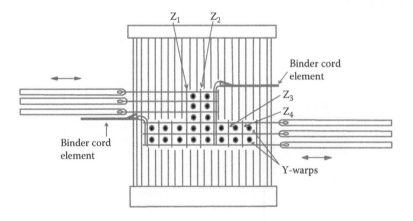

FIGURE 16.30
Principle of development of profiled section of 3-D woven fabric, US patent 5,085,252 of Mohamed and Zhang.

of the product and therefore having the maximum number of columns, the picking elements are inserted from the right side with much greater throw. Tips of the corresponding loops of weft are secured by a separate set of binder cords positioned differently than for the central bit.

The principle of operation of the binder cord stitching element for binding folded segments of weft yarns securely on one surface of the fabric is illustrated in Figure 16.31. A reciprocating latch needle catches the binder cord, which is brought across its path by means of the eye of a needle around which the weft yarns get folded during the return journey of the weft-insertion elements. When the two needles retract from the fabric edge, the binder cords get looped through the stack of weft loops along the fabric edge and form a chain stitch by being drawn through an old loop held by the latch needle.

Khokar (2001) introduced the concept of dual-directional shedding within the matrix of Y-warp yarns so that proper interlacement of warp with X- and Z-weft yarns can take place for forming true woven 3-D fabrics. One principle of such a system is illustrated in Figure 16.32. A 4 × 4 matrix of Y-warp yarns is drawn through a network of plates, one plate for each alternate row and column. Thus, in the figure, two plates are aligned along rows and two along columns. Each plate is equipped with suitable holes and slots at locations matching the position of yarns of the warp matrix for enabling respective yarns to be drawn in through these openings. Plates that are aligned along the X-direction are equipped with slots oriented along the Z-direction, and those aligned along the Z-direction have slots in the X-direction. The crossing plates are, moreover, so arranged that slots along the X-direction superimpose on slots along the Z-direction. A yarn drawn in through such superimposed slots can therefore be moved along both X- and Z-directions guided by movement of the corresponding plates. On the other hand, yarns

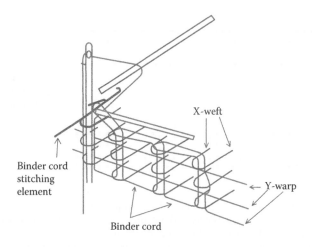

FIGURE 16.31
Principle of selvedge formation with binder cords, US patent 5,085,252 of Mohamed and Zhang.

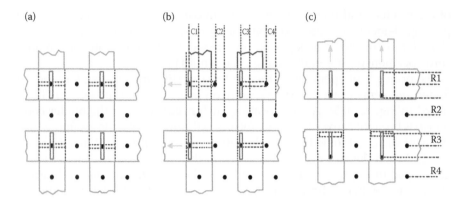

FIGURE 16.32
Dual-directional shedding by network of plates, US Patent 6,431,222 B1 by Khokar. (a) Yarn matrix without any shed, (b) shed formation along columns, and (c) shed formations along rows.

drawn in through holes can only be moved along either the X- or along the Z-direction. Owing to the nature of the arrangement of the crossing plates, one third of the yarns are left altogether free of any control, passing through the center of the gaps enclosed within neighboring plates. This arrangement thus breaks up the matrix of Z-warp yarns into three equal groups, one group capable of being moved along both X- and Y-directions, another group capable of being moved along either the X- or Y-direction, and a third group cannot be moved at all. The stationary matrix generated by free warp yarns forms the reference network about which yarns of the other two groups can be moved for creating different types of sheds. Figure 15.32b and 15.32c

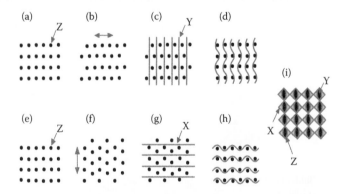

FIGURE 16.33
Sequence of shed formation and pick insertion by linear displacement method, US Patent 6,431,222 B1 by Khokar. (a) Yarn matrix without any shed, (b) shed formation along columns, (c) insertion of Y-picks, (d) closing of sheds and interlacement of Z-warps with Y-picks, (e) yarn matrix cleared of Y-picks and without any shed, (f) shed formation along rows, (g) insertion of X-picks, (h) closing of shed and Interlacement of Z-warps with X-picks, and (i) the interlaced sectional view of Z-warps, Y-picks and X-picks.

demonstrate how simple displacement of plates results in sheds along yarns within every column and row. Detailed illustrations of sequential shed formation and pick insertion resulting in formation of a woven network of 3-D fabric are shown in Figure 16.33.

Khokar (2002b) developed another method of dual-directional shedding based on angular and linear displacements of heald shafts (Figure 16.34). Each shaft (2) is equipped with healds (3), which are basically thin lamellar strips having openings at their tips and bases, namely eyes (4) and guides (5). An array of such shafts is mounted on frames S, which provide proper spatial location and necessary freedom for displacement of these shafts. The matrix of Y-warps is split up into active (Y_1) and passive (Y_2) yarns as shown in Figure 16.35 in such a way that each passive yarn is surrounded

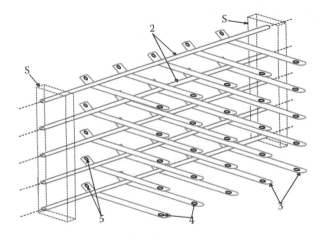

FIGURE 16.34
Modified dual-directional shedding by linear and angular displacement method, US Patent 6,431,222 B1 by Khokar.

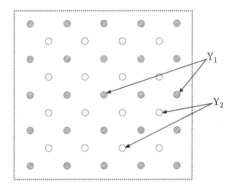

FIGURE 16.35
Arrangement of active and passive yarns in warp matrix, US Patent 6,431,222 B1 by Khokar.

by four active yarns. Each active warp yarn is drawn through a guide and eye of a particular heald commensurate to its position in the overall setup. The making up of this warp matrix is different from that shown in Figure 16.32. In this modified setup, the rows and columns of active yarns are distinctly separated from those of passive yarns, permitting a relative movement between the two. Accordingly, a lateral displacement of the rows of active yarns with respect to those of passive yarns creates column-wise shed openings, and a rotation of heald shafts around their axes shifts the columns of active yarns with respect to those of passive ones creating row-wise shed openings. The extent and direction of such lateral and angular displacements can be varied, leading to different types of sheds. Formation of two different row-wise sheds with such a setup is depicted in Figure 16.36, and a typical spatial configuration of an active warp yarn Y_1 (hatched portion) against the framework of passive warp yarns Y_2 and X- and Z-weft yarns is illustrated in Figure 16.37. Projection on the XY-plane of a typical 3-D fabric woven in the manner described in the foregoing is illustrated in Figure 16.38. The active Y_1 yarns shown in the hatched style interlace with X-weft and passive Y_2 warp yarns trapping securely the circular sections of Z-weft yarns.

One additional advantage of interlacing yarns during their assembly is that binder cord systems for securing weft yarns at their looped portions generated during their return movement are not necessary. Over and above, one can of course cut out suitable shapes from a block of such a 3-D fabric without disintegrating the assembly. Mohamed and Bogdanovich (2009), on

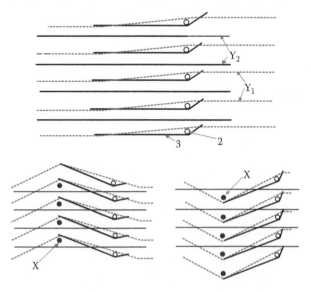

FIGURE 16.36
Sequence of shed formation and pick insertion by linear and angular displacement method, US Patent 6,431,222 B1 by Khokar.

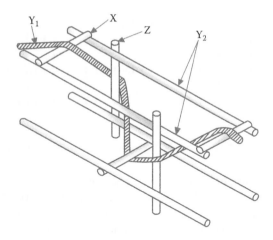

FIGURE 16.37
Schematic view of yarn interlacement in a 3-D fabric woven by linear and angular displacement method, US Patent 6,431,222 B1 by Khokar.

the other hand, point out the deleterious effect of yarn crimp caused precisely by yarn interlacements on functionality of such fabrics as reinforcing the component in the resultant composites. They also underline the possibility of damages that may be caused to certain kinds of materials, such as glass, carbon, ceramic, etc., by the shedding systems.

By suitably arranging the matrix of Y-warp yarns and working with independently moving weft insertion devices, it is possible to develop different cross-sectional shapes, some of which are illustrated in Figure 16.39.

FIGURE 16.38
Schematic plan view of a 3-D fabric woven by linear and angular displacement method, US Patent 6,431,222 B1 by Khokar.

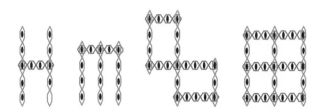

FIGURE 16.39
Some typical sections of 3-D woven fabric, US Patent 6,431,222 B1 by Khokar.

The 3-D woven (or noobed) fabric generated in a manner described by Fukuta et al.; Greenwood, Mohammed, and Zhang; or by Khokar exhibit load-bearing yarns along the X-, Y-, and Z-directions. However, such products do not exhibit a load-bearing element in bias directions along the XY- or YZ- or XZ-planes. As a result, the response of such materials to in-plane shear stresses are unsatisfactory, which indicates the necessity of building in bias threads along the three planes of a woven (or noobed) 3-D fabric.

Bilisik (2009) and Bilisik and Mohamed (2010) report about a development by Mohamed and Bilisik (USP application 5465760 of 1995) of a 3-D fabric-formation system, which incorporates layers of bias yarns along XY-planes interspersed with layers of matrix of Y-warp yarns. These yarns are bound together by X- and Z-weft yarns in the manner described by Mohamed and Zhang (1992). This particular system involving bias warp yarn insertion has been termed as tube carrier weaving. The conceptual development related to this process is illustrated in Figure 16.40. It is observed that a 4 × 4 matrix of Y-warp yarns is sandwiched between two layers of bias warp yarn assemblies. One of these assemblies is located above the 4 × 4 matrix, and the other is located below. Each assembly is made up of one layer of bias (+) and one layer of bias (–) warp yarns. The bias yarns in each assembly are supplied

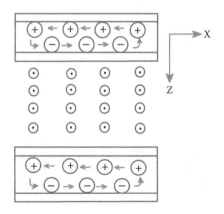

FIGURE 16.40
Principle of bias yarn displacement by tube carrier method in multiaxial 3-D weaving.

from tubes of warp yarns, which are mounted in a suitable carrier box. Each box resembles a carousel—somewhat similar to that illustrated in Figure 16.3. Thus, each warp tube dwells at one of the bias layers for a while as it is moved in steps from one edge to the other, and finally, after reaching the edge of the carousel, it is moved to the other bias layer. During each step-wise displacement of tubes in the framework, X-wefts in loop form and a pair of Z-wefts are inserted in a manner similar to that illustrated in Figures 16.27 and 16.28. However, no X-weft is inserted between layers of bias (+) and bias (–) yarns. A schematic view of the resultant product is illustrated in Figure 16.41 wherein a layer of bias yarns arranged along the XY-plane is clearly visible. To differentiate such 3-D products from those having load-bearing elements only along the three principal directions, the authors term these as multiaxial 3-D fabrics.

To circumvent the complexity of the tube carrier process, which requires a complex operation with the warp tubes for insertion of bias yarns as well as for enhancing versatility, a tube rapier 3-D fabric-formation process was also developed as reported by Bilisik (2010b). In this system, all yarns of the Y-warp matrix are drawn through tubes that are mounted on rows of flat metallic strips termed as rapiers by the author. A typical setup of such a rapier strip with Y-warp yarns drawn in through respective tubes is shown in Figure 16.42. Conceptually, there is a degree of similarity of such rapiers with the strips employed by Khokar (2001) in developing dual-directional shedding within the warp matrix. The rapiers can be moved laterally in both directions, displacing in the process warp yarns controlled by them. The row of warp matrix meant to serve as bias yarns is displaced continuously in one

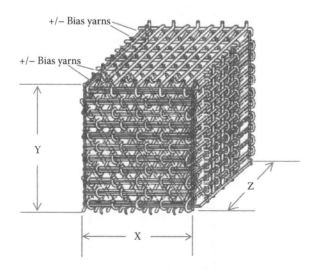

FIGURE 16.41
Unit cell of multiaxial 3-D woven fabric. (From Bilisik, K., *The Journal of Textile Institute*, 101, 5, 380–388, 2010a.)

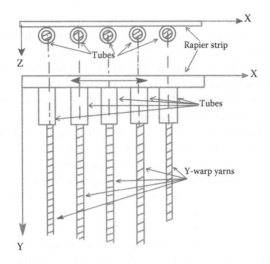

FIGURE 16.42
Principle of bias yarn displacement by tube rapier method in multiaxial 3-D weaving.

direction by means of the relevant rapier until one member of the row reaches the fabric edge. The direction of the rapier is then reversed and the process continued. The rapiers guiding the stationary warp yarns are not moved at all. Evidently, a bias yarn row can only cover a part of the fabric width in the X-direction in order that it can be moved laterally and still remain within the confines of the fabric edges. An illustrative trace of a layer of four bias yarns along the XY-plane is shown in Figure 16.43 by continuous lines. Such a configuration imparts a degree of nonuniformity in the mass distribution of bias yarns in the resultant product. If two complimentary layers of bias yarn, that

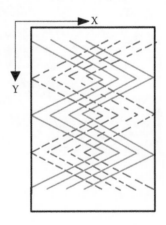

FIGURE 16.43
Trace of bias yarns in multiaxial 3-D fabric woven by tube rapier method.

is, two layers moving in opposite directions, are arranged in the immediate neighborhood of each other, then this nonuniformity is accentuated further as a very high concentration of bias yarns at the center is surrounded by zones with lower concentration and zones devoid of any bias yarn at all. Such a situation is illustrated in Figure 16.43 by superimposing traces of bias (+) and bias (−) yarns represented by continuous and discontinuous lines, respectively. The bias yarn layers generated by the tube carrier system, on the other hand, exhibit no such nonuniformity.

Warp yarn control by the tube rapier system enjoys an additional degree of freedom with respect to the tube carrier system in the sense that any row in the warp yarn matrix can be treated as a bias yarn layer, provided the number of yarns in that row is adjusted to satisfy the limitations imposed by constraints on the magnitude of lateral displacement.

16.4.3.2 Formation of Three-Dimensional Cylindrical Woven Products

Three-dimensional fabrics with rectangular sections or with sections based on a rectangular warp matrix can be converted to a class of composite materials that can be designed to withstand primarily tensile and bending strains. Torsional strains, on the other hand, are more conveniently negotiated by bodies with cylindrical or annular sections. To this end, a logical extension of the 3-D woven (or noobed) fabric-formation process led to systems designed for cylindrical or, more appropriately, annular constructions.

The change over from a rectangular 3-D fabric-formation system to a cylindrical system necessarily demands switching over from a Cartesian XYZ to a cylindrical $r\theta Z$ system whereby the symbol r stands for the radial, θ the circumferential, and Z the axial direction. To weave a corresponding product, then one needs radial, circumferential, and axial yarns.

Yasui et al. (1992) patented a system for weaving a 3-D cylindrical fabric with a number of axial Z-yarns stuffed at the core in a solid cylindrical format, the two ends of which are held firmly under tension by two discs suitably separated by a certain distance equaling the length of the targeted product. Around this core of Z-yarns an array of circumferential θ yarns are wrapped in many layers. While these θ yarns are wrapped around the core, they are simultaneously woven with radially arranged r yarns, during which a part of the r yarns are converted to segments of additional Z-yarns. During this process of part conversion of r yarns into segments of Z-yarns, interlacements that are typical of woven constructions occur with θ yarns. Hence, these radial yarns are more appropriately termed as rZ-yarns. Such yarns also travel back and forth along the radii of the cylinder, thus binding the structure firmly at the outer surface. The technique developed for the weaving of θ and rZ yarns forms the essence of the invention, which however, would not be gone into in detail here. From descriptions of the process provided in the patent document, it appears that this is a batch process, suitable for producing pieces of 3-D cylindrical fabrics the length of which is determined by the distance set between disks gripping the bundle of Z-yarns held at the core.

Bilisik (2000) invented a process for continuous production of 3-D annular fabric that can assume suitable shapes aided by appropriate mandrels, similar to those used in braiding processes. Indeed, the hand-driven rig devised by Bilisik resembles, in some respect, that of a 3-D braiding system. The rig itself was assembled in vertical tiers of basic elements—a schematic view of which is depicted in Figure 16.44—quite similar to the two-step braiding system depicted in Figure 16.24. The axial yarns that are supplied from stationary packages placed near the machine base are fed through a disk-shaped guide plate into tubes arranged in a polar matrix form on a bed plate and represented by small circles in Figure 16.45, showing a plan view of a bed plate along with a side view of the resultant construction.

The bed plate is formed in the shape of an annular ring with a prominent hole at the center. Besides the matrix of tubes, a bed plate contains a number of concentric rings and radial tracks that are suitably located in gaps between the tubes. Each ring carries a number of yarn carriers, indicated by the X in the figures, and each radial track supports only one yarn carrier, indicated by small circles with arrowheads, located either at the outermost or the innermost periphery of the bed plate. Concentric rings can be displaced in $\pm\theta$ directions, imparting in the process the necessary displacement to the yarn carriers supported by them while yarn carriers mounted on radial tracks are

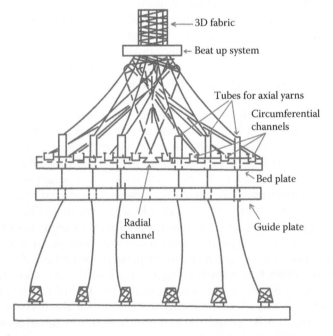

FIGURE 16.44

Schematic view of device for multiaxial weaving of a cylindrical 3-D fabric.

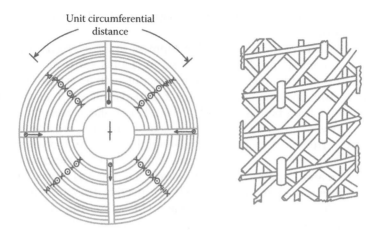

FIGURE 16.45
Arrangement of yarn carriers on the bed plate of a radial crossing system and a view of the resultant multiaxial woven fabric.

imparted movements directly. A suitable sequence of movements of various yarn carriers in each cycle results in their assemblage along $r\theta$ plane near fabric formation zone adjacent to a beat up plate located conveniently above the bed plate.

The basic principle of fabric formation with this system involves rotation of each ring through one carrier distance, causing the circumferential yarns to move through gaps between layers of axial yarns. This is followed by movement of the carriers of radial yarns from one extreme periphery of the annular ring to the other, locking in the process the three sets of yarns in their instantaneously displaced state. In the subsequent cycle, the circumferential yarns undergo another phase of displacement along the same θ direction, which is followed by the return journey of the radial yarn carriers. As a result, the radial yarns fold back across the outer/inner surfaces formed by the assembly of circumferential and axial yarns, thus imparting firmness to the respective inner and outer surfaces of the annular ring. Typical appearances of such folded segments of radial yarns on a fabric's outer surface are illustrated in Figure 16.45 adjacent to the respective view of the bed plate.

A closer scrutiny of the bed plate in Figure 16.45 reveals that the radial carriers are arranged alternately near the two extreme peripheries of the bed plate. Thus, the radial carriers cross each other while being moved from one end to the other. This arrangement can be modified such that all radial carriers are simultaneously located either near the inner periphery or near the outer one so that they move simultaneously either inward toward the center of the bed plate or outward away from the center. The former method is classified as radial-crossing weaving, and the latter is termed as radial in-out weaving. The resultant effect on fabric construction is evidenced by differences in the arrangement of folded segments of radial yarns on the two respective fabric surfaces.

A unique feature of the patented development involves the ability of constructing annular 3-D products incorporating ± bias yarns along its inner and outer surfaces. In effect then, the product of such a system can exhibit yarns in five directions, namely along r, θ, Z, and also along ± bias directions. Bilisik classifies such fabrics as multiaxial 3-D circular woven fabrics. The bias yarn carriers, indicated by + or − signs in the figures, are also mounted on concentric rings just as the circumferential yarns are. However, a bias yarn carrier ring can be made to move in either a clockwise or counterclockwise direction and, depending on the desired bias angle, is moved by a multiple of circumferential unit distance in the respective direction. In the fabric construction depicted in Figure 16.45, a pair of bias yarns moving in opposite directions are clearly visible near the fabric surface. It is also observed that, between the pair of bias yarns, no axial yarn passes through. Accordingly, the bias yarn rings are set either near the outermost periphery or near the innermost periphery of the bed plate, and no gap is provided between a pair of such rings for axial yarn tubes. The desired phase lag/lead vis-à-vis circumferential yarn rings is imparted to the bias yarn rings at the beginning of the fabric-formation process, and regular cyclic step-wise displacement in the desired direction ensures that this phase lag/lead is maintained between respective yarns in the resultant fabric.

One may form such a 3-D fabric without any bias yarns whatsoever or with either a single bias yarn layer or with a pair of such bias yarn layers near either the inner surface or near the outer surface or near both surfaces. One may also choose either the radial in-out method or the radial crossing method for assembling the yarns. Evidently, a large number of fabric constructions are possible for producing circular 3-D fabrics after the method developed by Bilisik (2000).

16.5 Key Technological Concerns for 3-D Fabric-Formation Systems

Even though various principles have been worked out for formation of 3-D knitted, braided, woven, and noobed fabrics, certain issues of concern need to be addressed before these materials find a wide range of commercial applications. Some of these will be gone into in the following.

The principal target of these textile materials is the composites sector in which properties, such as high energy absorption, good impact resistance, good formability (Cherif et al., 2012), good through-thickness stiffness, strength, and fatigue resistance to flexural and torsional strains (Mouritz et al., 1999) along with a certain degree of isotropicity and low density are prime requirements. However, the developed 3-D products do not fully satisfy all the desired criteria. Moreover, commercial applications usually call for well-established design strategies that relate functional properties of a product with its structural parameters and material properties as well as production systems that are efficient and reliable. Even in this respect, a considerable amount of development is still called for.

Conventional knitted spacer fabrics have been found eminently suitable for applications requiring high compressional resilience at a high volume and low bulk density but are structurally unsuitable for high-performance composite applications although developmental efforts at introducing multiaxial reinforcing elements on modern flat knitting machines have been reported in the literature. Composites produced from conventional spacer fabrics do, however, have the potential for replacing honeycomb panel structures employed in vehicles, cabins, and buildings (Abounaim et al., 2009).

Braided 3-D fabrics are, by and large, handicapped by limited width, a low rate of production, and an absence of straight load-bearing yarns in more than one principal directions. The solution worked out by Mungalov and Bogdanovich (2004) addresses the problem of limited width as the system permits assembling as many modules as necessary, and the six-step braiding system can hold together straight yarn segments in the X-, Y-, and Z-directions by means of braiding yarns. However, these are two distinctly different systems, and a braiding system that combines both these attributes would definitely make a more useful product. Moreover, the braiding yarns that crisscross the thickness direction of the product are sure to deflect axial yarns from their straight path unless the tension in all yarns is carefully controlled. Nonetheless, some readily available and highly automated 3-D braiding systems have enabled successful penetration of 3-D braided materials into the medical sector, such as in the form of tendons and ligaments, space and rocket propulsions, and the transportation industry (Bilisik, 2013), the fabrication of such products in small sections being cheap and not labor intensive.

Mouritz et al. (1999) list a number of issues affecting extensive application of 3-D woven fabrics in very demanding applications, such as in aerospace, and suggest that, with the current state of development, in respect to improved impact energy absorption capacity, 3-D woven fabrics may satisfy many less demanding applications, such as in the automobile sector, shipping containers, coal trucks, chemical containers, etc.

Control of yarn tension is one common problem encountered in forming fabrics with 3-D braiding and weaving systems. Studies on the structure property relationship of textile fabrics formed by an assemblage of yarns—be they woven, knitted, braided, or nonwoven—underline the importance of maintaining yarn tension at a desired level and within defined limits for forming fabrics of required properties. Incorporation of requisite tension control mechanisms in the systems described in the foregoing involving three to five sets of yarns, some of which assume tortuous paths during and after their assemblage, is a challenging but highly necessary measure. In this respect, Mohamed and Bogdanovich (2009) report successful weaving of a product with absolutely straight yarns in all three principal directions on a fully automated, computer-controlled 3-D weaving system developed by Mohamed and Zhang (1992) that features suitable tension control systems. However, the system referred to does not involve weaving with bias yarns, which, while promising to improve shear properties of the product, pose serious challenges with respect to tension control and the desired alignment of constituent yarns.

Woven and noobed fabric-formation systems invariably need a suitable beating up station, involving elements designed to push the interlocking yarns to the fell of the cloth after each cycle of operation. As all yarns, barring those of stationary warp matrix, need to be firmly pushed to a specific XZ (or $r\theta$) plane and held there until the succeeding layer of interlocking yarns is assembled and starts getting pushed in, multiaxial fingers that can penetrate into the appropriate part of the fabric-formation zone and move normal to the XZ (or $r\theta$) plane are usually employed for this purpose. These fingers are, later on, withdrawn from the fabric-formation zone to provide required free space for subsequent displacement of the interlocking yarns. Evidently, the force that each of these fingers need to apply to yarns in their path would vary with yarn material, yarn tension, and instantaneous yarn path. Moreover, the higher the number of yarns per unit volume of the 3-D fabric, the less would be the space available for such beating elements, and the more would be the force required for the beat up process. Understanding the exact geometry of yarns assembled at the beating up zone is also an important prerequisite for designing effective beating up systems. The presence of bias yarns adds to the complexity of the issues involved (Bilisik, 2012).

A suitable take up is a common requirement for all 3-D fabric-formation systems. In the event of fabric formation on a mandrel, the latter is moved at a desired rate in the required direction to accomplish this function. Otherwise, for solid or hollow rectangular or cylindrical products as well as for products with diverse shapes of sections, gripping the product securely without, in any way, distorting the fabric-formation zone and moving the fabric fell at a specific and constant rate constitutes a critical component of the fabric-formation process. Indeed, the rate and manner in which the fell of the cloth is moved away from the fabric-formation zone after each cycle of operation affects fiber volume fraction and orientation of yarns in the product, two very critical parameters for the composite into which the 3-D fabric is subsequently converted (Bilisik, 2012).

The distance between neighboring warp yarns arranged in a matrix format is, to a degree, influenced by the space needed by carriers of other sets of yarns to move through the gaps between rows and columns of the matrix. In a conventional 2-D weaving process, the neighboring warp yarns move away for providing space required by the weft yarn carrier and then close in again. However, in the 3-D systems developed so far, this inter-yarn space, governed by a dimension of yarn carriers, continues to persist in the final product although it gets partly reduced by the overall narrowing of the warp matrix at the beat up zone, caused by tension in the interlacing yarns. Owing to this system-based constraint, ensuring a reasonably high fiber volume fraction in composites can become challenging. This problem becomes more acute with systems that require carriers to be preloaded with a certain length of yarn, increasing in the process the size of the carriers and thereby the gaps between neighboring warp yarns.

References

Abounaim M, Hoffmann G, Diestel O, and Cherif C (2009). Development of flat knitted spacer fabrics for composites using hybrid yarns and investigation of two dimensional mechanical properties, *Textile Research Journal*, 79, No. 7, 596–610.

Bilisik A K (2000). Multiaxis three dimensional (3D) circular woven fabric, US Patent 6129122.

Bilisik A K (2009). New method of weaving multiaxis three dimensional flat woven fabric: Feasibility of prototype tube carrier weaving, *Fibres and Textiles in Eastern Europe*, 17, No. 6(77), 63–69.

Bilisik K (2010a). Dimensional stability of multiaxis 3D-woven carbon preforms, *The Journal of Textile Institute*, 101, No. 5, 380–388.

Bilisik K (2010b). Multiaxis 3D weaving: Comparison of developed two weaving methods (tube rapier weaving versus tube carrier weaving) and effects of bias yarn paths to the preform properties, *Fibres and Polymers*, 11, No. 1, 104–114.

Bilisik K (2010c). Multiaxis three dimensional circular woven preforms – "radial crossing weaving" and "radial in-out weaving": Preliminary investigation of feasibility of weaving and methods, *Journal of the Textile Institute*, 101, No. 11, 967–987.

Bilisik K (2012). Multiaxis three dimensional weaving for composites: A review, *Textile Research Journal*, 82, No. 7, 725–743.

Bilisik K (2013). Three dimensional braiding for composites: A review, *Textile Research Journal*, 83, 1414–1436.

Bilisik K, and Mohamed M H (2010). Multiaxis three dimensional flat woven preforms – Tube carrier weaving, *Textile Research Journal*, 80, 696–711.

Cherif C, Krzywinski S, Diestel O, Schulz C, Lin H, Klug P, and Trumper W (2012). Development of a process chain for the realization of multilayer weft knitted fabrics showing complex 2D/3D geometries for composite applications, *Textile Research Journal*, 82, No. 12, 1195–1210.

Dow N F (1969). Triaxial fabric, US Patent 3446251.

Dow N F, and Tranfield G (1970). Preliminary investigations of feasibility of weaving triaxial fabrics, *Textile Research Journal*, 40, 986–998.

Florentine R A (1982). Apparatus for weaving a three dimensional article, US Patent 4,312,261.

Fukuta K, Nagatsuka Y, Tsuburaya S, Miyashita R, Sekiguti J, Aoki E, and Sasahara M (1974). Three dimensional fabric and loom construction for the production thereof, US Patent 3,834,424.

Greenwood K (1974). Loom, US Patent 3,818,951.

Iida S, Chikaji O, and Ito T (1995). Multiaxial fabric with triaxial and quartaxial portions, US Patent 5,472,020.

Khokar N (2001). 3-D weaving: Theory and practice, *Journal of the Textile Institute, 92,* Part 1, No. 2, 193–207.

Khokar N (2002a). Noobing: A nonwoven 3D fabric forming process explained, *Journal of the Textile Institute, 93,* Part 1, No. 1, 52–74.

Khokar N (2002b). Network like woven 3D fabric material, US Patent 6,431,222 B1.

Ko F K (1987). Braiding, In *Engineered Materials Handbook, Vol. 1, Composites,* ASM International, Metals Park, Ohio, 519–528, CRC Press.

Kostar T D, and Chou T W (1994). Process simulation and fabrication of advanced multistep three dimensional braided preforms, *Journal of Materials Science, 29,* 2159–2167.

McConnel R F, and Popper P (1988). Complex shaped braided structures, US Patent 4,719,837.

Mohamed M H, and Bogdanovich A E (2009). Comparative analysis of different 3D weaving processes, machines and products, *Proceedings of the 17th International Conference on Composite Materials (ICCM-17),* July 27–31, Edinburgh, UK.

Mohamed M H, and Zhang Z H (1992). Method of forming variable cross sectional shaped three dimensional fabrics, US Patent 5,085,252.

Mouritz A P, Bannister M K, Falzon P J, and Leong K H (1999). Review of applications for advanced three dimensional fibre textile composites, *Composites Part A, 30,* 1445–1461.

Mungalov D, and Bogdanovich A (2004). Complex shape 3D braided composite preforms: Structural shapes for marine and aerospace, *SAMPE Journal, 40,* No. 3, 7–21.

Scardino F L, and Ko F K (1981). Triaxial woven fabrics: Behavior under tensile, shear and burst deformation, *Textile Research Journal, 51,* 80–89.

Uozumi T (1995). Braid structure body, US Patent 5,438,904.

Wunner R (1989). Apparatus for laying transverse weft threads for a warp knitting machine, US Patent 4,872,323.

Yasui Y, Anahara M, and Omori H (1992). Three dimensional fabric and method for making the same, US Patent 5,091,246.

Further Reading

Gokarneshan N, and Alagirusamy R (2009). Weaving of 3D fabrics: A critical appreciation of the developments, *Textile Progress, 41,* No. 1, 1–58.

Index

Page numbers followed by f and t indicate figures and tables, respectively.

Printed and bound by CPI Group (UK) Ltd, Croydon, CR0 4YY

22/10/2024

01777640-0007